The Philosophy of Science
A Collection of Essays

Series Editor

Lawrence Sklar
University of Michigan

A GARLAND SERIES
READINGS IN PHILOSOPHY
ROBERT NOZICK, *ADVISOR*
HARVARD UNIVERSITY

Series Contents

Probability and Confirmation

Edited with an introduction by

Lawrence Sklar
University of Michigan

Routledge
Taylor & Francis Group

NEW YORK AND LONDON

First published by Garland Publishing, Inc.

This edition published 2013 by Routledge

Routledge
Taylor & Francis Group
711 Third Avenue
New York
NY 10017

Routledge
Taylor & Francis Group
2 Park Square
Milton Park, Abingdon
Oxon OX14 4RN

Routledge is an imprint of the Taylor & Francis Group, an informa business

Library of Congress Cataloging-in-Publication Data

Philosophy of science : a collection of essays / edited with an
    introduction by Lawrence Sklar.
        p.   cm.
    "A Garland series"—Ser. t.p.
    Includes bibliographical references.
    Contents: 1. Explanation, law, and cause — 2. The nature of
scientific theory — 3. Theory reduction and theory change —
4. Probability and confirmation — 5. Bayesian and non-inductive
methods — 6. The philosophy of physics.
    ISBN 0-8153-2700-5 (v.1 : alk. paper) — ISBN 0-8153-2701-3
(v. 2 : alk. paper) — ISBN 0-8153-2702-1 (v. 3 : alk. paper) —
ISBN 0-8153-2703-X (v. 4 : alk. paper) — ISBN 0-8153-3492-3
(v. 5 : alk. paper) — ISBN 0-8153-3493-1 (v. 6 : alk. paper)
    1. Science—Philosophy. 2. Science—Methodology. 3. Physics—
Philosophy. I. Sklar, Lawrence.

    Q175.P51227 1999
    501—dc21   99-40012
                                                                CIP

SET ISBN   9780815326991
POD ISBN   9780415870719
    Vol1   9780815327004
    Vol2   9780815327011
    Vol3   9780815327028
    Vol4   9780815327035
    Vol5   9780815334927
    Vol6   9780815334934

Contents

Introduction

Probabilistic concepts first arose in the attempt to formulate reasonable betting rules for playing games of chance. It was only gradually realized that probabilistic and statistical notions have a vital role to play in the formulation of our most important theories about the world, and in our understanding of how one can reasonably assess those theories when the question at hand is whether or not to accept theories into our body of belief.

The first major "serious" application of probabilistic concepts was for the most part, in the studies of human social behavior, where such issues as inferring the predictable overall behavior of a society, say in economic terms, while knowing that the individuals in that society can act in different and unpredictable ways, was being explored. The development of the formal study of statistical theory was largely driven by a need to get a scientific grasp of how probabilities function in such contexts, and in contexts of the practical application of probabilistic methods.

In the late nineteenth and early twentieth centuries, probabilistic and statistical notions entered profoundly into fundamental physical theories as well. First they became essentially embedded in the theory that attempts to derive the macroscopic behavior of matter in transformations involving heat, that is to say the behavior described by thermodynamics, from the dynamical theory that describes the behavior of the many, many microscopic constituents (the molecules) that make up the macroscopic matter in question. In kinetic theory, and then in what is called statistical mechanics, especially as these sciences were developed by Maxwell, Boltzmann and Gibbs, probabilistic notions played a fundamental role in the theory. The development of quantum theory later opened up a new branch of fundamental physical theory where, once again, probabilistic notions were an essential component of the fundamental laws of the theory.

Probability was also seen to play an essential role in the methodologist's attempt to understand the dynamics of rational belief. In order to understand how our cognitive states are related to the world we are trying to understand, predict and control, it soon becomes apparent that merely invoking the simple notions of believing that something is the case or disbelieving that it is the case wouldn't do. Instead one needed a subtler apparatus, one that allowed for such notions as "partial belief," or "degrees of belief." And a number of important considerations indicated that it is the notion of probability that is ideally suited to provide the needed formal apparatus to understand such notions.

The formal, mathematical theory of probability achieved its formulation in the early twentieth century. It turned out to be a theory amazingly simple in its basic postulates, but with an extraordinary rich body of consequences derivable from these simple first principles.

But what is the nature of the magnitude of which probability theory is the formal theory? What in the world is probability? Here many views have contended. For some probability is the frequency with which some designated outcome arises in a set of happenings. Or, more generally, probability is the proportion with which some such outcome occurs. Or, perhaps, probability may be taken to be the limiting value of such frequencies or proportions when the sample space is allowed to grow without limit. For still others probabilities are thought of as a kind of dispositional property of a trial situation. These are versions of what is often called "objective" theories of probability. They take probabilities to be measures of some feature of the objective world.

For others, probabilities measure, rather, the logical relations propositions bear to one another. In this way probabilities resemble the logical notion of one proposition implying or entailing another, the fundamental concept of deductive logic. For these thinkers probability is a kind of "partial entailment." Such "logical" theories of probability immediately lead to attempts at constructing a formal probability logic to parallel the formal development of ordinary deductive logic.

For still others, probability is a measure of the "degree of belief" an agent has in some outcome whose true nature is still unknown. Or, perhaps, probability is the partial degree of belief the agent has in the truth of some proposition. Sometimes this "subjective" notion of probability is associated closely with the agent's rational behavior in the face of risky, betting type, situations. In other theories probability is related simply to a kind of "partial belief" on the part of the agent in purely cognitive situations. Profound results have been obtained that are designed to show why such degrees of belief on the part of an agent ought to conform to the usual laws of probability. Some of these results depend upon constraints upon rational betting, some on constraints on the rational ordering of our beliefs in terms of an ordinal notion of believability. In some subtle theories rational constraints on an agent's behavior are used to derive probabilistic laws for the agent's partial beliefs and, at the same time, measures of the subjective desirability of outcomes for the agent.

Some philosophers would argue that only one of the various notions of probability just outlined captures all there is to say about what probabilities are. Others would allow for more than one "correct" interpretation of probability. One vital question that arises is how the various interpretations of probability might be related to one another.

It is these questions concerning the nature of probability, that is to say the interpretive problem for probability theory, to which the readings in Section A are devoted.

We accept scientific hypotheses and theories. Usually we think of this acceptance as "belief." Believing in the truth of the scientific claims, we use them in our attempts to predict the future of the world and to control the world. But by what right do we claim that our beliefs in these hypotheses are correct? If the legitimacy of accepting the scientific claims is challenged, how could we defend our holding to the beliefs as reasonable or rational? Here again we have an area of philosophical inquiry as old as science and philosophy themselves.

For many centuries, the model of the justification for scientific belief was highly influenced by the early discovery of geometry in its traditional form of axioms, proofs, and theorems. It seemed to the great Greek philosophers, especially to Plato and to Aristotle, that geometry provided the ideal model of what every science ought to be like. And what was this model? There were the basic truths of science, expressed in its axioms. These axioms seemed so "obviously" true, that it was easy to convince oneself that their truth was apparent to some kind of "pure reason" as soon as their meaning was understood. And the warrant for believing in the truth of all the other propositions of science would then follow, since these other claims of science followed from the self-evidently true axiomatic truths by as kind of reasoning, deductive logic, that could also be seen, by pure reason, to carry one only to new truths from truths as premises. This "rationalist" model of science dominated methodological thought through the great rationalist philosopher-scientists of the Scientific Revolution, such as Descartes and Leibniz.

Most contemporary philosophers of science, however, are doubtful that the rationalist method has such scope as to include all of science, if, indeed, it is even applicable to pure mathematics itself. For most methodologists, the reason we have for taking our scientific hypotheses to truly describe the world, if there is any reason for doing so at all, must rest upon the support these hypotheses receive from observation and experiment. But how does such evidential support arise and what form does it take?

One important version of the general view that the support for rational believability in scientific claims rests on observation and experiment is the inductivist view. From the perspective of the inductivist, there are propositions whose reasonable believability rests upon the fact that they report the immediate results of observation. It is sensory awareness itself that justifies holding as true propositions about what that awareness is like. From this we get to truths about particular occurrences.

But the hypotheses of science that matter most to us are general in nature. They state putative laws of nature, and, in some cases, they even go beyond the conceptual realm appropriate for dealing with the results of observation since they posit theoretical entities and properties. How can known truths about particular occurrences provide the evidential support needed to warrant beliefs in generalities and in theories?

It is often argued that our warrant for believing generalities and theories rests upon a kind of cumulative support they receive from the truth of the particular instances predicted by them. Such an overall view may be called inductivist. We generalize our rational belief from belief that certain particular occurrences have happened, into belief that such-and-such regularities must always be the case.

But ever since Hume, who in his way was one of the prime exponents of the inductive method, questions about just what induction is, and how it itself can be justified as a method for rationally arriving at truths, have plagued philosophers. Just exactly what does inductive method consist of? And given the nature of the method, by what right do we assume that it is a guide to reasonable belief or to truth?

Some contemporary philosophy of science is devoted to trying to characterize the basic rules of generalizing from particular truths to scientific hypotheses, with all their generality and reference to the theoretical. One such proposal is to extend orthodox deductive logic by the formalization of a "logic" of induction. Other proposals look for general rules characterizing when specific experimental outcomes do or do not provide confirmatory evidence for general theories. Such studies of confirmation seek to outline criteria of adequacy any formal definition of a confirming instance for a theory would have to meet, and to find formal notions satisfying these criteria.

Other areas of methodological exploration consist in attempts at answering Hume's skeptical contention that there could be no good reason for us to ever come to the conclusion that results arrived at by means of inductive inference had any genuine warrant. That is they try to "justify" induction as a rational means for arriving at warranted belief and truth. Some propose to find a full justification for inductive reasoning. Others suggest weaker "vindications" of inductive methods might be possible. Still others argue that induction itself provides the standards of rationaltiy in the cases where it is applied.

One early attempt at characterizing the nature of induction was Mill's famous discussion of scientific method. There is no doubt that Mill put his finger on important aspects of scientific reasoning in his discussion of the methods of similarity, difference, and concomitant variation. But shortly after Mill first proposed these methods Whewell, influenced by Kant's philosophy, pointed out that the methods required that one already had at one's disposal the "correct" means of characterizing phenomena. Without knowing what features of things were and were not relevant to the questions at hand, Mill's methods were empty of genuine content.

The problems initially raised by Whewell have once more become major questions for anyone trying to understand methods of science broadly construed as inductive. What are the fundamental categories by which we classify things? If we expect the future to be like the past, in what respects do we have this expectation? Such questions give rise to the so-called issue of what the "natural kinds" may be taken to be in the world. That is, what are the appropriate features by which things ought to be categorized and classified, especially when issues of similarity are in question that are central to the issues of how to project our knowledge of particulars into claims about the truth of scientific generalizations.

These are the subjects taken up in the readings from Section B.

Probability and Confirmation

VII

TRUTH AND PROBABILITY (1926)

To say of what is that it is not, or of what is not that it is, is false, while to say of what is that it is and of what is not that it is not is true.—*Aristotle.*

When several hypotheses are presented to our mind which we believe to be mutually exclusive and exhaustive, but about which we know nothing further, we distribute our belief equally among them . . . . This being admitted as an account of the way in which we *actually do* distribute our belief in simple cases, the whole of the subsequent theory follows as a deduction of the way in which we must distribute it in complex cases *if we would be consistent.—W. F. Donkin.*

The object of reasoning is to find out, from the consideration of what we already know, something else which we do not know. Consequently, reasoning is good if it be such as to give a true conclusion from true premises, and not otherwise.—*C. S. Peirce.*

Truth can never be told so as to be understood, and not be believed.—*W. Blake.*

FOREWORD

In this essay the Theory of Probability is taken as a branch of logic, the logic of partial belief and inconclusive argument; but there is no intention of implying that this is the only or even the most important aspect of the subject. Probability is of fundamental importance not only in logic but also in statistical and physical science, and we cannot be sure beforehand that the most useful interpretation of it in logic will be appropriate in physics also. Indeed the general difference of opinion between statisticians who for the most part adopt the frequency theory of probability and logicians who mostly reject it renders it likely that the two schools are really discussing different things, and that the word 'probability' is used by logicians in one sense and by statisticians in another. The conclusions we shall come to as to the meaning of probability in logic must not, therefore, be taken as prejudging its meaning in physics.[1]

[1] [A final chapter, on probability in science, was designed but not written.—ED.]

CONTENTS

(1) THE FREQUENCY THEORY

In the hope of avoiding some purely verbal controversies, I propose to begin by making some admissions in favour of the frequency theory. In the first place this theory must be conceded to have a firm basis in ordinary language, which often uses 'probability' practically as a synonym for proportion; for example, if we say that the probability of recovery from smallpox is three-quarters, we mean, I think, simply that that is the proportion of smallpox cases which r cover. Secondly, if we start with what is called the calculus of probabilities, regarding it first as a branch of pure mathematics, and then looking round for some interpretation of the formulae which shall show that our axioms are consistent and our subject not entirely useless, then much the simplest and least controversial interpretation of the calculus is one in terms of frequencies. This is true not only of the ordinary mathematics of probability, but also of the symbolic calculus developed by Mr. Keynes; for if in his a/h, a and h are taken to be not propositions but propositional functions or class-concepts which define finite classes, and a/h is taken to mean the proportion of members of h which are also members of a, then all his propositions become arithmetical truisms.

4

Besides these two inevitable admissions, there is a third and more important one, which I am prepared to make temporarily although it does not express my real opinion. It is this. Suppose we start with the mathematical calculus, and ask, not as before what interpretation of it is most convenient to the pure mathematicism, but what interpretation gives results of greatest value to science in general, then it may be that the answer is again an interpretation in terms of frequency; that probability as it is used in statistical theories, especially in statistical mechanics—the kind of probability whose logarithm is the entropy—is really a ratio between the numbers of two classes, or the limit of such a ratio. I do not myself believe this, but I am willing for the present to concede to the frequency theory that probability as used in modern science is really the same as frequency.

But, supposing all this admitted, it still remains the case that we have the authority both of ordinary language and of many great thinkers for discussing under the heading of probability what appears to be quite a different subject, the logic of partial belief. It may be that, as some supporters of the frequency theory have maintained, the logic of partial belief will be found in the end to be merely the study of frequencies, either because partial belief is definable as, or by reference to, some sort of frequency, or because it can only be the subject of logical treatment when it is grounded on experienced frequencies. Whether these contentions are valid can, however, only be decided as a result of our investigation into partial belief, so that I propose to ignore the frequency theory for the present and begin an inquiry into the logic of partial belief. In this, I think, it will be most convenient if, instead of straight away developing my own theory I begin by examining the views of Mr Keynes, which are so well known and in essentials so widely accepted that readers probably feel that there is no ground for re-opening the subject *de novo* until they have been disposed of.

(2) MR KEYNES' THEORY

Mr Keynes[1] starts from the supposition that we make probable inferences for which we claim objective validity ; we proceed from full belief in one proposition to partial belief in another, and we claim that this procedure is objectively right, so that if another man in similar circumstances entertained a different degree of belief, he would be wrong in doing so. Mr Keynes accounts for this by supposing that between any two propositions, taken as premiss and conclusion, there holds one and only one relation of a certain sort called probability relations ; and that if, in any given case, the relation is that of degree a, from full belief in the premiss, we should, if we were rational, proceed to a belief of degree a in the conclusion.

Before criticising this view, I may perhaps be allowed to point out an obvious and easily corrected defect in the statement of it. When it is said that the degree of the probability relation is the same as the degree of belief which it justifies, it seems to be presupposed that both probability relations, on the one hand, and degrees of belief on the other can be naturally expressed in terms of numbers, and then that the number expressing or measuring the probability relation is the same as that expressing the appropriate degree of belief. But if, as Mr. Keynes holds, these things are not always expressible by numbers, then we cannot give his statement that the degree of the one is the same as the degree of the other such a simple interpretation, but must suppose him to mean only that there is a one-one correspondence between probability relations and the degrees of belief which

[1] J. M. Keynes, *A Treatise on Probability* (1921).

6

they justify. This correspondence must clearly preserve the relations of greater and less, and so make the manifold of probability relations and that of degrees of belief similar in Mr Russell's sense. I think it is a pity that Mr Keynes did not see this clearly, because the exactitude of this correspondence would have provided quite as worthy material for his scepticism as did the numerical measurement of probability relations. Indeed some of his arguments against their numerical measurement appear to apply quite equally well against their exact correspondence with degrees of belief ; for instance, he argues that if rates of insurance correspond to subjective, i.e. actual, degrees of belief, these are not rationally determined, and we cannot infer that probability relations can be similarly measured. It might be argued that the true conclusion in such a case was not that, as Mr Keynes thinks, to the non-numerical probability relation corresponds a non-numerical degree of rational belief, but that degrees of belief, which were always numerical, did not correspond one to one with the probability relations justifying them. For it is, I suppose, conceivable that degrees of belief could be measured by a psychogalvanometer or some such instrument, and Mr Keynes would hardly wish it to follow that probability relations could all be derivatively measured with the measures of the beliefs which they justify.

But let us now return to a more fundamental criticism of Mr. Keynes' views, which is the obvious one that there really do not seem to be any such things as the probability relations he describes. He supposes that, at any rate in certain cases, they can be perceived ; but speaking for myself I feel confident that this is not true. I do not perceive them, and if I am to be persuaded that they exist it must be by argument ; moreover I shrewdly suspect that others do not perceive them either, because they are able to come to so very little agreement as to which of them relates any two given propositions.

7

All we appear to know about them are certain general propositions, the laws of addition and multiplication ; it is as if
everyone knew the laws of geometry but no one could tell
whether any given object were round or square ; and I find
it hard to imagine how so large a body of general knowledge
can be combined with so slender a stock of particular facts.
It is true that about some particular cases there is agreement,
but these somehow paradoxically are always immensely complicated ; we all agree that the probability of a coin coming
down heads is $\frac{1}{2}$, but we can none of us say exactly what is
the evidence which forms the other term for the probability
relation about which we are then judging. If, on the other
hand, we take the simplest possible pairs of propositions such
as ' This is red ' and ' That is blue ' or ' This is red ' and
' That is red ', whose logical relations should surely be
easiest to see, no one, I think, pretends to be sure what is the
probability relation which connects them. Or, perhaps, they
may claim to see the relation but they will not be able to say
anything about it with certainty, to state if it is more or
less than $\frac{1}{3}$, or so on. They may, of course, say that it is
incomparable with any numerical relation, but a relation
about which so little can be truly said will be of little scientific
use and it will be hard to convince a sceptic of its
existence. Besides this view is really rather paradoxical ;
for any believer in induction must admit that between ' This is
red ' as conclusion and ' This is round ', together with a billion
propositions of the form ' a is round and red ' as evidence,
there is a finite probability relation ; and it is hard to suppose
that as we accumulate instances there is suddenly a point,
say after 233 instances, at which the probability relation
becomes finite and so comparable with some numerical relations.

It seems to me that if we take the two propositions ' a
is red ', ' b is red ', we cannot really discern more than four

8

simple logical relations between them ; namely identity of form, identity of predicate, diversity of subject, and logical independence of import. If anyone were to ask me what probability one gave to the other, I should not try to answer by contemplating the propositions and trying to discern a logical relation between them, I should, rather, try to imagine that one of them was all that I knew, and to guess what degree of confidence I should then have in the other. If I were able to do this, I might no doubt still not be content with it but might say ' This is what I should think, but, of course, I am only a fool ' and proceed to consider what a wise man would think and call that the degree of probability. This kind of self-criticism I shall discuss later when developing my own theory ; all that I want to remark here is that no one estimating a degree of probability simply contemplates the two propositions supposed to be related by it ; he always considers *inter alia* his own actual or hypothetical degree of belief. This remark seems to me to be borne out by observation of my own behaviour ; and to be the only way of accounting for the fact that we can all give estimates of probability in cases taken from actual life, but are quite unable to do so in the logically simplest cases in which, were probability a logical relation, it would be easiest to discern.

Another argument against Mr Keynes' theory can, I think, be drawn from his inability to adhere to it consistently even in discussing first principles. There is a passage in his chapter on the measurement of probabilities which reads as follows :—

" Probability is, *vide* Chapter II (§ 12), relative in a sense to the principles of *human* reason. The degree of probability, which it is rational for *us* to entertain, does not presume perfect logical insight, and is relative in part to the secondary propositions which we in fact know ; and it is not dependent upon whether more perfect logical insight

9

is or is not conceivable. It is the degree of probability to which those logical processes lead, of which our minds are capable ; or, in the language of Chapter II, which those secondary propositions justify, which we in fact know. If we do not take this view of probability, if we do not limit it in this way and make it, to this extent, relative to human powers, we are altogether adrift in the unknown ; for we cannot ever know what degree of probability would be justified by the perception of logical relations which we are, and must always be, incapable of comprehending." [1]

This passage seems to me quite unreconcilable with the view which Mr Keynes adopts everywhere except in this and another similar passage. For he generally holds that the degree of belief which we are justified in placing in the conclusion of an argument is determined by what relation of probability unites that conclusion to our premises. There is only one such relation and consequently only one relevant true secondary proposition, which, of course, we may or may not know, but which is necessarily independent of the human mind. If we do not know it, we do not know it and cannot tell how far we ought to believe the conclusion. But often, he supposes, we do know it ; probability relations are not ones which we are incapable of comprehending. But on this view of the matter the passage quoted above has no meaning : the relations which justify probable beliefs are probability relations, and it is nonsense to speak of them being justified by logical relations which we are, and must always be, incapable of comprehending.

The significance of the passage for our present purpose lies in the fact that it seems to presuppose a different view of probability, in which indefinable probability relations play no part, but in which the degree of rational belief depends on a variety of logical relations. For instance, there might be between the premiss and conclusion the relation

[1] p. 32, his italics.

that the premiss was the logical product of a thousand instances of a generalization of which the conclusion was one other instance, and this relation, which is not an indefinable probability relation but definable in terms of ordinary logic and so easily recognizable, might justify a certain degree of belief in the conclusion on the part of one who believed the premiss. We should thus have a variety of ordinary logical relations justifying the same or different degrees of belief. To say that the probability of *a* given *h* was such-and-such would mean that between *a* and *h* was some relation justifying such-and-such a degree of belief. And on this view it would be a real point that the relation in question must not be one which the human mind is incapable of comprehending.

This second view of probability as depending on logical relations but not itself a new logical relation seems to me more plausible than Mr Keynes' usual theory; but this does not mean that I feel at all inclined to agree with it. It requires the somewhat obscure idea of a logical relation justifying a degree of belief, which I should not like to accept as indefinable because it does not seem to be at all a clear or simple notion. Also it is hard to say what logical relations justify what degrees of belief, and why; any decision as to this would be arbitrary, and would lead to a logic of probability consisting of a host of so-called 'necessary' facts, like formal logic on Mr Chadwick's view of logical constants.[1] Whereas I think it far better to seek an explanation of this 'necessity' after the model of the work of Mr Wittgenstein, which enables us to see clearly in what precise sense and why logical propositions are necessary, and in a general way why the system of formal logic consists of the propositions it does consist of, and what is their common characteristic. Just as natural science tries to explain and

[1] J. A. Chadwick, " Logical Constants," *Mind*, 1927.

11

account for the   facts of nature, so philosophy should
try, in a sense, to explain and account for the facts
of logic ; a task ignored by the philosophy which dismisses
these facts as being unaccountably and in an indefinable
sense ' necessary '.

Here I propose to conclude this criticism of Mr Keynes'
theory, not because there are not other respects in which it
seems open to objection, but because I hope that what I have
already said is enough to show that it is not so completely
satisfactory as to render futile any attempt to treat the
subject from a rather different point of view.

(3) DEGREES OF BELIEF

The subject of our inquiry is the logic of partial belief,
and I do not think we can carry it far unless we have
at least an approximate notion of what partial belief is, and
how, if at all, it can be measured.   It will not be very
enlightening to be told that in such circumstances it would
be rational to believe a proposition to the extent of
$\frac{2}{3}$, unless we know what sort of a belief in it that means.
We must therefore try to develop a purely psychological
method of measuring belief.   It is not enough to measure
probability ; in order to apportion correctly our belief to the
probability we must also be able to measure our belief.

It is a common view that belief and other psychological
variables are not measurable, and if this is true our inquiry
will be vain ; and so will the whole theory of probability
conceived as a logic of partial belief; for if the phrase ' a
belief two-thirds of certainty ' is meaningless, a calculus
whose sole object is to enjoin such beliefs will be meaningless
also.   Therefore unless we are prepared to give up the whole
thing as a bad job we are bound to hold that beliefs can to
some extent be measured.   If we were to follow the analogy

of Mr Keynes' treatment of probabilities we should say that some beliefs were measurable and some not ; but this does not seem to me likely to be a correct account of the matter ; I do not see how we can sharply divide beliefs into those which have a position in the numerical scale and those which have not. But I think beliefs do differ in measurability in the following two ways. First, some beliefs can be measured more accurately than others; and, secondly, the measurement of beliefs is almost certainly an ambiguous process leading to a variable answer depending on how exactly the measurement is conducted. The degree of a belief is in this respect like the time interval between two events ; before Einstein it was supposed that all the ordinary ways of measuring a time interval would lead to the same result if properly performed. Einstein showed that this was not the case ; and time interval can no longer be regarded as an exact notion, but must be discarded in all precise investigations. Nevertheless, time interval and the Newtonian system are sufficiently accurate for many purposes and easier to apply.

I shall try to argue later that the degree of a belief is just like a time interval ; it has no precise meaning unless we specify more exactly how it is to be measured. But for many purposes we can assume that the alternative ways of measuring it lead to the same result, although this is only approximately true. The resulting discrepancies are more glaring in connection with some beliefs than with others, and these therefore appear less measurable. Both these types of deficiency in measurability, due respectively to the difficulty in getting an exact enough measurement and to an important ambiguity in the definition of the measurement process, occur also in physics and so are not difficulties peculiar to our problem ; what is peculiar is that it is difficult to form any idea of how the measurement is to be conducted, how a unit is to be obtained, and so on.

13

Let us then consider what is implied in the measurement of beliefs. A satisfactory system must in the first place assign to any belief a magnitude or degree having a definite position in an order of magnitudes ; beliefs which are of the same degree as the same belief must be of the same degree as one another, and so on. Of course this cannot be accomplished without introducing a certain amount of hypothesis or fiction. Even in physics we cannot maintain that things that are equal to the same thing are equal to one another unless we take 'equal' not as meaning 'sensibly equal' but a fictitious or hypothetical relation. I do not want to discuss the metaphysics or epistemology of this process, but merely to remark that if it is allowable in physics it is allowable in psychology also. The logical simplicity characteristic of the relations dealt with in a science is never attained by nature alone without any admixture of fiction.

But to construct such an ordered series of degrees is not the whole of our task ; we have also to assign numbers to these degrees in some intelligible manner. We can of course easily explain that we denote full belief by 1, full belief in the contradictory by 0, and equal beliefs in the proposition and its contradictory by $\frac{1}{2}$. But it is not so easy to say what is meant by a belief $\frac{2}{3}$ of certainty, or a belief in the proposition being twice as strong as that in its contradictory. This is the harder part of the task, but it is absolutely necessary ; for we do calculate numerical probabilities, and if they are to correspond to degrees of belief we must discover some definite way of attaching numbers to degrees of belief. In physics we often attach numbers by discovering a physical process of addition [1] : the measure-numbers of lengths are not assigned arbitrarily subject only to the proviso that the greater length shall have the greater measure ; we determine them further by deciding on a

[1] See N. Campbell, *Physics The Elements* (1920), p. 277.

14

physical meaning for addition ;   the length got by putting together two given lengths must have for its measure the sum of their measures.   A system of measurement in which there is nothing corresponding to this is immediately recognized as arbitrary, for instance Mohs' scale of hardness [1] in which 10 is arbitrarily assigned to diamond, the hardest known material, 9 to the next hardest, and so on. We have therefore to find a process of addition for degrees of belief, or some substitute for this which will be equally adequate to determine a numerical scale.

Such is our problem ;   how are we to solve it ?   There are, I think, two ways in which we can begin.   We can, in the first place, suppose that the degree of a belief is something perceptible by its owner ;   for instance that beliefs differ in the intensity of a feeling by which they are accompanied, which might be called a belief-feeling or feeling of conviction, and that by the degree of belief we mean the intensity of this feeling.   This view would be very inconvenient, for it is not easy to ascribe numbers to the intensities fo feelings ; but apart from this it seems to me observably false, for the beliefs which we hold most strongly are often accompanied by practically no feeling at all ;   no one feels strongly about things he takes for granted.

We are driven therefore to the second supposition that the degree of a belief is a causal property of it, which we can express vaguely as the extent to which we are prepared to act on it.   This is a generalization of the well-known view, that the differentia of belief lies in its causal efficacy, which is discussed by Mr Russell in his *Analysis of Mind*. He there dismisses it for two reasons, one of which seems entirely to miss the point.   He argues that in the course of trains of thought we believe many things which do not lead to action.   This objection is however beside the mark, because

[1] Ibid., p. 271.

15

it is not asserted that a belief is an idea which does actually lead to action, but one which would lead to action in suitable circumstances ; just as a lump of arsenic is called poisonous not because it actually has killed or will kill anyone, but because it would kill anyone if he ate it.   Mr Russell's second argument is, however, more formidable.   He points out that it is not possible to suppose that beliefs differ from other ideas only in their effects, for if they were otherwise identical their effects would be identical also.   This is perfectly true, but it may still remain the case that the nature of the difference between the causes is entirely unknown or very vaguely known, and that what we want to talk about is the difference between the effects, which is readily observable and important.

As soon as we regard belief quantatively, this seems to me the only view we can take of it.   It could well be held that the difference between believing and not believing lies in the presence or absence of introspectible feelings.   But when we seek to know what is the difference between believing more firmly and believing less firmly, we can no longer regard it as consisting in having more or less of certain observable feelings ;   at least I personally cannot recognize any such feelings.   The difference seems to me to lie in how far we should act on these beliefs : this may depend on the degree of some feeling or feelings, but I do not know exactly what feelings and I do not see that it is indispensable that we should know.   Just the same thing is found in physics ; men found that a wire connecting plates of zinc and copper standing in acid deflected a magnetic needle in its neighbourhood. Accordingly as the needle was more or less deflected the wire was said to carry a larger or a smaller current.   The nature of this ' current ' could only be conjectured : what were observed and measured were simply its effects.

It will no doubt be objected that we know how strongly

we believe things, and that we can only know this if we can measure our belief by introspection. This does not seem to me necessarily true ; in many cases, I think, our judgment about the strength of our belief is really about how we should act in hypothetical circumstances. It will be answered that we can only tell how we should act by observing the present belief-feeling which determines how we should act ; but again I doubt the cogency of the argument. It is possible that what determines how we should act determines us also directly or indirectly to have a correct opinion as to how we should act, without its ever coming into consciousness.

Suppose, however, I am wrong about this and that we can decide by introspection the nature of belief, and measure its degree ; still, I shall argue, the kind of measurement of belief with which probability is concerned is not this kind but is a measurement of belief *qua* basis of action. This can I think be shown in two ways. First, by considering the scale of probabilities between 0 and 1, and the sort of way we use it, we shall find that it is very appropriate to the measurement of belief as a basis of action, but in no way related to the measurement of an introspected feeling. For the units in terms of which such feelings or sensations are measured are always, I think, differences which are just perceptible : there is no other way of obtaining units. But I see no ground for supposing that the interval between a belief of degree $\frac{1}{3}$ and one of degree $\frac{1}{2}$ consists of as many just perceptible changes as does that between one of $\frac{2}{3}$ and one of $\frac{5}{6}$, or that a scale based on just perceptible differences would have any simple relation to the theory of probability. On the other hand the probability of $\frac{1}{3}$ is clearly related to the kind of belief which would lead to a bet of 2 to 1, and it will be shown below how to generalize this relation so as to apply to action in general. Secondly, the quantitative aspects of beliefs as the basis of action are evidently more important than the intensities of belief-feelings.

17

The latter are no doubt interesting, but may be very variable from individual to individual, and their practical interest is entirely due to their position as the hypothetical causes of beliefs *qua* bases of action.

It is possible that some one will say that the extent to which we should act on a belief in suitable circumstances is a hypothetical thing, and therefore not capable of measurement. But to say this is merely to reveal ignorance of the physical sciences which constantly deal with and measure hypothetical quantities ; for instance, the electric intensity at a given point is the force which would act on a unit charge if it were placed at the point.

Let us now try to find a method of measuring beliefs as bases of possible actions. It is clear that we are concerned with dispositional rather than with actualized beliefs ; that is to say, not with beliefs at the moment when we are thinking of them, but with beliefs like my belief that the earth is round, which I rarely think of, but which would guide my action in any case to which it was relevant.

The old-established way of measuring a person's belief is to propose a bet, and see what are the lowest odds which he will accept. This method I regard as fundamentally sound ; but it suffers from being insufficiently general, and from being necessarily inexact. It is inexact partly because of the diminishing marginal utility of money, partly because the person may have a special eagerness or reluctance to bet, because he either enjoys or dislikes excitement or for any other reason, e.g. to make a book. The difficulty is like that of separating two different co-operating forces. Besides, the proposal of a bet may inevitably alter his state of opinion ; just as we could not always measure electric intensity by actually introducing a charge and seeing what force it was subject to, because the introduction of the charge would change the distribution to be measured.

In order therefore to construct a theory of quantities of belief which shall be both general and more exact, I propose to take as a basis a general psychological theory, which is now universally discarded, but nevertheless comes, I think, fairly close to the truth in the sort of cases with which we are most concerned. I mean the theory that we act in the way we think most likely to realize the objects of our desires, so that a person's actions are completely determined by his desires and opinions. This theory cannot be made adequate to all the facts, but it seems to me a useful approximation to the truth particularly in the case of our self-conscious or professional life, and it is presupposed in a great deal of our thought. It is a simple theory and one which many psychologists would obviously like to preserve by introducing unconscious desires and unconscious opinions in order to bring it more into harmony with the facts. How far such fictions can achieve the required result I do not attempt to judge : I only claim for what follows approximate truth, or truth in relation to this artificial system of psychology, which like Newtonian mechanics can, I think, still be profitably used even though it is known to be false.

It must be observed that this theory is not to be identified with the psychology of the Utilitarians, in which pleasure had a dominating position. The theory I propose to adopt is that we seek things which we want, which may be our own or other people's pleasure, or anything else whatever, and our actions are such as we think most likely to realize these goods. But this is not a precise statement, for a precise statement of the theory can only be made after we have introduced the notion of quantity of belief.

Let us call the things a person ultimately desires ' goods ', and let us at first assume that they are numerically measurable and additive. That is to say that if he prefers for its own sake an hour's swimming to an hour's reading, he will prefer two hours' swimming to one hour's swimming and one hour's reading. This is of course absurd in the given case but this may only be because swimming and reading are not ultimate goods, and because we cannot imagine a second hour's swimming precisely similar to the first, owing to fatigue, etc.

Let us begin by supposing that our subject has no doubts about anything, but certain opinions about all propositions. Then we can say that he will always choose the course of action which will lead in his opinion to the greatest sum of good.

It should be emphasized that in this essay good and bad are never to be understood in any ethical sense but simply as denoting that to which a given person feels desire and aversion.

The question then arises how we are to modify this simple system to take account of varying degrees of certainty in his beliefs. I suggest that we introduce as a law of psychology that his behaviour is governed by what is called the mathematical expectation ; that is to say that, if p is a proposition about which he is doubtful, any goods or bads for whose realization p is in his view a necessary and sufficient condition enter into his calculations multiplied by the same fraction, which is called the ' degree of his belief in p '. We thus define degree of belief in a way which presupposes the use of the mathematical expectation.

We can put this in a different way. Suppose his degree of belief in p is $\frac{m}{n}$; then his action is such as he would choose it to be if he had to repeat it exactly n times, in m of which p was true, and in the others false. [Here it may be necessary to suppose that in each of the n times he had no memory of the previous ones.]

This can also be taken as a definition of the degree of belief, and can easily be seen to be equivalent to the previous definition. Let us give an instance of the sort of case which might occur. I am at a cross-roads and do not know the way ; but I rather think one of the two ways is right. I propose therefore to go that way but keep my eyes open for someone to ask ; if now I see someone half a mile away over the fields, whether I turn aside to ask* him will depend on the relative inconvenience of going out of my way to cross the fields or of continuing on the wrong road if it is the wrong road. But it will also depend on how confident I am that I am right ; and clearly the more confident I am of this the less distance I should be willing to go from the road to check my opinion. I propose therefore to use the distance I would be prepared to go to ask, as a measure of the confidence of my opinion ; and what I have said above explains how this is to be done. We can set it out as follows : suppose the disadvantage of going x yards to ask is $f(x)$, the advantage of arriving at the right destination is r, that of arriving at the wrong one w. Then if I should just be willing to go a distance d to ask, the degree of my belief that I am on the right road is given by

$$p = 1 - \frac{f(d)}{r - w}.$$

For such an action is one it would just pay me to take, if I had to act in the same way n times, in np of which I was on the right way but in the others not.

For the total good resulting from not asking each time

$$= npr + n(1 - p)w$$
$$= nw + np(r - w),$$

that resulting from asking at distance x each time

$$= nr - nf(x). \qquad \text{[I now always go right.]}$$

This is greater than the preceding expression, provided

$$f(x) < (r - w)(1 - p),$$

∴ the critical distance d is connected with p, the degree of belief, by the relation $f(d) = (r - w)(1 - p)$

$$\text{or } p = 1 - \frac{f(d)}{r - w} \qquad \text{as asserted above.}$$

21

It is easy to see that this way of measuring beliefs gives results agreeing with ordinary ideas; at any rate to the extent that full belief is denoted by 1, full belief in the contradictory by 0, and equal belief in the two by $\frac{1}{2}$. Further, it allows validity to betting as means of measuring beliefs. By proposing a bet on p we give the subject a possible course of action from which so much extra good will result to him if p is true and so much extra bad if p is false. Supposing the bet to be in goods and bads instead of in money, he will take a bet at any better odds than those corresponding to his state of belief; in fact his state of belief is measured by the odds he will just take; but this is vitiated, as already explained, by love or hatred of excitement, and by the fact that the bet is in money and not in goods and bads. Since it is universally agreed that money has a diminishing marginal utility, if money bets are to be used, it is evident that they should be for as small stakes as possible. But then again the measurement is spoiled by introducing the new factor of reluctance to bother about trifles.

Let us now discard the assumption that goods are additive and immediately measurable, and try to work out a system with as few asumptions as possible. To begin with we shall suppose, as before, that our subject has certain beliefs about everything; then he will act so that what he believes to be the total consequences of his action will be the best possible. If then we had the power of the Almighty, and could persuade our subject of our power, we could, by offering him options, discover how he placed in order of merit all possible courses of the world. In this way all possible worlds would be put in an order of value, but we should have no definite way of representing them by numbers. There would be no meaning in the assertion that the difference in value between α and β was equal to that between γ and δ. [Here and elsewhere we use Greek letters to represent the different possible totalities of events between which our subject chooses—the ultimate organic unities.]

Suppose next that the subject is capable of doubt ; then we could test his degree of belief in different propositions by making him offers of the following kind. Would you rather have world a in any event; or world β if p is true, and world γ if p is false? If, then, he were certain that p was true, he would simply compare a and β and choose between them as if no conditions were attached ; but if he were doubtful his choice would not be decided so simply. I propose to lay down axioms and definitions concerning the principles governing choices of this kind. This is, of course, a very schematic version of the situation in real life, but it is, I think, easier to consider it in this form.

There is first a difficulty which must be dealt with ; the propositions like p in the above case which are used as conditions in the options offered may be such that their truth or falsity is an object of desire to the subject. This will be found to complicate the problem, and we have to assume that there are propositions for which this is not the case, which we shall call ethically neutral. More precisely an atomic proposition p is called ethically neutral if two possible worlds differing only in regard to the truth of p are always of equal value ; and a non-atomic proposition p is called ethically neutral if all its atomic truth-arguments [1] are ethically neutral.

We begin by defining belief of degree $\frac{1}{2}$ in an ethically neutral proposition. The subject is said to have belief of degree $\frac{1}{2}$ in such a proposition p if he has no preference between the options (1) a if p is true, β if p is false, and (2) a if p is false, β if p is true, but has a preference between a and β simply. We suppose by an axiom that if this is true of any

---

[1] I assume here Wittgenstein's theory of propositions ; it would probably be possible to give an equivalent definition in terms of any other theory.

one pair α, β it is true of all such pairs.[1]   This comes roughly to defining belief of degree $\frac{1}{2}$ as such a degree of belief as leads to indifference between betting one way and betting the other for the same stakes.

Belief of degree $\frac{1}{2}$ as thus defined can be used to measure values numerically in the following way.   We have to explain what is meant by the difference in value between α and β being equal to that between γ and δ; and we define this to mean that, if p is an ethically neutral proposition believed to degree $\frac{1}{2}$, the subject has no preference between the options (1) α if p is true, δ if p is false, and (2) β if p is true, γ if p is false.

This definition can form the basis of a system of measuring values in the following way :—

Let us call any set of all worlds equally preferable to a given world a value : we suppose that if world α is preferable to β any world with the same value as α is preferable to any world with the same value as β and shall say that the value of α is greater than that of β.   This relation ' greater than ' orders values in a series.   We shall use α henceforth both for the world and its value.

Axioms.

(1) There is an ethically neutral proposition p believed to degree $\frac{1}{2}$.

(2) If p, q are such propositions and the option

α if p, δ if not-p is equivalent to β if p, γ if not-p

then α if q, δ if not-q is equivalent to β if q, γ if not-q.

Def.   In the above case we say $\alpha\beta = \gamma\delta$.

Theorems.   If $\alpha\beta = \gamma\delta$,

$$\text{then } \beta\alpha = \delta\gamma, \ \alpha\gamma = \beta\delta, \ \gamma\alpha = \delta\beta.$$

---

[1] α and β must be supposed so far undefined as to be compatible with both p and not-p.

(2a) If $a\beta = \gamma\delta$, then $a > \beta$ is equivalent to $\gamma > \delta$

and $a = \beta$ is equivalent to $\gamma = \delta$.

(3) If option A is equivalent to option B and B to C then A to C.

Theorem. If $a\beta = \gamma\delta$ and $\beta\eta = \zeta\gamma$,

then $a\eta = \zeta\delta$.

(4) If $a\beta = \gamma\delta$, $\gamma\delta = \eta\zeta$, then $a\beta = \eta\zeta$.

(5) (a, β, γ). E ! $(\imath x)$ $(ax = \beta\gamma)$.

(6) (a, β). E ! $(\imath x)$ $(ax = x\beta)$.

(7) Axiom of continuity :—Any progression has a limit (ordinal).

(8) Axiom of Archimedes.

These axioms enable the values to be correlated one-one with real numbers so that if a^1 corresponds to a, etc.

$$a\beta = \gamma\delta \; . \; \equiv \; . \; a^1 - \beta^1 = \gamma^1 - \delta^1.$$

Henceforth we use a for the correlated real number a^1 also.

Having thus defined a way of measuring value we can now derive a way of measuring belief in general. If the option of a for certain is indifferent with that of β if p is true and γ if p is false,[1] we can define the subject's degree of belief in p as the ratio of the difference between a and γ to that between β and γ ; which we must suppose the same for all a's, β's and γ's that satisfy the conditions. This amounts roughly

[1] Here β must include the truth of p, γ its falsity ; p need no longer be ethically neutral. But we have to assume that there is a world with any assigned value in which p is true, and one in which p is false.

to defining the degree of belief in p by the odds at which the subject would bet on p, the bet being conducted in terms of differences of value as defined. The definition only applies to partial belief and does not include certain beliefs ; for belief of degree 1 in p, a for certain is indifferent with a if p and any β if not-p.

We are also able to define a very useful new idea—' the degree of belief in p given q '. This does not mean the degree of belief in ' If p then q ', or that in ' p entails q ', or that which the subject would have in p if he knew q, or that which he ought to have. It roughly expresses the odds at which he would now bet on p, the bet only to be valid if q is true. Such conditional bets were often made in the eighteenth century.

The degree of belief in p given q is measured thus. Suppose the subject indifferent between the options (1) a if q true, β if q false, (2) γ if p true and q true, δ if p false and q true, β if q false. Then the degree of his belief in p given q is the ratio of the difference between a and δ to that between γ and δ, which we must suppose the same for any a, β, γ, δ which satisfy the given conditions. This is not the same as the degree to which he would believe p, if he believed q for certain ; for knowledge of q might for psychological reasons profoundly alter his whole system of beliefs.

Each of our definitions has been accompanied by an axiom of consistency, and in so far as this is false, the notion of the corresponding degree of belief becomes invalid. This bears some analogy to the situation in regard to simultaneity discussed above.

I have not worked out the mathematical logic of this in detail, because this would, I think, be rather like working out to seven places of decimals a result only valid to two. My logic cannot be regarded as giving more than the sort of way it might work.

From these definitions and axioms it is possible to prove the fundamental laws of probable belief (degrees of belief lie between 0 and 1) :

(1) Degree of belief in p + degree of belief in $\bar{p} = 1$.

(2) Degree of belief in p given q + degree of belief in $\bar{p}$ given $q = 1$.

(3) Degree of belief in (p and q) = degree of belief in p × degree of belief in q given p.

(4) Degree of belief in (p and q) + degree of belief in (p and $\bar{q}$) = degree of belief in p.

The first two are immediate. (3) is proved as follows.

Let degree of belief in $p = x$, that in q given $p = y$.

Then ξ for certain $\equiv \xi + (1 - x)t$ if p true, $\xi - xt$ if p false, for any t.

$\xi + (1 - x)\, t$ if p true $\equiv$

$$\begin{cases} \xi + (1 - x)\, t + (1 - y)\, u \text{ if `}p \text{ and } q\text{ ' true,} \\ \xi + (1 - x)\, t - yu \text{ if } p \text{ true } q \text{ false ;} \qquad \text{for any } u. \end{cases}$$

Choose u so that $\xi + (1 - x)\, t - yu = \xi - xt$,

i.e. let $u = t/y\ (y \neq 0)$

Then ξ for certain $\equiv$

$$\begin{cases} \xi + (1 - x)\, t + (1 - y)\, t/y \text{ if } p \text{ and } q \text{ true} \\ \xi - xt \text{ otherwise,} \end{cases}$$

$\therefore$ degree of belief in `p and q' $= \dfrac{xt}{t + (1 - y)\, t/y} = xy. \ (t \neq 0)$

If $y = 0$, take $t = 0$.

27

Then ξ for certain $\equiv \xi$ if p true, ξ if p false

$$\equiv \xi + u \text{ if } p \text{ true, } q \text{ true ; } \xi \text{ if } p \text{ false, } q$$
$$\text{false ; } \xi \text{ if } p \text{ false}$$

$$\equiv \xi + u, pq \text{ true ; } \xi, pq \text{ false}$$

$\therefore$ degree of belief in $pq = 0$.

(4) follows from (2), (3) as follows :—

Degree of belief in $pq =$ that in $p \times$ that in q given p, by (3). Similarly degree of belief in $p\bar{q} =$ that in $p \times$ that in $\bar{q}$ given p $\therefore$ sum $=$ degree of belief in p, by (2).

These are the laws of probability, which we have proved to be necessarily true of any consistent set of degrees of belief. Any definite set of degrees of belief which broke them would be inconsistent in the sense that it violated the laws of preference between options, such as that preferability is a transitive asymmetrical relation, and that if a is preferable to β, β for certain cannot be preferable to a if p, β if not-p. If anyone's mental condition violated these laws, his choice would depend on the precise form in which the options were offered him, which would be absurd. He could have a book made against him by a cunning better and would then stand to lose in any event.

We find, therefore, that a precise account of the nature of partial belief reveals that the laws of probability are laws of consistency, an extension to partial beliefs of formal logic, the logic of consistency. They do not depend for their meaning on any degree of belief in a proposition being uniquely determined as the rational one ; they merely distinguish those sets of beliefs which obey them as consistent ones.

Having any definite degree of belief implies a certain measure of consistency, namely willingness to bet on a given proposition at the same odds for any stake, the stakes being measured

28

in terms of ultimate values. Having degrees of belief obeying the laws of probability implies a further measure of consistency, namely such a consistency between the odds acceptable on different propositions as shall prevent a book being made against you.

Some concluding remarks on this section may not be out of place. First, it is based fundamentally on betting, but this will not seem unreasonable when it is seen that all our lives we are in a sense betting. Whenever we go to the station we are betting that a train will really run, and if we had not a sufficient degree of belief in this we should decline the bet and stay at home. The options God gives us are always conditional on our guessing whether a certain proposition is true. Secondly, it is based throughout on the idea of mathematical expectation ; the dissatisfaction often felt with this idea is due mainly to the inaccurate measurement of goods. Clearly mathematical expectations in terms of money are not proper guides to conduct. It should be remembered, in judging my system, that in it value is actually defined by means of mathematical expectation in the case of beliefs of degree $\frac{1}{2}$, and so may be expected to be scaled suitably for the valid application of the mathematical expectation in the case of other degrees of belief also.

Thirdly, nothing has been said about degrees of belief when the number of alternatives is infinite. About this I have nothing useful to say, except that I doubt if the mind is capable of contemplating more than a finite number of alternatives. It can consider questions to which an infinite number of answers are possible, but in order to consider the answers it must lump them into a finite number of groups. The difficulty becomes practically relevant when discussing induction, but even then there seems to me no need to introduce it. We can discuss whether past experience gives a high probability to the sun's rising to-morrow without bothering about what probability it gives to the sun's rising each morning for evermore. For this reason I cannot but feel that Mr Ritchie's discussion of the problem [1] is unsatisfactory; it is true that we can agree that inductive generalizations need have no finite probability, but particular expectations entertained on inductive grounds undoubtedly do have a high numerical probability in the minds of all of us. We all are more certain that the sun will rise to-morrow than that I shall not throw 12 with two dice first time, i.e. we have a belief of higher degree than $\frac{35}{36}$ in it. If induction ever needs a logical justification it is in connection with the probability of an event like this.

(4) THE LOGIC OF CONSISTENCY

We may agree that in some sense it is the business of logic to tell us what we ought to think; but the interpretation of this statement raises considerable difficulties. It may be said that we ought to think what is true, but in that sense we are told what to think by the whole of science and not merely by logic. Nor, in this sense, can any justification be found for partial belief; the ideally best thing is that we should have beliefs of degree 1 in all true propositions and beliefs of degree 0 in all false propositions. But this is too high a standard to expect of mortal men, and we must agree that some degree of doubt or even of error may be humanly speaking justified.

------

[1] A. D. Ritchie, "Induction and Probability," *Mind*, 1926, p. 318. 'The conclusion of the foregoing discussion may be simply put. If the problem of induction be stated to be "How can inductive generalizations acquire a large numerical probability?" then this is a pseudo-problem, because the answer is "They cannot". This answer is not, however, a denial of the validity of induction but is a direct consequence of the nature of probability. It still leaves untouched the real problem of induction which is "How can the probability of an induction be increased?" and it leaves standing the whole of Keynes' discussion on this point.'

30

Many logicians, I suppose, would accept as an account of their science the opening words of Mr Keynes' *Treatise on Probability* : " Part of our knowledge we obtain direct ; and part by argument. The Theory of Probability is concerned with that part which we obtain by argument, and it treats of the different degrees in which the results so obtained are conclusive or inconclusive." Where Mr Keynes says ' the Theory of Probability ', others would say Logic. It is held, that is to say, that our opinions can be divided into those we hold immediately as a result of perception or memory, and those which we derive from the former by argument. It is the business of Logic to accept the former class and criticize merely the derivation of the second class from them.

Logic as the science of argument and inference is traditionally and rightly divided into deductive and inductive ; but the difference and relation between these two divisions of the subject can be conceived in extremely different ways. According to Mr Keynes valid deductive and inductive arguments are fundamentally alike ; both are justified by logical relations between premiss and conclusion which differ only in degree. This position. as I have already explained, I cannot accept. I do not see what these inconclusive logical relations can be or how they can justify partial beliefs. In the case of conclusive logical arguments I can accept the account of their validity which has been given by many authorities, and can be found substantially the same in Kant, De Morgan, Peirce and Wittgenstein. All these authors agree that the conclusion of a formally valid argument is contained in its premisses ; that to deny the conclusion while accepting the premisses would be self-contradictory ; that a formal deduction does not increase our knowledge, but only brings out clearly what we already know in another form ; and that we are bound to accept its validity on pain of being inconsistent with ourselves. The logical relation which justifies the inference is that the sense or import of the conclusion is contained in that of the premisses. But in the case of an inductive argument this does not happen in the least; it is impossible to represent it as resembling a deductive argument and merely weaker in degree ; it is absurd to say that the sense of the conclusion is partially contained in that of the premisses. We could accept the premisses and utterly reject the conclusion without any sort of inconsistency or contradiction.

It seems to me, therefore, that we can divide arguments into two radically different kinds, which we can distinguish in the words of Peirce as (1) ' explicative, analytic, or deductive ' and (2) ' amplifiative, synthetic, or (loosely speaking) inductive '.[1] Arguments of the second type are from an important point of view much closer to memories and perceptions than to deductive arguments. We can regard perception, memory and induction as the three fundamental ways of acquiring knowledge ; deduction on the other hand is merely a method of arranging our knowledge and eliminating inconsistencies or contradictions.

Logic must then fall very definitely into two parts: (excluding analytic logic, the theory of terms and propositions) we have the lesser logic, which is the logic of consistency, or formal logic ; and the larger logic, which is the logic of discovery, or inductive logic.

What we have now to observe is that this distinction in no way coincides with the distinction between certain and partial beliefs ; we have seen that there is a theory of consistency in partial beliefs just as much as of consistency in certain beliefs, although for various reasons the former is not so important as the latter. The theory of probability is in fact a generalization of formal logic ; but in the process

---

[1] C. S. Peirce, *Chance Love and Logic*, p. 92.

of generalization one of the most important aspects of formal logic is destroyed. If p and $\bar{q}$ are inconsistent so that q follows logically from p, that p implies q is what is called by Wittgenstein a ' tautology ' and can be regarded as a degenerate case of a true proposition not involving the idea of consistency. This enables us to regard (not altogether correctly) formal logic including mathematics as an objective science consisting of objectively necessary propositions. It thus gives us not merely the ἀνάγκη λέγειν, that if we assert p we are bound in consistency to assert q also, but also the ἀνάγκη εἶναι, that if p is true, so must q be. But when we extend formal logic to include partial beliefs this direct objective interpetation is lost ; if we believe pq to the extent of $\frac{1}{3}$, and $p\bar{q}$ to the extent of $\frac{1}{3}$, we are bound in consistency to believe p also to the extent of $\frac{2}{3}$. This is the ἀνάγκη λέγειν ; but we cannot say that if pq is $\frac{1}{3}$ true and $p\bar{q}$ $\frac{1}{3}$ true, p also must be $\frac{2}{3}$ true, for such a statement would be sheer nonsense. There is no corresponding ἀνάγκη εἶναι. Hence, unlike the calculus of consistent full belief, the calculus of objective partial belief cannot be immediately interpreted as a body of objective tautology.

This is, however, possible in a roundabout way ; we saw at the beginning of this essay that the calculus of probabilities could be interpreted in terms of class-ratios ; we have now found that it can also be interpreted as a calculus of consistent partial belief. It is natural, therefore, that we should expect some intimate connection between these two interpretations, some explanation of the possibility of applying the same mathematical calculus to two such different sets of phenomena. Nor is an explanation difficult to find ; there are many connections between partial beliefs and frequencies. For instance, experienced frequencies often lead to corresponding partial beliefs, and partial beliefs lead to the expectation of corresponding frequencies in accordance with Bernouilli's

Theorem. But neither of these is exactly the connection we want ; a partial belief cannot in general be connected uniquely with any actual frequency, for the connection is always made by taking the proposition in question as an instance of a propositional function. What propositional function we choose is to some extent arbitrary and the corresponding frequency will vary considerably with our choice. The pretensions of some exponents of the frequency theory that partial belief means full belief in a frequency proposition cannot be sustained. But we found that the very idea of partial belief involves reference to a hypothetical or ideal frequency ; supposing goods to be additive, belief of degree $\frac{m}{n}$ is the sort of belief which leads to the action which would be best if repeated n times in m of which the proposition is true ; or we can say more briefly that it is the kind of belief most appropriate to a number of hypothetical occasions otherwise identical in a proportion $\frac{m}{n}$ of which the proposition in question is true. It is this connection between partial belief and frequency which enables us to use the calculus of frequencies as a calculus of consistent partial belief. And in a sense we may say that the two interpretations are the objective and subjective aspects of the same inner meaning, just as formal logic can be interpreted objectively as a body of tautology and subjectively as the laws of consistent thought.

We shall, I think, find that this view of the calculus of probability removes various difficulties that have hitherto been found perplexing. In the first place it gives us a clear justification for the axioms of the calculus, which on such a system as Mr Keynes' is entirely wanting. For now it is easily seen that if partial beliefs are consistent they will obey these axioms, but it is utterly obscure why Mr Keynes' mysterious logical relations should obey them.[1] We should be so curiously ignorant of the instances of these relations, and so curiously knowledgeable about their general laws.

Secondly, the Principle of Indifference can now be altogether dispensed with ; we do not regard it as belonging to formal logic to say what should be a man's expectation of drawing a white or a black ball from an urn ; his original expectations may within the limits of consistency be any he likes ; all we have to point out is that if he has certain expectations he is bound in consistency to have certain others. This is simply bringing probability into line with ordinary formal logic, which does not criticize premises but merely declares that certain conclusions are the only ones consistent with them. To be able to turn the Principle of Indifference out of formal logic is a great advantage ; for it is fairly clearly impossible to lay down purely logical conditions for its validity, as is attempted by Mr Keynes. I do not want to discuss this question in detail, because it leads to hair-splitting and arbitrary distinctions which could be discussed for ever. But anyone who tries to decide by Mr Keynes' methods what are the proper alternatives to regard as equally probable in molecular mechanics, e.g. in Gibbs' phase-space, will soon be convinced that it is a matter of physics rather than pure logic. By using the multiplication formula, as it is used in inverse probability, we can on Mr Keynes' theory reduce all probabilities to quotients of *a priori* probabilities ; it is therefore in regard to these latter that the Principle of Indifference is of primary importance ; but here the question is obviously not one of formal logic. How can we on merely

---

[1] It appears in Mr Keynes' system as if the principal axioms—the laws of addition and multiplication—were nothing but definitions. This is merely a logical mistake ; his definitions are formally invalid unless corresponding axioms are presupposed. Thus his definition of multiplication presupposes the law that if the probability of a given bh is equal to that of c given dh, and the probability of b given h is equal to that of d given k, then will the probabilities of ab given h and of cd given k be equal.

logical grounds divide the spectrum into equally probable bands ?

A third difficulty which is removed by our theory is the one which is presented to Mr Keynes' theory by the following case. I think I perceive or remember something but am not sure ; this would seem to give me some ground for believing it, contrary to Mr Keynes' theory, by which the degree of belief in it which it would be rational for me to have is that given by the probability relation between the proposition in question and the things I know for certain. He cannot justify a probable belief founded not on argument but on direct inspection. In our view there would be nothing contrary to formal logic in such a belief ; whether it would be reasonable would depend on what I have called the larger logic which will be the subject of the next section ; we shall there see that there is no objection to such a possibility, with which Mr Keynes' method of justifying probable belief solely by relation to certain knowledge is quite unable to cope.

(5) THE LOGIC OF TRUTH

The validity of the distinction between the logic of consistency and the logic of truth has been often disputed ; it has been contended on the one hand that logical consistency is only a kind of factual consistency ; that if a belief in p is inconsistent with one in q, that simply means that p and q are not both true, and that this is a necessary or logical fact. I believe myself that this difficulty can be met by Wittgenstein's theory of tautology, according to which if a belief in p is inconsistent with one in q, that p and q are not both true is not a fact but a tautology. But I do not propose to discuss this question further here.

From the other side it is contended that formal logic or the logic of consistency is the whole of logic, and inductive logic either nonsense or part of natural science. This contention, which would I suppose be made by Wittgenstein, I feel more difficulty in meeting. But I think it would be a pity, out of deference to authority, to give up trying to say anything useful about induction.

Let us therefore go back to the general conception of logic as the science of rational thought. We found that the most generally accepted parts of logic, namely, formal logic, mathematics and the calculus of probabilities, are all concerned simply to ensure that our beliefs are not self-contradictory. We put before ourselves the standard of consistency and construct these elaborate rules to ensure its observance. But this is obviously not enough ; we want our beliefs to be consistent not merely with one another but also with the facts [1] : nor is it even clear that consistency is always advantageous ; it may well be better to be sometimes right than never right. Nor when we wish to be consistent are we always able to be : there are mathematical propositions whose truth or falsity cannot as yet be decided. Yet it may humanly speaking be right to entertain a certain degree of belief in them on inductive or other grounds : a logic which proposes to justify such a degree of belief must be prepared actually to go against formal logic ; for to a formal truth formal logic can only assign a belief of degree 1. We could prove in Mr Keynes' system that its probability is 1 on any evidence. This point seems to me to show particularly clearly that human logic or the logic of truth, which tells men how they should think, is not merely independent of but sometimes actually incompatible with formal logic.

In spite of this nearly all philosophical thought about human logic and especially induction has tried to reduce it in some way

[1] Cf. Kant : ' Denn obgleich eine Erkenntnis der logischen Form völlig gemäss sein möchte, dass ist sich selbst nicht widerspräche, so kann sie doch noch immer dem Gegenstande widersprechen.' *Kritik der reinen Vernunft*, First Edition, p. 59.

to formal logic. Not that it is supposed, except by a very few, that consistency will of itself lead to truth ; but consistency combined with observation and memory is frequently credited with this power.

Since an observation changes (in degree at least) my opinion about the fact observed, some of my degrees of belief after the observation are necessarily inconsistent with those I had before. We have therefore to explain how exactly the observation should modify my degrees of belief ; obviously if p is the fact observed, my degree of belief in q after the observation should be equal to my degree of belief in q given p before, or by the multiplication law to the quotient of my degree of belief in pq by my degree of belief in p. When my degrees of belief change in this way we can say that they have been changed consistently by my observation.

By using this definition, or on Mr Keynes' system simply by using the multiplication law, we can take my present degrees of belief, and by considering the totality of my observations, discover from what initial degrees of belief my present ones would have arisen by this process of consistent change. My present degrees of belief can then be considered logically justified if the corresponding initial degrees of belief are logically justified. But to ask what initial degrees of belief are justified, or in Mr Keynes' system what are the absolutely *a priori* probabilities, seems to me a meaningless question ; and even if it had a meaning I do not see how it could be answered.

If we actually applied this process to a human being, found out, that is to say, on what *a priori* probabilities his present opinions could be based, we should obviously find them to be ones determined by natural selection, with a general tendency to give a higher probability to the simpler alternatives. But, as I say, I cannot see what could be meant by asking whether these degrees of belief were logically justified. Obviously the best thing would be to know for certain in advance what was true and what false, and therefore if any one system of initial beliefs is to receive the philosopher's approbation it should be this one. But clearly this would not be accepted by thinkers of the school I am criticising. Another alternative is to apportion initial probabilities on the purely formal system expounded by Wittgenstein, but as this gives no justification for induction it cannot give us the human logic which we are looking for.

Let us therefore try to get an idea of a human logic which shall not attempt to be reducible to formal logic. Logic, we may agree, is concerned not with what men actually believe, but what they ought to believe, or what it would be reasonable to believe. What then, we must ask, is meant by saying that it is reasonable for a man to have such and such a degree of belief in a proposition ? Let us consider possible alternatives.

First, it sometimes means something explicable in terms of formal logic : this possibility for reasons already explained we may dismiss. Secondly, it sometimes means simply that were I in his place (and not e.g. drunk) I should have such a degree of belief. Thirdly, it sometimes means that if his mind worked according to certain rules, which we may roughly call ' scientific method ', he would have such a degree of belief. But fourthly it need mean none of these things ; for men have not always believed in scientific method, and just as we ask ' But am I necessarily reasonable ', we can also ask ' But is the scientist necessarily reasonable ? ' In this ultimate meaning it seems to me that we can identify reasonable opinion with the opinion of an ideal person in similar circumstances. What, however, would this ideal person's opinion be ? As has previously been remarked, the highest ideal would be always to have a true opinion and be certain of it ; but this ideal is more suited to God than to man.[1]

We have therefore to consider the human mind and what is the most we can ask of it.[2]   The human m‡nd works essentially according to general rules or habits ; a process of thought not proceeding according to some rule would simply be a random sequence of ideas ; whenever we infer *A* from *B* we do so in virtue of some relation between them. We can therefore state the problem of the ideal as " What habits in a general sense would it be best for the human mind to have ? "   This is a large and vague question which could hardly be answered unless the possibilities were first limited by a fairly definite conception of human nature.   We could imagine some very useful habits unlike those possessed by any men.   [It must be explained that I use habit in the most general possible sense to mean simply rule or law of behaviour, including instinct :   I do not wish to distinguish acquired

---

[1] [Earlier draft of matter of preceding paragraph in some ways better.—F.P.R.

What is meant by saying that a degree of belief is reasonable ? First and often that it is what I should entertain if I had the opinions of the person in question at the time but was otherwise as I am now, e.g. not drunk.   But sometimes we go beyond this and ask : ' Am I reasonable ? '   This may mean, do I conform to certain enumerable standards which we call scientific method, and which we value on account of those who practise them and the success they achieve.   In this sense to be reasonable means to think like a scientist, or to be guided only by ratiocination and induction or something of the sort (i.e. reasonable means reflective).   Thirdly, we may go to the root of why we admire the scientist and criticize not primarily an individual opinion but a mental habit as being conducive or otherwise to the discovery of truth or to entertaining such degrees of belief as will be most useful.   (To include habits of doubt or partial belief.)   Then we can criticize an opinion according to the habit which produced it. This is clearly right because it all depends on this habit ; it would not be reasonable to get the right conclusion to a syllogism by remembering vaguely that you leave out a term which is common to both premisses.

We use reasonable in sense 1 when we say of an argument óf a scientist this does not seem to me reasonable ;   in sense 2 when we *contrast* reason and superstition or instinct ;   in sense 3 when we *estimate* the value of new methods of thought such as soothsaying.]

[2] What follows to the end of the section is almost entirely based on the writings of C. S. Peirce. [Especially his "Illustrations of the Logic of Science ", *Popular Science Monthly*, 1877 and 1878, reprinted in *Chance Love and Logic* (1923).]

rules or habits in the narrow sense from innate rules or instincts, but propose to call them all habits alike.] A completely general criticism of the human mind is therefore bound to be vague and futile, but something useful can be said if we limit the subject in the following way.

Let us take a habit of forming opinion in a certain way ; e.g. the habit of proceeding from the opinion that a toadstool is yellow to the opinion that it is unwholesome. Then we can accept the fact that the person has a habit of this sort, and ask merely what degree of opinion that the toadstool is unwholesome it would be best for him to entertain when he sees it ; i.e. granting that he is going to think always in the same way about all yellow toadstools, we can ask what degree of confidence it would be best for him to have that they are unwholesome. And the answer is that it will in general be best for his degree of belief that a yellow toadstool is unwholesome to be equal to the proportion of yellow toadstools which are in fact unwholesome. (This follows from the meaning of degree of belief.) This conclusion is necessarily vague in regard to the spatio-temporal range of toadstools which it includes, but hardly vaguer than the question which it answers. (Cf. density at a point of gas composed of molecules.)

Let us put it in another way : whenever I make an inference, I do so according to some rule or habit. An inference is not completely given when we are given the premiss and conclusion ; we require also to be given the relation between them in virtue of which the inference is made. The mind works by general laws ; therefore if it infers q from p, this will generally be because q is an instance of a function ϕx and p the corresponding instance of a function ψx such that the mind would always infer ϕx from ψx. When therefore we criticize not opinions but the processes by which they are formed, the rule of the inference determines for us a range to which the frequency theory can be applied. The rule of the inference

41

may be narrow, as when seeing lightning I expect thunder, or wide, as when considering 99 instances of a generalization which I have observed to be true I conclude that the 100th is true also. In the first case the habit which determines the process is 'After lightning expect thunder'; the degree of expectation which it would be best for this habit to produce is equal to the proportion of cases of lightning which are actually followed by thunder. In the second case the habit is the more general one of inferring from 99 observed instances of a certain sort of generalization that the 100th instance is true also; the degree of belief it would be best for this habit to produce is equal to the proportion of all cases of 99 instances of a generalization being true, in which the 100th is true also.

Thus given a single opinion, we can only praise or blame it on the ground of truth or falsity: given a habit of a certain form, we can praise or blame it accordingly as the degree of belief it produces is near or far from the actual proportion in which the habit leads to truth. We can then praise or blame opinions derivatively from our praise or blame of the habits that produce them.

This account can be applied not only to habits of inference but also to habits of observation and memory; when we have a certain feeling in connection with an image we think the image represents something which actually happened to us, but we may not be sure about it; the degree of direct confidence in our memory varies. If we ask what is the best degree of confidence to place in a certain specific memory feeling, the answer must depend on how often when that feeling occurs the event whose image it attaches to has actually taken place.

Among the habits of the human mind a position of peculiar importance is occupied by induction. Since the time of Hume a great deal has been written about the justification for inductive inference. Hume showed that it could not

42

be reduced to deductive inference or justified by formal logic. So far as it goes his demonstration seems to me final ; and the suggestion of Mr Keynes that it can be got round by regarding induction as a form of probable inference cannot in my view be maintained. But to suppose that the situation which results from this is a scandal to philosophy is, I think, a mistake.

We are all convinced by inductive arguments, and our conviction is reasonable because the world is so constituted that inductive arguments lead on the whole to true opinions. We are not, therefore, able to help trusting induction, nor if we could help it do we see any reason why we should, because we believe it to be a reliable process. It is true that if any one has not the habit of induction, we cannot prove to him that he is wrong ; but there is nothing peculiar in that. If a man doubts his memory or his perception we cannot prove to him that they are trustworthy ; to ask for such a thing to be proved is to cry for the moon, and the same is true of induction. It is one of the ultimate sources of knowledge just as memory is : no one regards it as a scandal to philosophy that there is no proof that the world did not begin two minutes ago and that all our memories are not illusory.

We all agree that a man who did not make inductions would be unreasonable : the question is only what this means. In my view it does not mean that the man would in any way sin against formal logic or formal probability ; but that he had not got a very useful habit, without which he would be very much worse off, in the sense of being much less likely [1] to have true opinions.

This is a kind of pragmatism : we judge mental habits by whether they work, i.e. whether the opinions they lead

[1] ' Likely ' here simply means that I am not sure of this, but only have a certain degree of belief in it.

to are for the most part true, or more often true than those
which alternative habits would lead to.

Induction is such a useful habit, and so to adopt it is
reasonable. All that philosophy can do is to analyse it,
determine the degree of its utility, and find on what characteristics of nature this depends. An indispensable means for
investigating these problems is induction itself, without which
we should be helpless. In this circle lies nothing vicious.
It is only through memory that we can determine the degree
of accuracy of memory ; for if we make experiments to determine this effect, they will be useless unless we remember them.

Let us consider in the light of the preceding discussion
what sort of subject is inductive or human logic—the logic
of truth. Its business is to consider methods of thought,
and discover what degree of confidence should be placed in
them, i.e. in what proportion of cases they lead to truth.
In this investigation it can only be distinguished from the
natural sciences by the greater generality of its problems.
It has to consider the relative validity of different types of
scientific procedure, such as the search for a causal law
by Mill's Methods, and the modern mathematical methods
like the *a priori* arguments used in discovering the Theory
of Relativity. The proper plan of such a subject is to be
found in Mill [1] ; I do not mean the details of his Methods
or even his use of the Law of Causality. But his way of
treating the subject as a body of inductions about inductions,
the Law of Causality governing lesser laws and being itself
proved by induction by simple enumeration. The different
scientific methods that can be used are in the last resort
judged by induction by simple enumeration ; we choose
the simplest law that fits the facts, but unless we found that
laws so obtained also fitted facts other than those they were
made to fit, we should discard this procedure for some other.

[1] Cf. also the account of ' general rules ' in the Chapter ' Of Un-
philosophical Probability ' in Hume's *Treatise*.

44

Degrees of Belief

. . . the wise man follows many things probable that he has not grasped nor perceived nor assented to but that possess verisimilitude.

Cicero
Academica

Rational belief comes in various strengths, a fact that is most dramatic when grounds for belief are equivocal and the stakes involved are high: e.g., in medicine, jurisprudence, and serious gambling. This did not escape the ancients.[1] Signs which are evidence but not conclusive evidence are discussed by Aristotle in the *Rhetoric*, and by Cicero in many places. When the point is pressed, the gradations become finer and finer, and less and less seems absolutely certain. Careful epistemologists from Carneades in the second century B.C. to J. L. Austin and Wilfred Sellars have concluded that even the immediate evidence of the senses should not bestow absolute certainty. A reasonable theory of rational action and decision must take into account degrees of belief.

The idea of justifying the probability calculus as embodying laws of static coherence for degrees of belief goes back to Ramsey's famous essay "Truth and Probability."[2] Ramsey had the idea that qualitative constraints could lead to a representation theorem for probability. And he had the idea of a Dutch book theorem: a theorem to the effect that if probabilities are taken as betting quotients then someone who violates the laws of the probability calculus would be susceptible to a system of bets, each of which he considers fair or favorable, such that he would suffer a net loss no matter what happened. Since the leading ideas of the Dutch book theorems are remarkably simple, we will discuss them first.

DUTCH BOOK THEOREMS

What do you consider a fair price for a wager which pays $1 if p is true; nothing otherwise? To keep you honest, we can rely on the wisdom of Solomon: You set the price, but I decide whether you buy the bet from me or I buy the bet from you. The price you judge to be fair for this bet we will take[3] as your personal probability for p.

Certain features of personal probability follow immediately: (I) You would be mad to have a personal probability greater than one or less than zero, for in the first case I could sell you the bet for more than you could possibly win, and in the second case I could buy it from you for less than nothing, i.e., you would pay me to take it off your hands. (II) You had best give a tautology, p or not-p, personal probability one and a contradiction, p and not-p, personal probability zero, for a bet on a tautology is a bet that you must win and a bet on a contradiction is a bet that you must lose.

Another property follows almost as quickly. Suppose that you buy (sell) a bet that pays $1 if p but nothing otherwise for $.25 and that you buy (sell) a similar bet on q for $.50 and that p is logically incompatible with q. Then you have, in effect, bought (sold) a bet which pays $1 if p or q but nothing otherwise for $.75. If you are to be consistent, your personal probability for p or q had better be $.75, i.e., probability (p) + probability (q). If it is lower, I could buy a bet on p or q from you at less than $.75 and sell it back to you piecemeal (as bets on p and on q) for $.75. If it is higher, you could be embarrassed by the converse transaction.[4]

That's almost all there is to it. If we pause to survey this simple argument we see that what is basic is the consistency condition that you evaluate a betting arrangement independently of how it is described (e.g., as a bet on p or q or as a system of bets consisting of a bet on p and a bet on q). Ramsey puts it this way:

> If anyone's mental condition violated these laws, his choice would depend on the precise form in which the option were offered him, which would be absurd. He could then have book made against him by a cunning bettor and would then stand to lose in any event.[5]

The cunning bettor is simply a dramatic device—the Dutch book a striking corollary—to emphasize the underlying issue of coherence.

I said that the properties just given a pragmatic justification—(i) probability is non-negative, (ii) probability of a tautology is one, (iii) probability is finitely additive—are *almost* all there is to probability. The standard measure–theoretic treatment of probability due to Kolmogoroff assumes a stronger form of additivity: countable additivity.

Let us assume that the propositions about which we have degrees of belief are closed not only under finite truth functions but also under countable disjunction and conjunction. Let us assume that the probability values are real numbers. We say that our probability function is *countably additive* iff (iv) for a countable disjunction of pairwise mutually incompatible propositions, p_1 or p_2 or . . . ,

$$\Pr(p_1 \text{ or } p_2 \text{ or } \ldots) = \Sigma_i \Pr(p_i)$$

[where $\Sigma_i \Pr(p_i)$ is the limit of the sequence of partial sums $\Pr(p_1) + \Pr(p_2) + \ldots + \Pr(p_n)$]. Countable additivity is the property which allows the standard measure–theoretic treatment of probability density functions.

Although mathematically convenient, countable additivity has been a matter of some controversy in the foundations of personal probability. De Finetti has opposed it, arguing that a rational agent should be able to believe that the tickets in a denumerably infinite lottery are equally likely to win.

Assuming that probability is real-valued, this is impossible if probability is countably additive for then if the probability of a given ticket winning is greater than zero, the probability that some ticket wins would have to be greater than one. But if the probability of a given ticket winning is zero, then by countable additivity the probability that some ticket wins would have to be zero. Weakening the condition to finite additivity allows that for each ticket, the probability that it wins is zero, while the probability that some ticket wins is one.

Notice, however, that de Finetti's agent leaves himself open to a Dutch book. Bet him $101 against $½ that ticket one wins; $101 against ½^n that the nth ticket wins. You will be assured of a net winning of at least $100 no matter what happens. Indeed, as more than one person has noticed,[6] the coherence argument works just as well for countable additivity as for finite additivity provided that probability is real-valued and we allow countably infinite systems of bets.

Do I ever take on an infinite number of bets? Suppose that I bet that Achilles will drop out of the race. Let it be a very favorable bet: I gain one unit of value if he drops out and lose nothing otherwise. This is a betting arrangement that is tantamount to a countable number of bets: the bet of nothing against one that he drops out in the first half, the bet of nothing against one that he drops out in the first half of the remaining distance, and so forth. Let p_1 be the proposition that he drops out in the first half,[7] p_2 be the proposition that he drops out at the first half of the remaining distance, and so forth. Then the proposition that he drops out is the countable disjunction, p_1 or p_2 or . . . . (You can't drop out on the finish line. If you reach the finish line, you've finished.) Countable additivity of probability tells us that the probability that he drops out should be equal to $\Sigma_i Pr(p_i)$. We will countenance the consideration of countable disjunctions and countable systems of bets as in the foregoing, and will feel free to use the standard theory of countably-additive real-valued probability.[8]

In the Kolmogoroff version of that theory, conditional probability on a condition with non-zero probability is introduced as:

$$Pr(q \text{ given } p) = Pr(p \text{ \& } q)/Pr(p) \qquad [Pr(p) \neq 0]$$

But the question may arise as to whether this is the appropriate definition. De Finetti provides an answer of a piece with the rationale for the unconditional probabilities by introducing the notion of a conditional bet. A bet on q conditional on p is called off if p is false. If p is true, it is won or lost depending on the truth value of q. Such a betting arrangement can again be described as the upshot of separate bets on p *and* q and against p with the consequence that coherence requires this definition of conditional probability.[9]

So the *statics* of coherence leads us to model rational degrees of belief as countably additive probability measures with conditional probability for conditions of positive probability being defined in the usual way. The question arises as to whether there is any analogous argument for *changing* degrees of belief. Ramsey strongly suggests that he believes that there is such an argument for updating by conditionalization on the evidence:

> Since an observation changes (in degree at least) my opinion about the fact observed, some of my degrees of belief after the observation are necessarily inconsistent with those I had before. We have therefore to explain exactly how the observation should modify my degrees of belief; obviously if p is the fact observed, my degree of belief in q after the observation should be equal to my degree of belief in q given p before. . . . When my degrees of belief change in this way we can say that they have been changed consistently by my observation.[10]

Ramsey, however, does not explicitly set out any such argument.

Despite some skeptical doubts about the possibility of such an argument, Dutch book arguments for conditionalization have been given.[11] Suppose that I am about to learn whether a certain proposition, p, is true or false and that I have a rule or disposition to change my degrees of belief in a certain way if I learn that p is true. Prior to finding out whether or not p is true, *my rule or disposition to change my beliefs in a certain way upon learning p is tantamount to having a set of ratios for bets conditional on p.* (Someone can achieve a betting arrangement for a bet on q conditional on p with me just by reserving a sum that he will bet on q with me *after* I change my degrees of belief if p turns out true, and which he will not bet at all if p turns out false.) But we also know from de Finetti's observation that my prior degrees of belief commit me to betting ratios for conditional bets in a different way, with those betting ratios being reflected in the prior conditional probabilities. For these conditional betting ratios to coincide, the degrees of belief that I would move to upon learning p must coincide with those gotten by conditionalization on p.

This observation yields a Dutch book theorem as a corollary. If someone does not change his degrees of belief by conditionalization, then someone who knows how he does change his degrees of belief can exploit the different betting ratios to make a Dutch book conditional on p, which can be turned into an unconditional Dutch book by making an appropriately small sidebet against p.

In this diachronic coherence argument it is again not the cunning opponent that is fundamental but rather the requirement that different ways in which we can be committed to the same conditional bet receive the same evaluation.

The argument does not show that *all* rational changes of belief should be by conditionalization. There might be no proposition within the domain of the agent's degrees of belief that expresses what is learned; and in such a case nothing appropriate to conditionalize on. Only when such a proposition is at hand and is learned with certainty does diachronic coherence require updating by conditionalization.

If degrees of belief have real numerical values which are tied to betting in the way indicated, we have a rationale for the standard Bayesian assumptions that degrees of belief should be represented by a countably additive probability measure which is updated by conditionalization upon learning a proposition with positive prior probability. But our degrees of belief do not in general have precise numerical values. What can we say if we weaken our framework and only assume comparative degrees of belief, or preferences instead of expected utilities?

REPRESENTATION THEOREMS

A relational structure is said to have a *representation* in some domain of mathematical objects if there is some mapping of the objects and relations of the relational structure onto the mathematical objects and their relations which preserves structure. For example, if my relational structure contains three objects, *Sam*, *Tom*, and *Sue*, and specifies the relation of being *at least as tall as* such that Sam and Sue are at least as tall as each other and neither is at least as tall as Sam and such that the relation is a total order, then one numerical representation would map Sam and Sue both onto 67, Tom onto 72, and *at least as tall* onto *greater than or equal to*. Obviously, there are many other numerical representations as well. A rich enough structure may determine a unique representation, or one that is unique up to a tight group of transformations, but in general representations are underdetermined.

Ramsey had the idea of starting with the qualitative ordering of preference on gambles and extracting a probability-utility representation, an assignment of probabilities and utilities such that the expected utilities that they give order the gambles in the same way as the original ordering. Again, the leading ideas are remarkably simple.

Suppose that you prefer champagne to beer with a Chinese meal but are indifferent between the gambles: (a) Champagne if this coin comes up heads, beer otherwise; (b) Beer if this coin comes up heads, champagne otherwise. Then we might reasonably infer that you take it to be no more likely that the coin at issue come up heads than that it come up tails. Likewise, if there are six goods such that your preferences totally order them with no ties, and you are indifferent between all gambles which assign exactly one of them to each of the faces of a certain die (i.e., your preferences are invariant under permutations of the die faces), then we may reasonably conclude that you take each face of the die to be equiprobable.

The foregoing interpretation depends on the outcome of the flip of the coin or the toss of the die itself not mattering to you. It has no value for you in and of itself, nor does it in any way affect the value of the goods which do matter. Ramsey calls such propositions as "The coin comes up heads" *ethically neutral*. Your ethically neutral propositions are identifiable from your preference ranking in that you are indifferent between payoffs consisting of maximal collections of goods (possible worlds, if you please) which differ only with respect to ethically neutral propositions.

A rich enough set of preferences leads to a set of ethically neutral propositions with arbitrary probabilities. These, in turn, can be used to determine a utility scale. Suppose that your most preferred payoff, TOP, is arbitrarily given utility of 1 and your least preferred alternative, BOTTOM, is given utility of zero. Then a wager *TOP if p, otherwise BOTTOM*, where p is ethically neutral will have a utility equal to the probability of p, as will any wager ranked equally with it. Anything preferred to it will have a greater utility; anything to which it is preferred will have a lesser utility. In this way, the probability scale on ethically neutral propositions determines a utility scale on all gambles. Given these utilities one can solve for the probabilities of propositions which are not ethically neutral.

The foregoing sketch of Ramsey's leading ideas is here only to give the uninitiated reader some idea of one natural path from preferences to probability and utility. There are many different ways to such representation theorems, among which are those of Savage, Bolker, and Luce and Krantz.[12] Some representation theorems get only a finitely additive probability, but it is possible to get a representation theorem that yields a countably additive probability (e.g., Bolker) by imposing a continuity condition on preferences.

In all the representation theorems available, the conditions put on preferences to arrive at the representation fall into two groups: the first being fairly intuitive consistency conditions on preferences (e.g., Nothing is preferred to itself. If A is preferred to B and B to C, then A is preferred to C, etc.), and the second being structural conditions which guarantee that the preferences are over a rich enough domain to peg the probabilities and utilities (e.g., that there be enough of the appropriate gambles involving ethically neutral propositions).

Structural richness properties guarantee the uniqueness properties of the representation. (In most approaches the probabilities are determined uniquely and the utilities are determined up to choice of origin and unit; Bolker's result is slightly different.) Thus, on Ramsey's approach, we would like there to be enough gambles on ethically neutral propositions so that we could subdivide the world as finely as we please into equiprobable ethically neutral propositions.

With regard to a similar condition put forward by de Finetti, Savage has this to say:

> It might fairly be objected that such a postulate would be fla-
> grantly *ad hoc*. On the other hand, such a postulate could be
> made relatively acceptable by observing that it will obtain if, for
> example, in all the world there is a coin that the person is firmly
> convinced is fair, that is, a coin such that any finite sequence of
> heads and tails is for him no more probable than any other se-
> quence of the same length; though such a coin is, to be sure, a
> considerable idealization.[13]

What are we to make of such simple and attractive idealizations?

We do not, I trust, want to maintain that any rational person must believe that there is such a coin, or indeed that any rational person have a very rich structure of preferences at all. Rather, I believe that we should follow Brian Ellis[14] in interpreting structural richness conditions as *embeddibility* conditions. A rational system of preferences need not be rich, but it should be able to grow. It should not preclude expansion, e.g., by addition of a gamble contingent on an independent flip of a fair coin. If we are willing to say this, then by iterated expansion we have as a condition of rationality embeddibility in the sort of rich system of preferences which gives us the representation theorem. Then rational systems of preferences will have a probability-utility representation.

What we lose, by taking this viewpoint, is the uniqueness of the representation. A system of preferences over a modest domain which is embeddible in a rich system of preferences will typically be embeddible in more than one such system. It will therefore have not one probability-utility representation but rather a set of probability-utility representations. This is, I think, a virtue rather than a defect of the embeddibility approach. It gives a more realistic picture of a decision maker—one that does not assume that people have unique precise probability assignments—while at the same time licensing the analytical techniques of the probability calculus and utility theory.

HIGHER-ORDER DEGREES OF BELIEF

It is hardly in dispute that people have beliefs about their beliefs. Thus, if we distinguish degrees of belief, we should not shrink from saying that people have degrees of belief about their degrees of belief. Nevertheless, the founding fathers of the theory of personal probability are strangely reticent about extending that theory to probabilities of a higher order. Ramsey does not consider the possibility. De Finetti rejects it. Savage toys with the idea but decides against it. This reticence is, I think, ill-founded.

That the mathematics is at hand to treat probabilities of probabilities should not be in question. The mathematical treatment of probability as a random variable is familiar from standard examples of uncertain chances. And as we will see in the next chapter, ordinary first-order probabilities may be tantamount to such a set-up. De Finetti's representation theorem for exchangeable sequences is, among other things, a consistency proof for taking probability as a random variable. Some writers who claim that dealing with probabilities of probabilities leads to inconsistency prove, on examination, to be simply confused.[15]

One might however hold that although formally consistent, a theory of higher-order *personal* probabilities is, in some way, *philosophically* incoherent. This appears to be de Finetti's position. He adopts an *emotive* theory of probability attribution:

> Any assertion concerning probabilities of events is merely the expression of somebody's opinion and not itself an event. There is no meaning, therefore, in asking whether such an assertion is true or false or more or less probable.

> Speaking of unknown probabilities must be forbidden as meaningless.[16]

If probability attribution statements are merely ways of evincing degrees of belief, they do not express genuine propositions and are not capable themselves of standing as objects of belief.

De Finetti's positivism stands in sharp contrast to Ramsey's pragmatism:

> There are, I think, two ways in which we can begin. We can, in the first place, suppose that a degree of belief is something perceptible by its owner; for instance that beliefs differ in the intensity of a feeling by which they are accompanied, which might be called a belief-feeling of conviction, and that by degree of belief we mean the intensity of this feeling. This view . . . seems to me observably false, for the beliefs we hold most strongly are often accompanied by practically no feeling at all.

> We are driven therefore to the second supposition that the degree of belief is a causal property of it, which we can express vaguely as the extent to which we are prepared to act on it.

> The kind of measurement of belief with which probability is concerned is . . . a measurement of belief *qua* basis of action.[17]

For Ramsey, then, a probability attribution is a theoretical claim. It is evident that on Ramsey's conception of personal probability, higher-order personal probabilities are permitted (and indeed required).

Even from de Finetti's viewpoint, the situation is more favorable to a theory of higher-order personal probabilities than might at first appear. For a given person and time there must *be*, after all, a proposition to the effect that that person has the degree of belief that he might evince by uttering a certain probability attribution statement. De Finetti grants as much:

> The situation is different, of course, if we are concerned not with the assertion itself but with whether "someone holds or expresses such an opinion or acts according to it," for this is a real event or proposition.[18]

With this, de Finetti grants the existence of propositions on which a theory of higher-order personal probabilities can be built, but never follows up this possibility.

Perhaps this is because of another sort of philosophical objection to higher-order personal probabilities which, I think, is akin to the former in philosophical presupposition, though not in substance. Higher-order personal probabilities are well-defined, all right—so this line goes—but they are trivial; they only take on the values zero and one. According to this story, personal probabilities—if they exist at all—are directly open to introspection so that one should be certain about their values. If my degree of belief in p is x, then my degree of belief that my degree of belief in p is x will be one; and my degree of belief that my degree of belief is unequal to x will be zero. Put so baldly, the objection seems a bit silly, but I will discuss it because I think that something like it often hovers in the background of discussions of personal probability. But first I would like to point out that this objection has much narrower scope than the previous one. According to this view, it is perfectly all right to postulate nontrivial personal probabilities about personal probabilities if they are my probabilities now about your probabilities now, or my probabilities now about my probabilities yesterday, or tomorrow, or whenever the results of the experiments we are running come in. What become trivial, according to this view, are my probabilities now about my probabilities (that I am introspecting) now.

But even this seems bad psychology and bad epistemology. And if we follow Ramsey in focusing on degrees of belief *qua* basis of action rather than the intensity of feeling notion, the objection vanishes entirely. For in this sense of belief it is entirely possible for a person not to *know his own mind* with certainty.[19]

Given that the conception of higher-order personal probabilities is philosophically legitimate and non-trivial, the question remains as to whether they are of any special interest. Savage's brief discussion in *The Foundations of Statistics* is along these lines:

> To approach the matter in a somewhat different way, there seem to be some probability relations about which we feel relatively "sure" as compared with others. When our opinions, as reflected in real or envisaged action, are inconsistent, we sacrifice the unsure opinions to the sure ones. The notion of "sure" and "unsure" introduced is vague, and my complaint is precisely that neither the theory of personal probability, as it is developed in this book, nor any other device known to me renders the notion less vague. There is some temptation to introduce probabilities of the second order so that a person would find himself saying such things as "the probability that B is more probable than C is greater than the probability that F is more probable than G." But such a program seems to meet insurmountable difficulties. The first of these—pointed out to me by Max Woodbury—is this. If the primary probability of an event B were a random variable b with respect to the secondary probability, then B would have a "composite" probability by which I mean the (secondary) expectation of b. Composite probability would then play the allegedly villainous role that secondary probability was intended to obviate, and nothing would have been accomplished.
>
> Again, once second order probabilities have been introduced, the introduction of an endless hierarchy seems inescapable. Such a hierarchy seems very difficult to interpret, and it seems to make the theory less realistic, not more.[20]

In this passage, Savage seems to have two rather different motivations in mind for higher-order probabilities. The first is the consideration that he begins with—that there is a second-order aspect to our beliefs, i.e., "sureness" about our first-order beliefs, which is not adequately reflected in the first-order probability distribution alone. The second is the idea that second-order distributions might be a tool for representing vague, fuzzy, or ill-defined first-order degrees of belief with greater psychological realism than a first-order account alone would provide. The second motivation is implicit in the discussion of the insuperable difficulties, and is made explicit in a footnote in the second edition.[21]

I think that it is very important to carefully distinguish these two lines of thought. Savage's "insuperable difficulties" are serious objections against the suggestion that second-order distributions provide a good mathematical representation of vague, fuzzy, or ill-defined first-order beliefs. Indeed, an apparatus of second-order distributions presumes more structure rather than less. But however we wish to treat vague or fuzzy first-order degrees of belief, we shall, given beliefs about beliefs, wish to treat vague or fuzzy second-order degrees of belief as well.

In particular, the treatment of imprecise first-order degrees of belief as represented by a set of precise representations is not in competition with second-order probabilities; it is aimed at a different problem. If we return to Savage's first motivation, we find that the "insuperable objections" are no longer objections at all. The extra structure of higher-order probabilities is just what is wanted. That two second-order distributions can have the same mean but different variance gives us a representation of the phenomenon with which Savage broached the discussion: Two people may have the same first-order probabilities but different degrees of sureness about them.

If you can be more or less sure about probabilities, you can also be more or less happy about them. That is, there can be value associated with propositions of probability. In particular, it is possible for there to be genuine disutility of risk. On a first-order approach apparent disutility of risk must be explained as an illusion, generated by the diminishing marginal utility of money.

For example, I prefer no bet to wagering $1,000 on the flip of a coin because the loss of $1,000 would hurt more than the gain of $1,000 would help. This explanation of the apparent disutility of risk, due to Daniel Bernoulli,[22] no doubt describes a real phenomenon, but many from Bernoulli's time to ours have not been able to believe that it is the whole story.[23] Now it is certainly coherent to ascribe a negative value to "butterflies in the stomach" and if an agent believes that the second bet will certainly lead to them, its real consequences are either "Win $1,000 with gastrointestinal upset" or "Lose $1,000 with gastrointestinal upset." So the second bet can in this way have a smaller expected utility than the first even if the marginal utility of money is constant. In like manner, if there is a proposition (or a family of propositions) in our language specifying riskiness (or "ambiguity") of a wager, it is coherent to attach negative or positive utility to riskiness (or "ambiguity") according to one's temperament. We can have such propositions if we have probability as a random variable, with the operative sense of risk depending on the exact interpretation of that random variable. Extending the theory of personal probabilities to higher-order probabilities thus gives one a richer framework within which it is possible, within the constraints of coherence, to give a responsive answer to one of the most persistent criticisms of that theory.

There is one further point in the passage from Savage that invites comment. Savage speculates that the notion of sureness may give some insight into probability change, and indeed the introduction of higher-order degrees of belief does that as well. Our consideration of probability change so far has focused on the rationale for conditionalization; for taking our new probabilities as our old probabilities conditional on the evidence. This gives the evidence probability one. But the assumption that every observation can be interpreted as conferring certainty on some observational proposition appears to lead to an unacceptable epistemology of the given.

There is, however, another way in which conditionalization can be brought to bear via higher-order degrees of belief.

When our epistemic inputs give each first-order proposition a less-than-certain probability, they may nevertheless sometimes be plausibly characterizable as learning (for certain) *that* a proposition has a less-than-certain probability. We can then conditionalize on such a probabilistic proposition.

Rules that appear to be *ad hoc* extensions of conditionalization in a first-order setting can be evaluated as special cases of conditionalization within the framework of higher-order probabilities. For instance, suppose observation raises the probability of a p from .6 to .9. Following Jeffrey[24] we say that observation changes my degrees of belief by *probability kinematics* on the partition [p, $\sim p$], just in case the probabilities conditional on p and on $\sim p$ remain the same. From the second-order point of view, let us assume that what we learned was that $pr(p) = .9$, and conditionalize on that proposition. It can then be shown that second-order conditionalization yields first-order probability kinematics under a simple hypothesis of conditional independence.[25] That is that for every first-order proposition, q, $PR(q$ given p & $pr(p)$ = a) = $PR(q$ given p) and $PR(q$ given $\sim p$ & $pr(p)$ = a) = PR (q given $\sim p$). This result can be generalized, and the general approach can be extended to analyze other first-order generalizations of conditionalization.[26] Once higher-order degrees of belief are in hand we can conditionalize on a rich variety of propositions: the proposition that the probability of p is within a certain interval; the proposition that the expectation of any random variable is within a certain interval; conjunctions of the foregoing types of proposition; and so forth. The theory of conditionalization can therefore become a much more comprehensive and powerful theory of rational belief change.

Once we have gone this far, there is no reason why not to consider beliefs about beliefs about beliefs, and so on as far as we please. But there is no reason to view this Bayesian hierarchy as vicious. It provides a very rich framework within which to work. For any practical problem we can utilize as much of it as we need.

In the nearly thirty years since the publication of *The*

Foundations of Statistics, statisticians *have* worked in hierarchical models. I. J. Good[27] considers mixed hierarchies which contain physical and other notions of probability as well as degrees of belief. In comments on a review article by Good, DeGroot remarks:

> I strongly agree with Professor Good about the importance and usefulness of hierarchical models. However, I do not see any need to consider physical logical and subjective probabilities. Subjective probabilities would seem to be enough; they unify the theory and are usually convenient to use.[28]

The implied dispensability of physical probabilities is a reference to de Finetti's theorem and generalizations thereof that we will discuss in the next chapter. For more details about statistics and Bayesian hierarchies, I will have to refer the reader to the literature.[29] Enough has perhaps been said to show how a hierarchy allows a Bayesian a much more inclusive theory of rational belief change than might at first seem possible.

CONCLUSION

Here then, we have the full Bayesian framework of rational degrees of belief together with the leading ideas for its justification. I would not want to hold the framework above criticism, but its foundations compare favorably with most of what passes as epistemology. Careful criticism is more likely to call for modification or generalization than for junking the whole approach. (We will discuss the possibility of some modifications in the chapter on causal decision theory.) For now, let us provisionally adopt the framework and pursue the questions of empiricism from within it.

Learning from Experience

Aptly did Plato call natural science the science of the probable.[1]

Simplicius
in Arist. Phys.

. . . inductive reasoning is reduced in this way to Hume's justification.

Bruno de Finetti
"On the Condition of Partial Exchangeability"

COIN FLIPS AND THE UNIFORMITY OF NATURE

Consideration of induction in a non-probabilistic setting might lead one to conclude that scientific inductive logic must presume that nature is *uniform* with the future resembling the past and the unexamined resembling the examined, rather than *random* like the outcome of independent flips of a coin,

---

This chapter is largely drawn from my "Presuppositions of Induction," forthcoming in a festschrift for Arthur Burks (ed. M. Salmon), and read at the University of Western Ontario, the University of Chicago, Princeton University, the University of Southern California, and the 1982 Pacific Division Meetings of the American Philosophical Association. The research on which it is based was partially supported by National Science Foundation grant SES-8007884.

or willfully *perverse* like a clever opponent in a zero-sum game.

A closer examination of objective uniformity reveals unexpected complications. Postulation of uniformity means nothing unless we specify the respects in which uniformity is postulated. Uniformity is relative to a system of classification. This basic insight was already there, submerged in the chorus line, with Hume and Mill.[2] Nelson Goodman brought it to center stage with such glamor[3] that it dazzled the eye.

There are complications even if we abstract away Goodman's problem. Suppose that we consider a sequence of outcomes of the same experiment, assuming that the notion of "same experiment" and the outcome space have been chosen with an eye toward projectibility. The question arises as to what we are to count as a uniform sequence; or better, as a sequence with some degree of uniformity. One is naturally led to the consideration of *random* sequences as the really non-uniform ones.

What are the random sequences? Well, the random process consisting of a sequence of independent and identically distributed trials (like the coin tosses) should have a high probability[4] of producing random sequences. But this still leaves open a variety of possible definitions. There is an extensive mathematical literature on the characterization of random sequences; it is by no means a trivial question either mathematically or philosophically. Without pursuing the matter in detail, I want to note a fact that is invariant over questions of fine tuning the analysis. It is that random sequences *must* have a limiting relative frequency.[5] This is a rather spicy revelation in view of Reichenbach's taking the existence of limiting relative frequencies as the principle of the uniformity of nature. The most chaotic and disordered alternative to uniformity that we can find *entails* uniformity-in-the-sense-of-Reichenbach! The naive intuition of a dichotomy between randomness and order needs to be qualified. Randomness is indeed a kind of disorder, but it carries with it of necessity a kind of statistical order in the large. If randomness is taken as a standard of non-uniformity, then postulation of this type of non-uniformity is postulation of the existence of limiting relative frequencies.

I haven't said anything yet about the third *prima facie* possibility of nature as a willfully perverse opponent. Certainly the world could be so arranged that its creatures with fixed finite intelligence could be frustrated at every turn. Put in this way there is simply no answer to Hume's problem. It is, however, worth noting that if the capabilities of nature's opponents knew no bounds, then the best a perverse nature could do is adopt a random strategy.

Now what I have been building up to is that the whole question takes on a quite different complexion when considered in the context of rational degrees of belief.

Consider one of the oldest illustrations in the business[6]: A biased coin is to be flipped a number of times. There are two hypotheses about the physical probabilities or chances which characterize the process: the hypothesis that the coin is biased two to one in favor of heads and the tosses are independent (BH); and the hypothesis that the coin is biased two to one in favor of tails and the tosses are independent (BT).[7] Suppose that we believe that these are the only two possibilities with non-negligible probability and that they are equally likely. We assign them each epistemic probability ½. Then it is reasonable that we take as our epistemic probabilities of outcomes an average of their physical probabilities on each hypothesis, with the weights of the average (in this case equal) being the epistemic probabilities of the corresponding hypotheses being correct. So the epistemic probability of heads on toss one is $(½)(⅓) + (½)(⅔) = ½$, likewise for the epistemic probability of heads on toss two. But the epistemic probability of heads on toss two *conditional* on toss one is not ½, but a bit more. Intuitively, this is because the information that the coin comes up heads on toss one supports the hypothesis of bias toward heads a bit more strongly than the hypothesis of bias toward tails. Our elementary example has led us to a philosophical conclusion of fundamental importance. *We can have learning from experience without uniformity of nature.*

In our epistemic probability distribution the proposition

(heads on toss one) has positive statistical relevance to the proposition (heads on toss two) even though we are sure that the tosses are really (*vis-à-vis* the true physical probabilities) independent. Mathematically, this positive statistical relevance is an artifact of the averaging (mixture) of possible chance distributions to come up with our rational degrees of belief. The conditions for learning from experience are here not created by knowledge, but rather by ignorance. Here we are learning from experience not by presupposing that nature is uniform, but rather in the teeth of the conviction that it is not uniform, by virtue of our uncertainty about *how* it is not uniform.

Of course, one might say that we have here uniformity of nature at the level of chance. But I hope that my opening remarks about randomness have cast some doubt on the possibility of belief that nature is not uniform at any level. The discussion of generalized representation theorems later in this chapter will show that belief in uniformity *for some sense of chance or other* is almost inescapable.

With regard to the coin-flip example, it is also worth pointing out that the epistemic probabilities can be represented as a mixture (weighted average) in more than one way. Consider the *relative frequency* distributions corresponding to different possible relative frequencies of heads. Let us construct them so that the outcome sequences giving the same relative frequencies to heads are equally probable.[8] For simplicity, consider the case in which there are only two tosses. Then the relative frequency distributions corresponding to the realizable relative frequencies in such a sequence are:

Relative Frequency of Heads:	1	½	0
Pr (Heads on 1 & Heads on 2)	1	0	0
Pr (Heads on 1 & Tails on 2)	0	½	0
Pr (Tails on 1 & Heads on 2)	0	½	0
Pr (Tails on 1 & Heads on 2)	0	0	1

Each of the columns represents a possible relative frequency distribution. If we average these relative frequency distribu-

tions using the epistemic probabilities of the relative frequencies as weights[9] we get back our epistemic distribution.
Note here that the relative *frequency* distributions into which
we factor the epistemic probability distribution are radically
different from the *chance* distributions in terms of which it
was represented previously. In the chance distributions, the
tosses were independent. But consider the relative frequency distribution corresponding to the relative frequency
of heads being ½. In this distribution, the probability of heads
on toss 2 is ½, but the probability of heads on toss two conditional on heads on toss one is zero! We have here *negative*
statistical relevance with a vengeance. Such distributions
(hypergeometric) have been taken as a model[10] for counterinductive reasoning. That is, if I was sure that the relative
frequency of heads was ½ in a sequence of two tosses, observation of a head on toss one would lead me to drop the
probability of a head on toss two from ½ to zero. Obviously
the same sort of effect can be illustrated with respect to any
finite number of tosses.

We have, in our example, an illustration of three different
conceptions of probability: epistemic probability; physical
propensity or chance; and relative frequency. Our epistemic
probability distributions could be viewed either as a mixture
of possible relative frequency distributions or as a mixture
of possible propensity distributions, even though the propensity distributions made the trials independent while the
relative frequency distributions made an outcome on a trial
negatively relevant to the same outcome on another trial.
The interaction of these three conceptions of probability lies
at the heart of inductive reasoning.

DE FINETTI'S THEOREM

In the preceding section, we took a weighted average to get
an epistemic distribution. If we start with the epistemic distribution, we say that it has a *representation as a mixture of*
probability distributions. In general, such representations need
not be unique. We just saw how this distribution has two

quite different representations as mixtures: the mixture of chance distributions and the mixture of relative frequency distributions.

Each of the distributions in the preceding paragraph has the property of being invariant under finite permutations of trials [Pr(Heads on 1 & Tails on 2) = Pr(Tails on 1 & Heads on 2)]. Probability measures with this property are said to make the trials *exchangeable*. Any independent sequence of trials is exchangeable and any mixture[11] of exchangeable sequences (i.e., of probability measures that make the sequences exchangeable) is exchangeable. De Finetti used the concept of exchangeability to prove a representation theorem. It is most neatly put if we consider infinite sequences of trials and assume countable additivity of probability. Then de Finetti's theorem is that an exchangeable sequence of trials has a unique representation as a mixture of independent sequences of trials.[12]

De Finetti's theorem establishes a broad domain for the sort of model of learning from experience that we looked at in miniature in the last section. If our degrees of belief are exchangeable, they act as if they were averages of independent chance distributions. We learn from experience just as before. If we observe a sequence of outcomes and conditionalize on that observation, then the subsequence of remaining outcomes is still exchangeable, but the weights of the independent sequences into which it factors will have shifted as a result of our observation. As the sequence of observed outcomes grows, then with probability (degree of belief) approaching one, the weight of the average will become concentrated on one of the independent sequences.[13] Positive statistical relevance between like outcomes of different trials will arise from the averaging, except in the extreme case of a degenerate average where we start out with our degrees of belief already concentrated on one independent sequence. In the case of infinite sequences of trials, de Finetti appears to have replaced the presupposition of the *objective* uniformity of nature with the modest *subjective* condition of exchangeability. Thus, we have de Finetti's sanguine remarks about having resolved the problem of induction along Humean lines. How far de Finetti's theorem goes toward resolving the problem of induction remains to considered. But it cannot be denied that de Finetti's ideas put the problem in a dramatically new light.

The second important thing that de Finetti's thorem does for us is to establish the relationship between the conceptions of degree of belief, relative frequency, and chance in the cases under consideration. Any probability for a finite sequence can be represented as a mixture of those distributions gotten by conditionalizing out on statements of relative frequency. Exchangeability of the original distribution assures that in the factors so obtained all outcome sequence with positive probability are equally probable. Decomposition of finite exchangeable sequences into relative frequency probabilities in this way yields hypergeometric distributions (the "counterinductive," sampling from an urn without replacement type distributions that were the relative frequency distributions of the last section). Every finite exchangeable sequence has a unique representation in this way as a mixture of hypergeometric sequences. As the length of the finite sequence approaches infinity the hypergeometric distribution approaches independence. The factors of the de Finetti representation are gotten by conditionalizing on statements of relative frequency for finite subsequences and passing to the limit.

If we believe that our infinite exchangeable sequence is governed by unknown propensities which make the trials independent, then by the uniqueness clause of de Finetti's theorem we cannot represent our degrees of belief in two materially distinct ways: as a mixture of relative frequency probabilities and as a mixture of physical propensities. That is, the sum of possibilities of disagreement between propensity and relative frequency must have degree of belief zero. It comes to the same thing, whether we think of our sequence as governed by a mixture of propensities or chances, by a mixture of relative frequency probabilities, or simply by an exchangeable degree of belief. De Finetti regards this as a way of dispensing with metaphysically dubious propensities. With regard to our degree-of-belief probabilities, the postulation of propensities pulls no weight.

De Finetti's analysis of the situation is so beautiful that we would almost like to forget that it is a special case. But what if the number of trials is finite? What if the epistemic probabilities are not exchangeable? How far can the bold insights of de Finetti's analysis be generalized?

FINITE EXCHANGEABLE SEQUENCES

In the first section of this chapter, an example was given of a finite exchangeable sequence that could be represented as a mixture in two ways: as a mixture of hypergeometric sequences and as a mixture of independent sequences. Every finite exchangeable sequence has a unique representation as a mixture of hypergeometric sequences, with the elements of the mixture gotten from the original probability by conditioning on statements of relative frequency, and the weights being the respective original probabilities of those relative frequency statements. (E.g., fixing a sequence of coin tosses at length n, we can take the appropriate statements to be: no heads, 1 head, . . . n heads.) Not every finite exchangeable sequence has a representation as a mixture of independent sequences. For instance, consider the hypergeometric distribution in the two-toss case corresponding to one head and one tail.[14]

Those finite exchangeable sequences which do admit a representation as a mixture of independent sequences are of special interest because they are ones which permit learning from experience in the way illustrated in section 1. Under what conditions do we have representability as a mixture of independent sequences in the finite case, or a good approximation to it? There is an illuminating treatment of this question in recent work by Diaconis, and Diaconis and Freedman.[15] They analyze how a finite exchangeable sequence of length r which can be extended to a longer finite exchangeable sequence of length k approximates a mixture of independent sequences, with the error in the approximation going to zero as k approaches infinity. What is important for the finite case is that the analysis yields a bound on the error for finite k which shows that the error disappears fairly fast.[16] We may think it highly likely that an experiment will only be repeated a finite number of times but not be absolutely sure. If we are sure that it will only be repeated a finite number of times, we may be unsure as to how many. Even if we are sure that it can only be repeated r times, we may believe that it could have been repeated a greater number of times and thus have exchangeable beliefs about a sequence of length r that could be extended to exchangeable beliefs about a sequence of longer length k. Diaconis and Freedman take the example of a coin which is flipped only 10 times, where our beliefs are such that they can be extended to an exchangeable sequence of 1,000 flips. Here the error in the representation of the original sequence as a mixture of sequences of 10 independent flips must be no more than .02. There is considerable scope for the application of de Finetti's analysis to finite sequences.

What does this tell us about finite hypergeometric sequences as a model for counterinduction? It should be reemphasized at the onset that hypergeometric sequences are not, in themselves, perverse. If I know that an urn contains 50 red balls and 50 white balls and nothing else, and my experiment consists of random sampling from that urn without replacement, then the hypergeometric distribution is the one that I should adopt. If, after drawing 50 red balls, I change the probability that the 51st draw wil be red to zero, I will have learned from experience in a quite unexceptionable way, notwithstanding the negative statistical relevance of like outcomes on different trials. It is adopting such a distribution in less appropriate circumstances, say in the coin-flipping case, that is perverse.

Concentrating one's probability on a hypergeometric distribution in a case like the coin case, which is more plausibly a mixture of independent distributions, can be taken as a model of counterinductive behavior.[17] Finding a model for a counterinductive *method* may be another matter. Suppose someone believes there will be only 100 tosses of this coin and that exactly half will be heads, and adopts the corresponding hypergeometric distribution as his epistemic probability. What does he predict about the outcome of the 101st toss while the coin is spinning in the air? What probabilities does he assign after observing the 51st heads? There are falsifiable contingent assumptions which lie behind his assignment. They provide no clue as to how the assignment should be extended to a greater number of trials. One might require of a counterinductive *method* that it deal with any possible sequence of observations as input. Diaconis' result then shows just how tightly the counterinductivist is hemmed in by exchangeability.

(This is not to claim that in no sense is a model of counterinductive method possible. The counterinductivist might, at the moment of truth, simply leap to a different hypergeometric distribution consistent with the observations to date, e.g., the one corresponding to an urn with 51 red balls and 50 black balls.[18] This sort of counterinductive strategy would not respect extendability—life would be full of crises requiring leaps to entirely new probability distributions which are not extensions of the old[19] but the counterinductivist is, after all, rather an odd duck anyway.)

Scientific induction has its presuppositions. We do not really need the picture of "the counterinductivist" to dramatize the point. The difficulties encountered in painting the picture are none the less instructive. They emphasize the considerable role of mixing in the creation of statistical relevance in the finite case as well. The theoretical importance of mixtures of independent sequences leads, in the finite case, to a special appreciation of alternative ways of representing a sequence as a mixture. The canonical de Finetti representation as a mixture of relative frequency distributions is, in the finite case, a representation as a mixture of non-independent sequences. In many cases, the same mixture can naturally be represented as a mixture of independent chance sequences. Thus the study of learning from experience in the finite case tends to emphasize the distinction between relative frequency and chance.

STOCHASTIC PROCESSES AND PARTIAL EXCHANGEABILITY

In the theory of stochastic processes, the numbers which index the events are taken to convey information as to some physical order (often temporal order). Here, in the discrete case, we have repetitions of an experiment indexed by the positive integers, or a doubly infinite sequence indexed by the whole numbers (. . . -1, 0, $+1$ . . .), and in the continuous case indexed by a real "time" parameter. (The theory can be generalized so that the experiment is parametrized in several dimensions, in which case we have a *random field*. Random fields form the analytical framework for the study of the statistical mechanics of lattice gases.) Exchangeability in the epistemic distribution and independence in the objective distributions must be considered a special case. However, a great many physical stochastic processes which are not independent satisfy Markov's weakening of independence where the chances of an outcome on a trial depend only on the outcome of the preceding trial. (The Markov property is obviously connected with causality, and an appropriate Markov property in a random field where the parameters represent space and time would represent a probabilistic principle of locality of causation.) It is therefore of great interest whether the de Finetti analysis can be extended to Markov chains. Again, when we are dealing with arbitrary sequences of random variables, rather than just repetitions of an experiment, there is no reason to assume in general that the random variables in question should be exchangeable in the epistemic distribution, or independent in the possible objective distribution.

De Finetti appreciated the importance of these questions from the beginning. In 1937 he wrote: "To get from the case of exchangeability to other cases which are more general but still tractable, we must take up the case where we still encounter 'analogies' among the events under consideration, but without their attaining the limiting case of exchange-ability."[20] To this end, de Finetti introduced the notion of partial exchangeability:

> All the conclusions and formulas which hold for the case of ex-changeability are easily extended to the present case of *partially exchangeable* events which we could define by the same sym-metry condition, specifying that the events divide into a certain number of types 1, 2, . . . g, and that it is only events of the same type which will be treated as "interchangeable" for all probabi-listic purposes. Here again, for this definition to be satisfied quite generally, it need only be satisfied in a seemingly much more special case: it suffices that there be a unique probability ω n_1, n_2, . . . n_g that n given events all happen, where n_1, n_2 . . . , n_g belong respectively to the first, second, . . . gth types and $n_1 + n_2$ + . . . + $n_g = n$—and where this unique probability is indepen-dent of the choice of particular events from each type.[21]

De Finetti introduces this notion of partial exchangeability with a coin-tossing example, where several odd-shaped coins are tossed. Here it is not so plausible to expect exchange-ability in the sequence of all tosses, but it is plausible to expect exchangeability within the sequence of tosses of a given coin. The example can be varied to illustrate various degrees of subjective analogy between the different types of toss. In one limiting case, the coins may be thought to be much the same, with the upshot of this judgment being global exchangeability. As the other end of the spectrum, the coins may be thought so different that the results of tossing one give no information about the bias of the others.

It may not be apparent how this notion of partial ex-changeability is meant to be applied to the theory of stochas-tic processes. The leading idea, in the Markov case, is clear: "In the particular case of events occurring in chronological order, the division into classes may depend on the result of the previous trial; in such a case we have the Markov form of partial exchangeability."[22] That the probability of an out-come on toss 56 of a Markov chain depends on the outcome of toss 55 is not a reflection of any special status of toss 55 over and above its being the predecessor of toss 56. Thus, if we take the subsequence of tosses whose predecessor has a given outcome, renumber them 1, 2, . . . in order of occurrence (thus suppressing some information as to position in the original sequence), we should expect the resulting sequence to be exchangeable.

Even though the leading idea is clear, this case is more complicated than the case of the odd coins, and de Finetti's suggestion constitutes a broadening of the notion of partial exchangeability. To illustrate, consider an example of Diaconis and Freedman.[23] Instead of flipping a coin we flick a thumbtack across the floor, and we play it as it lays. The probability of its landing point up (U) or not may plausibly depend on the result of the previous trial. It would be a mistake to think that we could get an exchangeable subsequence of trials just by conditioning on a statement which posits a fixed outcome (say, point up) for each of their predecessors. Conditioning on the statement that tosses 1, 5, and 6 have predecessors that land point up makes it certain that toss 5 lands point up but cannot be expected to make up on 1 or up on 6 certain. Here the very act of dividing the trials into classes may presuppose information which destroys the exchangeability of the trials within those classes.

On the other hand, there is a generalization from de Finetti's alternative characterization of exchangeability. We can almost give a frequentist characterization of a sequence, so that this characterization can plausibly be held to determine uniquely the probability of that sequence. That is, we can specify the frequency of transitions from state to state (up to up, up to down, etc.) in the sequence. I said "almost" because to this, we must add the non-frequentist information of the initial state of the system.[24] Given that we believe that the probabilities are governed by a Markov law, sequences with the same initial state and the same transition count should be equiprobable.

Freedman, in 1962,[25] brought exchangeability and various forms of partial exchangeability together under the concept of a certain form of sufficient statistic. Let a statistic here be a function from sequences of length n into some set of values (which may be a set of vectors). We say that the statistic, T, is a summarizing statistic for the probability assignment, P, on sequences of length n; or alternatively, that the probability assignment is *partially exchangeable with respect to the statistic* iff $T(x) = T(y)$ implies $P(x) = P(y)$ where x and y are sequences of length n. As Diaconis and Freedman[26] point out, this covers all the intended cases of partial exchangeability. In the case of exchangeability, the summarizing statistic consists just of the frequencies (or relative frequencies) of outcomes. In the case of tossing several funny coins and analogous cases, the value of the summarizing statistic consists of a vector whose components are the frequencies (or relative frequencies) for each coin. In the Markov case, we have a vector whose components are the transition counts together with the outcome of the first trial.

Although these two applications are much in the spirit of de Finetti's original analysis, the notion of partial exchangeability with respect to a statistic is a very substantial generalization of the notion of partial exchangeability. In particular, it should be noted that there is nothing in the definition of a summarizing statistic that requires it to be or involve a frequency count. Freedman's treatment moves us very far toward the higher levels of generality that will be discussed in the subsequent sections.

If the statistic, T, is a *summarizing statistic* for two probability assignments P_1 and P_2, then, by definition, for any value of the statistic, the sequences in its inverse image are given the same value by P_1 and likewise for P_2. Those sequences will therefore be given the same value by any weighted average of P_1 and P_2. That is, partial exchangeability with respect to a fixed statistic T is a property of probability distributions preserved by mixing. If we consider the family of probability assignments (on a given set of sequences of length n) which are partially exchangeable with respect to T, it is closed under mixing.[27] Those probability measures which concentrate probability one on the set of sequences which are the inverse image of one value of the statistic are the extreme points of the set of measures, i.e., the ones that cannot be represented as a non-degenerate[28] mixture of distinct measures in the set. If the set contains a regular assignment (one which gives each possible outcome

non-zero probability), then all the extremal probability assignments can be recovered from it by conditioning on the
possible values of the statistic. We now have the following
generalization of de Finetti's theorem for finite sequences:

> Every probability assignment P which is partially exchangeable
> with respect to T is a unique mixture of the extreme measures P_i.
> The mixing weights are $w_i = P (x:T(x) = t_i)$. [Diaconis and Freed-
> man, 1980]

(For a discussion of the extreme measures in the finite case
for some of the special cases of partial exchangeability that
have been mentioned, see Diaconis and Freedman.)[29]

Freedman (1962) proves a general representation theorem
for partial exchangeability with respect to a statistic. Since
this concept is defined with respect to finite sequences, it is
necessary to find a way to relate the various summarizing
statistics for subsequences of an infinite sequence. For our
probability space with infinite sequence as its elements, let
us take as our new sense of summarizing statistic an infinite
sequence of summarizing statistics in the old sense. That is,
a statistic, U, in the new sense can be taken as a function
from the integers to statistics in the old sense for finite sequences, and a summarizing statistic in the new sense is one
such that for any n, U_n, is such that if two finite sequences
have the same value of U_n the set of infinite sequences that
has them as subsequences in a specified position must be
equiprobable. Freedman introduced the notion of an S-
structure to assure that the U_ns mesh together nicely. U has
an S-structure iff for any finite sequences A, A' of length n
and B, B' of length m, if $U_n(A) = U_n(A')$ and $U_m(B) = U_m(B')$
then $U_{n+m}(A \ @ \ B) = U_{n+m}(A' \ @ \ B')$, where $A \ @ \ B$ is the
result of concatenating the sequence A with the sequence
B. The summarizing statistics mentioned specifically in
this section (relative frequency, frequency within types, initial outcome together with transition counts) all have an
S-structure. Freedman then proves that a probability P is
partially exchangeable with respect to (is summarized by) a
statistic U with an S-structure if and only if it has a representation as a mixture of ergodic (metrically transitive) proba-

bilities which are partially exchangeable with respect to U. (I will postpone the discussion of ergodic measures until the next section.) He uses this result to attack the question of the characterization of mixtures of Markov chains. The answer when the stochastic process is stationary is in Freedman (1962). Diaconis and Freedman (1980b)[30] show that when a stochastic process is recurrent[31] it has a representation as a mixture of Markov chains if and only if it is partially exchangeable with respect to the statistic of initial state and transition count. The representation is unique.

In the simpler case of partial exchangeability (several different biased coins), a representation as a mixture of products of Bernoullian measures is given in de Finetti (1938),[32] Link (1980),[33] and Diaconis and Freedman (1980).[34] Link gives proofs of representation theorems for such "k-fold partial exchangeability" using Choquet's theorem as the main analytical tool. Diaconis and Freedman indicate how the proof of one of these theorems can be gotten by passing to the limit in the finite case.

On the other hand, when random processes are generalized to random fields, there are generalizations of de Finetti's theorem. This is an area of much current interest, largely because of the connections with statistical mechanics and thermodynamics. Two recent studies are Preston[35] and Georgii.[36]

It is clear that methods in the spirit of de Finetti can be used to give an analogous account of a more general class of cases than those treated in de Finetti's original theorem. The question as to how general that class in fact is has an answer which depends in part on ongoing mathematical research and in part on how generously we interpret "analogous" and "in the spirit of de Finetti." The consequences of a generous interpretation will be discussed in the next two sections.

INVARIANCE, RESILIENCY, AND ERGODICITY

Many of us would like to think of physical probabilities quite generally in the way that de Finetti thinks of the factors in his theorem; as an artifact of our representation of the epistemic probabilities as a mixture. What is needed to back up such a position is a general mathematical theory. Ideally such a theory should do two things for us. First, it should prove a unique representation theorem. That is, given degrees of belief about the outcome of an experiment (which might be required to be invariant provided that the basic experimental arrangement is fixed), we would like to have a unique representation of our degrees of belief as a mixture of physical probabilities. Second, we would like to have an account of learning from experience. This could be provided if we could prove some kind of law of large numbers for repetitions of an experiment. If we had such an account, we could show how to learn the correct physical probabilities from an experiment, and how it is possible to use the correct physical probabilities to predict the results of a sequence of experiments. Finally we, or at least I, would like to claim that the "objective physical probabilities" should have an objective *resiliency* or invariance under conditionalization, which is connected with the subjective invariance of the degrees of belief with which we started.

Let me formulate these desiderata for the mathematics of a subjectivist theory of chance precisely at a high level of generality. Let our degrees of belief be represented by a probability measure, P, on a standard Borel space (Ω, F, P),[37] where Ω is a set, F is a sigma-field of measurable subsets of Ω, and P is a probability measure on F. We will think of a point in Ω as specifying not only the outcome of an experiment but also the experimental arrangement and background conditions of the experiment. Although we may not know exactly what the physical probabilities are, we may know that some of the background conditions (e.g., time of experiment) are irrelevant to them. That is, the physical probabilities should be invariant under variation of some background conditions and so should the epistemic probability that is a mixture of the possible physical probabilities. We may typically expect that there may be some such symmetry or invariance which characterizes our epistemic probabilities. This may be expressed formally by introducing a transformation (e.g., shift in time of the experiment) which maps the prob-

ability space, Ω, into itself. (We should require that the transformation respect measurability, i.e., that the inverse image
of a measurable set be measurable.) We will then say that our
probability, P, is *invariant with respect to the transformation, T*, if for every measurable set A, a set which the transformation carries into it, $T^{-1}A$, has the same probability that
it does: $P\ (T^{-1}A) = P(A)$. Invariance with respect to a fixed
transformation is a property of probability measures that is
preserved under mixing.

We know that some background factors are irrelevant to
the physical probabilities, and this information manifests
itself as invariance of epistemic probabilities with respect to
some transformation, T. But we may not know the true values of the relevant variables, so we are uncertain about the
correct physical probabilities. We want to think of the possible physical probabilities as probabilities which are fixed
by a knowledge of *all* the relevant factors. The "all" is important. If we are uncertain as to some of them, we would
have our degrees of belief being a mixture of possible physical probabilities. We are thus led naturally to the following
characterization of the possible physical probability measures of which our epistemic probability measure is supposed to be a mixture:

I. *INVARIANCE*: A possible physical probability measure
             should be invariant with respect to T.
II. *RESILIENCY*: A possible physical probability measure
             should be such that it is not possible to
             move to a "truer" physical probability by
             conditionalizing on a further specification
             of *projectible experimental factors*.

The second condition, *resiliency*, is the condition that we
have *all* the relevant factors built into our notion of possible
physical probability.

Of course, no non-degenerate probability space will be such
that the probability measure will be invariant under conditionalization on *any* set, so the question of *projectibility* is
vital. A specification of *projectible experimental factors* is
represented by a set in our probability space that is *invariant*

with respect to the transformation (technically: a set A such that the probability of the symmetric difference between A and $T^{-1}(A)$ has probability zero). The idea is that projectible factors should be invariant under the transformation which represents repetitions of the experiment.[38] Resiliency then becomes II': *There is no invariant set which the measure in question gives probability other than zero or one.* Conditioning on a set of measure one would leave the probability unchanged; conditioning on a set of measure zero is undefined. If a probability satisfies both I and II', I will say it is *objectively resilient* with respect to T. The desired representation theorem is then: Every probability *invariant* with respect to T has a unique representation as a mixture of probabilities that are *objectively resilient* with respect to T.

We also want our physical probabilities to have a connection with frequency. For a probability measure that is *objectively resilient* with respect to T, the transformation, T, is to be thought of as a repetition of the experiment with all the relevant factors fixed. For each point, ω, in the probability space, Ω, we can consider the sequence, ω, $T\omega$, $T(T\omega)$, ... $T^n\omega$ ..., and we can consider the relative frequency of any measurable set, A, in any finite segment of that sequence, and the limiting relative frequency in the sequence provided that it exists.

Call this sequence the (positive) *orbit* of ω. The nicest generalization of the law of large numbers that we could imagine here is that for any measurable set A, the limiting relative frequency of A in the orbit of ω exists and equals the physical probability of A almost everywhere (i.e., for every ω in Ω excepting a set of measure zero).

The theory that I have been hypothesizing already exists, although it is not usually viewed in this way. It is none other than the theory of measure-preserving transformations—the modern form of ergodic theory.[39] Where I said that the probability measure, P, is invariant with respect to T, the more usual terminology is that T is measure preserving with respect to P. Where I said that P is *objectively resilient* with respect to T, it is more common to say that T is *ergodic* (or metrically transitive or indecomposable) with respect to P.

The fantasized generalization of the law of large numbers is nothing other than Birkhoff's celebrated pointwise ergodic theorem[40] (or rather a special case thereof). The fantasized representation theorem is a version of the Kryloff-Bogoliouboff theorem.[41]

Actually, Birkhoff established the existence of the limiting relative frequency almost everywhere in general, when T is measure preserving with respect to P, but not necessarily ergodic with respect to P. If the transformation is, in addition, ergodic with respect to P, then the limiting relative frequency is almost everywhere constant and equal to the probability.[42]

In light of the representation theorem, this should be understood as follows: If the probability P is *invariant* with respect to T, then the limiting relative frequencies will almost surely converge to a probability measure P' which is one of the probability measures *objectively resilient* with respect to T. In terms of the interpretation we are pursuing here, let P be the probability measure representing *your* degrees of belief. Let T be a transformation which leaves *your* degrees of belief invariant. It determines *your* conception of a repetition of the same experiment and *your* notion of projectibility and *your* notion of *chance*. Then the limiting relative frequencies under repetition of the experiment will almost surely (with respect to *your* degrees of belief) converge to one of *your* chance distributions. *By virtue of a symmetry in your degrees of belief you must act as if you believe in objectively resilient chances to which you believe relative frequencies will converge.*

We characterized your objectively resilient "chance" probability measures as ones which were invariant under conditionalization on projectible predicates. We can show that these are the extreme points of the convex set of invariant measures. That is, we can show that a probability measure is objectively resilient if and only if it is (a) invariant and (b) cannot be represented as a non-trivial mixture:

$$P = (a/a + b)P_1 + (b/a + b)P_2 \qquad [a, b > 0; P_1 \neq P_2]$$

of distinct invariant measures.

Proof: One direction is trivial. If P is not objectively resilient, then by condition II there is a projectible property, A, such that $P(A) \neq 1$ & $P(A) \neq 0$. Then P can be written as a non-trivial mixture of invariant probability measures, i.e., the measures gotten by conditionalizing on A and on $\sim A$. For the other direction, notice that if P is a non-trivial average of P_1 and P_2, and P is objectively resilient, then P_1 and P_2 must be also, since under these assumptions P gives a set probability 0 only if P_1 and P_2 give it probability 0; likewise for probability one. Now suppose for *reductio* that P is objectively resilient and a non-trivial average of the invariant measures P_1 and P_2. Then, by the previous observation, P_1 and P_2 are also objectively resilient. By hypothesis, P_1 and P_2 are distinct, so there is a measurable set, A, to which they assign different values. By the ergodic theorem, the limiting relative frequency of A almost everywhere exists and is equal to $P_1(A)$, $P_2(A)$ and $P(A)$. This contradicts the assumption that $P_1(A) \neq P_2(A)$.

Historically, ergodic theory arose out of statistical mechanics. Birkhoff replaced Boltzmann's unworkable conception of ergodicity with the hypothesis of metric indecomposability. Here the transformation, T, is thought of as a time transformation specified by the dynamical law of the system. Then, taking M to be Lebesgue measure on phase space (suitably coordinatized), we have *invariance* with respect to T by Liouville's theorem. The question is whether one can prove that the Lebesgue measure restricted to constant energy hypersurfaces is objectively resilient (ergodic, metrically indecomposable) with respect to T. This is the version of the ergodic question that is relevant post-Birkhoff. Where the Kryloff-Bogoliouboff theorem applies we know that the Liouville measure on phase space is representable as a mixture of objectively resilient ergodic measures. We also know that each of these ergodic measures must concentrate probability one on some subset of a constant energy hypersurface, because the Liouville measure restricted to a given constant energy hypersurface is again an invariant measure. But the possibility is left open by the general theory that several ergodic measures each assign probability one to a different

member of a partition of a given constant energy hypersurface. Whether this is so depends on the particular dynamical
system in question. Proving the ergodic theorem for a given
dynamical system requires proving that this is not so. This
is an exceptionally difficult mathematical problem for the
physical systems for which the theory was designed, and
even for idealized approximations of them.[43] Although the
mathematics of ergodic theory was born with the special
concerns of statistical mechanics in mind it is congenial to
the much more general interpretation that I have put on it.

The treatment of partial exchangeability for stochastic processes in the last section can be seen as a special case of the
theory of measure-preserving transformations. Consider a
stochastic process consisting of a doubly infinite sequence
of random variables, $\ldots f_{-2}, f_{-1}, f_0, f_1, f_2 \ldots$, on a common
probability space, S, and let Ω be the infinite product space
whose elements consist of doubly infinite sequences of
members of S, $\omega = (\omega_{-1}, \omega_0, \omega_1 \ldots)$. Let T be the *shift* transformation which takes $(\omega_{-1}, \omega_0, \omega_1, \ldots)$ to $(\omega_0, \omega_1, \omega_2 \ldots)$.
Any statement about the random variables can be rephrased
in terms of one of them and the shift transformation. To say
that the shift is *measure preserving* or that the measure is
invariant with respect to the shift is the same as saying that
the stochastic process is *stationary*.

Suppose that the stochastic process is Bernoullian; that is,
that the random variables are independent. Then the shift is
ergodic with respect to the product measure on the product
space, Ω. The Birkhoff ergodic theorem applied to this case
gives the strong law of large numbers for Bernoulli sequences.[44] The Bernoullian measure is not the only sort of
measure for which the shift transformation is ergodic. Certain Markov measures corresponding to certain Markov processes do as well.[45] So the Bernoullian measures are not the
only extreme measures corresponding to the shift transformation. However, if, instead of the shift, we consider the
group of transformations which take the sequence ω into
another ω' just in case one can be gotten from the other by a
finite permutation of its elements, the measures with respect

to which that group of transformations is measure preserving are the exchangeable ones, and the objectively resilient probability measures are the independent ones.

The de Finetti analysis of exchangeable sequences as mixtures of independent ones is a special case of the theorems of ergodic theory. Freedman[46] uses the ergodic representation theorem as a tool to show that any stochastic process that is partially exchangeable with respect to the statistic consisting of the transition counts and the outcome of the first trial is a mixture of stationary Markov chains. In the course of this investigation he proves the more general theorem concerning partial exchangeability with respect to a statistic to which I referred in the previous section.

At a higher level of generality, the theory of measure-preserving transformations itself *is* the generalization of de Finetti's analysis. The theory of stationary stochastic processes is a special case. We need not confine ourselves to stochastic processes and to the shift transformation. Rather we can, under modest regularity conditions, consider an arbitrary probability space and an arbitrary measure-preserving transformation (or group of measure preserving transformations) which leave the probability invariant. This is quite a generalization.

The key element of the whole representation is the symmetry in one's subjective-probability measure, the transformation or group of transformations which leave it invariant. Does every subjective-probability distribution have some such symmetry? There is a trivial affirmative answer. The identity map is a measure-preserving transformation. If this is the only measure-preserving transformation for a person's subjective probability, then he will count as a repetition of the experiment only one with the same experimental conditions and the same experimental results. His chance distributions will each be concentrated on one point in his probability space. His conception of chance will be degenerate, with the chances all being zero or one. Most believers will have more generous symmetries in their degrees of belief, and will act as if they had a more interesting conception of chance.

We think of our group of transformations as varying those factors which are irrelevant to fixing the physical probabilities. Then the objectively resilient (ergodic) probability measures are intuitively probabilities conditional on the relevant experimental variables being fixed in a certain definite way. This intuition is correct in the following sense: *physical probability or chance* as a random variable is here probability conditional on the sub-sigma algebra of *projectible experimental factors* (invariant sets).[47] The chances you believe to be present in a given experiment (a point, ω,) are determined by the totality of the projectible experimental factors (invariant sets) which are present in that experiment (which have ω as a member). In this way, the invariant sets represent one's individuation of the chance setup.

We have now come almost full circle to the sort of picture given in section 1 of this chapter, but the account has gained in detail. There *chance* as a random variable was probability conditional on a partition; here that idea is given its straightforward generalization to probability conditional on a sub-sigma algebra. Furthermore, we can say something useful about this sub-sigma algebra. It is the sub-sigma algebra of invariant sets with respect to the transformation or group of transformations determining chance. To give a representation in terms of a person's degrees of belief about chance, we find the appropriate measure preserving transformations from his degrees of belief about experiments and outcomes.

Thus the question "When is it rational to have degrees of belief about chances, or to behave as if one does?" is reduced to the question of the rationality of degrees of belief about less mysterious entities. This is what I mean by a *pragmatic reduction* of the concept of chance.

PROBABILISTIC PRESUPPOSITIONS OF INDUCTION

You can have your prior degrees of belief be in accordance with the probability calculus, conditionalize on your data and still refuse to learn from experience if your prior is con-

centrated on one extreme point. Or your state space for describing nature might be so pathological that the representation theorems fail because of failures in measurability. In
this sense, induction has its probabilistic presuppositions.

Our conception of these presuppositions is profoundly
transformed by the line of analysis initiated by de Finetti. I
count the theory of measure-preserving transformations as
part of this line of analysis. Invariance under a group of
measure-preserving transformations is a natural generalization of partial exchangeability, even though it was not at first
introduced with these concerns in mind.

The common theme of this line of analysis, at increasingly
higher levels of generality, has been that the precondition for
learning from experience is the adoption of an epistemic
probability measure which is a nontrivial mixture of extremal measures. The effect of conditioning on a finite number
of observations will, in general, be to remix the extremal
measures with a different weighting function. As the number
of observations goes to infinity, the weight almost surely con-
centrates on one of the extremal measures. We say that we
learn what the *chances* (propensities, physical probabilities)
are. In the special case of mixtures of independent sequences, this means that we have positive relevance of outcome types between trials; e.g., observation of a head makes
it more likely that the next trial will come up heads. We do
not have this type of learning from experience in general,
and do not want it. As we approach concentration on the
physical probabilities answering to the following Markov
transition matrix:

	S_1	S_2
S_1	.1	.9
S_2	.9	.1

we want just the opposite. Learning from experience is best
thought of as convergence to an extremal measure.

The ergodic theorem clarifies both the nature of the extremal measures and their connection with limiting relative
frequency. Once we fasten on a relevant notion of repetition
of an experiment by choice of a transformation which preserves our subjective probability measure, we find that the
extremal measures are objectively resilient and that they satisfy the form of the law of large numbers with respect to
repetitions of the experiment that finds expression in Birkhoff's ergodic theorem.

It is perhaps worthwhile underlining the differences between this analysis and customary applications of the strong
law of large numbers for independent trials. In the first place,
we are not assuming that you, the epistemic agent, know that
the trials are independent or even that your degrees of belief
exhibit any given symmetry. We assume only that your degrees of belief exhibit some symmetry or other, which is hardly
any assumption at all. In the second place, there is the question of your proper attitude to the proof that there is convergence to a limiting relative frequency except for a set of points
of measure zero. It is here inappropriate for you to ask the
standard question, "Why should I believe that the real situation is not in that set of measure zero?" The measure in
question is your degree of belief. You do believe that the real
situation is not in that set, with degree of belief one.

We see how difficult it is to not behave inductively and to
not believe in inductive success in the long run. We see how
difficult it is to act as if one were not uncertain about chance.
The interesting questions are not whether we do or should
behave inductively, but about how we do or should behave
inductively with respect to what conception of chance. These
questions reduce to questions about the symmetries that we
do or should have in our subjective degrees of belief.

The propensity interpretation of the calculus of probability, and the quantum theory

by

K. R. POPPER

IN this paper, I propose briefly to put forth and to explain the following theses, and to indicate the manner of their defence.

(1) The solution of the problem of interpreting probability theory is fundamental for the interpretation of quantum theory; for quantum theory is a probabilistic theory.

(2) The idea of a statistical interpretation is correct, but is lacking in clarity.

(3) As a consequence of this lack of clarity, the usual interpretation of probability in physics *oscillates* between two extremes: an *objective* purely statistical interpretation and a *subjective* interpretation in terms of our incomplete knowledge, or of the available information.

(4) In the orthodox Copenhagen interpretation of quantum theory we find the same oscillation between an objective and a subjective interpretation: *the famous intrusion of the observer into physics.*

(5) As opposed to all this, a revised or reformed statistical interpretation is here proposed. It is called the *propensity interpretation of probability.*

(6) The propensity interpretation is a purely objective interpretation. It eliminates the oscillations between objective and subjective interpretation, and with it the intrusion of the subject into physics.

(7) The idea of propensities is 'metaphysical', in exactly the same sense as forces or fields of forces are metaphysical.

(8) It is also 'metaphysical' in another sense: in the sense of providing a coherent programme for physical research.

These are my theses. I begin by explaining what I call the propensity interpretation of probability theory.[1]

Section 1. Objective and Subjective Interpretations of Probability.

Let us assume that we have two dice: one is a *regular* die of homogeneous material, the other is *loaded*, in such a way that in long sequences of throws the side marked '6' comes uppermost in about 1/4 of the throws. We say, in this case, that the probability of throwing a 6 is 1/4.

Now the following line of arguing seems attractive.

We ask what we *mean* by saying that the probability is 1/4; and we may arrive at the answer: What we *mean*, precisely, is that the relative frequency, or the statistical

---

[1] I have explained the propensity interpretation of probability and of quantum theory very briefly in my paper 'Three Views Concerning Human Knowledge', in *Contemporary British Philosophy*, edited by H. D. Lewis, 1956, p. 388. A full treatment of the propensity interpretation and of its repercussions upon quantum theory will be found in the *Postscript: After Twenty Years* to my *Logic of Scientific Discovery*, 1957.

frequency, of the results in long sequences is $1/4$. Thus probability is relative frequency in the long run. This is the statistical interpretation.

The statistical interpretation has been often criticized because of the difficulties of the phrase 'in the long run'. I will *not* discuss this question. Instead I will discuss the question of *the probability of a* SINGLE EVENT. This question is of importance in connexion with quantum theory because the ψ-function determines the probability of a *single electron* to take up a certain state, under certain conditions.

Thus we ask ourselves now what it *means* to say 'The probability of throwing 6 *with the next throw* of this loaded die is $1/4$.'

From the point of view of the *statistical interpretation*, this can only mean one thing: 'The next throw is a *member of a sequence* of throws, and the relative frequency within this sequence is $1/4$.'

At first sight, this answer seems satisfactory. But we can ask the following awkward question:

What if the sequence consists of throws of a *loaded* die, with one or two throws of a *regular* die occurring in between the others? Clearly, we shall say about the throws with the regular die that their probability is different from $1/4$, in spite of the fact that these throws are members of a sequence of throws with the frequency $1/4$.

This simple objection is of fundamental importance. It can be answered in various ways. I shall mention two of these answers, one leading to a *subjective interpretation*, the other to the *propensity interpretation*.

The first or subjective answer is this. 'You have assumed in your question', the subjectivist may address me, 'that *we know* that the one die is loaded, the other regular, and also that *we know* whether the one or the other is used at a certain place in the sequence of throws. In view of this information, we shall of course attribute the proper probabilities to the various single throws. For probability, as your own objection shows, is not simply a frequency in a sequence. Admittedly, observed frequencies are important as providing us with valuable *information*. But we must use *all* our information. The probability is our assessment, in the light *of all we know*, of reasonable betting odds. It is a measure which depends essentially upon our incomplete information, and *it is a measure of the incompleteness of our information*: if our information about the conditions under which the die will be thrown were sufficiently precise, then there would be no difficulty in predicting the result with certainty.'

This is the subjectivist's answer, and I shall take it as a characterization of the subjectivist position which I shall not discuss further in this paper, although I shall mention it in various places.[2]

Now what will the defender of an objective interpretation say to our fundamental objection? Most likely he will say (as I myself used to say for a long time) the following:

'To make a statement about probability is to propose a *hypothesis*. It is a hypothesis about frequencies in a sequence of events. In proposing this hypothesis, we can make use of all sorts of things—of past experience, or of inspiration: it does not matter *how we get* it; all that matters is *how we test it*. Now in the case mentioned, we all agree on the frequency hypothesis, and we all agree that the frequency of $1/4$

[2] I have discussed and criticized the subjectivist position very fully elsewhere (see the preceding footnote). The subjectivist interpretation of probability is a necessary consequence of determinism. Its retention within the quantum theory is a residue of a not yet fully eliminated determinist position.

90

will not be affected by having one or two throws with a regular die in between the throws with a loaded die. As to the regular throws, *if* we consider them merely as belonging to this sequence, we have to attribute to them, strange as it may sound, the probability of $1/4$, even though they are throws with a regular die. And if, on the other hand, we attribute to them the probability of $1/6$, then we do so because of the hypothesis that in *another* sequence—one of throws with the regular die—the frequency will be $1/6$.'

This is the objectivist's defence of the purely statistical interpretation, or of the frequency interpretation, and *as far as it goes* I still agree with it.

But I now think it strange that I did not press my question further. For it seems clear to me now that this answer of mine, or of the objectivist's, implies the following. In attributing probabilities to sequences, we consider as decisive the *conditions under which the sequence is produced*. In assuming that a sequence of throws of a loaded die will be different from a sequence of throws of a regular die, we attribute the probability to the *experimental conditions*. But this leads to the following result.

Even though probabilities may be said to be frequencies, we believe that these *frequencies will depend on the experimental arrangement.*

But with this, we come to a new version of the objective interpretation. It is as follows.

Every experimental arrangement is *liable to produce*, if we repeat the experiment very often, a sequence with frequencies which depend upon this particular experimental arrangement. These virtual frequencies may be called probabilities. But since the probabilities turn out to depend upon the experimental arrangement, they may be looked upon as *properties of this arrangement*. *They characterize the disposition, or the propensity*, of the experimental arrangement to give rise to certain characteristic frequencies *when the experiment is often repeated.*

Section 2. The Propensity Interpretation

We thus arrive at the propensity interpretation of probability.[3] It differs from the

[3] What we interpret is not a word, 'probability', and its 'meaning', but formal systems—the probability calculus (especially in its measure-theoretical form), and the formalism of quantum theory.

A formalized set of axioms for relative (or conditional) probability is the following. (The theory of real numbers is assumed.)
(A) $a, b \in K \rightarrow : p(a, b)$ is a real number, and
      (a) $((c)(c \in K \rightarrow p(a, c) = p(b, c))) \rightarrow (d)(d \in K \rightarrow p(d, a) = p(d, b))$
      (a') $p(a, a) = 1$
(B) $a, b \in K \rightarrow : ab \in K$; and $b, c, bc \in K \rightarrow p(a, bc)p(b, c) = p(ab, c) \leqslant p(a, c)$
(C) $a \in K \rightarrow : 'a \in K$, and $b, c \in K \rightarrow . p(b, c) \neq 1 \rightarrow p(a, c) + p('a, c) = 1$
(D) $(Ea)(Eb)(a, b \in K, $ and $p(a, b) \neq 1)$
(ab is here, of course, the meet or conjunction of a and b, and 'a the complement of a.)

This set is equivalent to a more concise set in which no numerical constant such as '1' occurs. It is obtained by retaining (a), omitting (a'), and replacing the second (operational) lines of (B) and (C), as well as (D), respectively, by the following lines ((B) is now made 'organic'):
      (b) $d \in K \rightarrow ((p(a, bc)p(b, c) = p(d, c) \rightarrow p(a, c) < p(d, c)) \rightarrow p(ab, c) < p(d, c))$
      (c) $d \in K \rightarrow (p(a, a) \neq p(b, c) \rightarrow p(a, c) + p('a, c) = p(d, d))$
      (d) $(Ea)(Eb)(a, b \in K, $ and $p(a, b) = p(b, b))$

This axiom system has the following properties: (1) if we define absolute probability by
$$p(a) = p(a, '(a' a)),$$
then we can say that $p(a, b)$ is defined even if $p(b) = 0$. In this respect, the system is a generalization of the systems known so far (except my systems published in *B.J.P.S.*, **6**, 1955, pp. 56f., and in *British Philosophy in the Mid-Century*, ed. C. A. Mace, 1957, p. 191). (2) It is not, as in other systems, tacitly assumed that the elements of K satisfy the postulates of Boolean algebra (this fact, on the contrary, can be proved). This is of advantage in connexion with the problem of interpretation, such as the problem whether propensities satisfy Boolean algebra: it becomes superfluous to postulate that they do.

F

purely statistical or frequency interpretation only in this—that it considers the probability as a characteristic property of the experimental arrangement rather than as a property of a sequence.

The main point of this change is that we now take as fundamental *the probability of the result of a single experiment*, with respect to its *conditions*, rather than the frequency of results in a sequence of experiments. Admittedly, if we wish to *test* a probability statement, we have to test an experimental sequence. But now the probability statement is not a statement *about* this sequence: it is a statement *about* certain properties of the experimental conditions, of the experimental set-up. (Mathematically, the change corresponds to the transition from the frequency-theory to the measure-theoretical approach.)

A statement about propensities may be compared with a statement about the strength of an electric field. We can test this statement only if we introduce a test body and measure the effect of the field upon this body. But the statement which we test speaks about the field rather than about the body. It speaks about certain *dispositional properties* of the field. And just as we can consider the field as physically real, so we can consider the propensities as physically real. They are *relational* properties of the experimental set-up. For example, the propensity 1/4 *is not a property of our loaded die*. This can be seen at once if we consider that in a very weak gravitational field, the load will have little effect—the propensity of throwing a 6 may decrease from 1/4 to very nearly 1/6. In a strong gravitational field, the load will be more effective and the same die will exhibit a propensity of 1/3 or 1/2. The tendency or disposition or propensity is therefore, as a relational property of the experimental set-up, something more abstract than, say, a Newtonian force with its simple rules of vectorial addition. *The propensity distribution attributes weights to all possible results of the experiment.* Clearly, it can be represented by a vector in the *space of possibilities*.

Section 3. Propensity and Quantum Theory

The main thing about the propensity interpretation is that *it takes the mystery out of quantum theory, while leaving probability and indeterminism in it*. It does so by pointing out that all the apparent mysteries would also involve thrown dice, or tossed pennies—*exactly* as they do electrons. In other words, it shows that quantum theory is a probability theory just as any theory of any other game of chance, such as the bagatelle board (pin board).

In our interpretation, Schrödinger's ψ-function determines the propensities of the states of the electron. We therefore have no 'dualism' of particles and waves. The electron is a particle, but its wave theory is a propensity theory which attributes weights to the electron's possible states. The waves in configuration space are waves of weights, or waves of propensities.

Let us consider Dirac's example of a photon and a polarizer. According to Dirac, we have to say that the photon is in both possible states at once, half in each; even although it is indivisible, and although we can find it, or observe it, in only one of its possible states.

We can translate this as follows. The theory describes, and gives weight to, all the possible states—in our case, two. The photon will be in one state only. The situation is exactly the same as with a tossed penny. Assume that we have tossed the penny, and that we are shortsighted and have to bend down before we can observe which side is upmost. The probability formalism tells us then that each of the possible states has a probability of 1/2. So we can say that the penny is half in one state, and half in the other. And when we bend down to observe it, the Copenhagen spirit will inspire the penny to make a quantum jump into one of its two *Eigen*-states. For nowadays a quantum jump is said by Heisenberg to be the same as a reduction of the wave packet. And by 'observing' the penny, we induce exactly what in Copenhagen is called a 'reduction of the wave packet'.

The famous two-slit experiment allows exactly the same analysis. If we shut one slit, we interfere with the possibilities, and therefore get a different ψ-function, and a different probability distribution of the possible results. *Every change in the experimental arrangement such as the shutting of a slit, will lead to a different distribution of weights to the possibilities* (just as will the shifting of a pin on a pin board). That is, we obtain a different ψ-function, determining a different distribution of the propensities.

There is nothing peculiar about the role of the observer: he does not come in at all. What 'interferes' with the ψ-function are only changes of experimental arrangements.

The opposite impression is due to an oscillation between an objective and a subjective interpretation of probability. It is the subjective interpretation which drags in our knowledge, and its changes, while we ought to speak only of experimental arrangements, and the results of experiments.

Section 4. Metaphysical Considerations

I have stressed that the propensities are not only as objective as the experimental arrangements but also *physically real*—in the sense in which forces, and fields of forces, are *physically real*. Nevertheless they are *not* pilot-waves in ordinary space, but weight functions of possibilities, that is to say, vectors in possibility space. (Bohm's 'quantum-mechanical potential' would become here a propensity to accelerate, rather than an accelerating force. This would give full weight to the Pauli-Einstein criticism of the pilot-wave theory of de Broglie and Bohm.) We are quite used to the fact that such abstract things as, for example, degrees of freedom, have a very real influence on our results, and are in so far something physically real. Or consider the fact that, compared with the mass of the sun, the masses of the planets are negligible, and that, compared with the masses of the planets, those of their moons are also negligible. This is an abstract, a relational fact, not attributable to any planet or to any point in space, but a relational property of the whole solar system. Nevertheless, there is every reason to believe that it is one of the 'causes' of the stability of the solar system. Thus abstract relational facts can be 'causes', and in that sense physically real.

It seems to me that by stressing that the ψ-function describes physical realities, we may be able to bridge the gap between those who rightly stress the statistical character of modern physics and those who, like Einstein and Schrödinger, insist that physics has to describe an objective physical reality. The two points of view are incompatible on the subjectivist assumption that statistical laws describe our own imperfect state of knowledge. They become compatible if only we realize that these statistical laws describe propensities, that is to say, objective relational propertie of the physical world.

Beyond this, the propensity interpretation seems to offer a new metaphysical inter pretation of physics (and incidentally also of biology and psychology). For we car say that all physical (and psychological) properties are dispositional. That a surfac is coloured red means that it has the disposition to reflect light of a certain wav length. That a beam of light has a certain wave length means that it is disposed t behave in a certain manner if surfaces of various colours, or prisms, or spectographs or slotted screens, etc., are put in its way.

Aristotle put the propensities as potentialities *into* the things. Newton's was th first *relational* theory of physical dispositions and his gravitational theory led, almos inevitably, to a theory of fields of forces. I believe that the propensity interpretatior of probability may take this development one step further.

Rudolf Carnap

STATISTICAL AND
INDUCTIVE PROBABILITY

If you ask a scientist whether the term "probability" as used in science has always the same meaning, you will find a curious situation. Practically everyone will say that there is only one scientific meaning; but when you ask that it be stated, two different answers will come forth. The majority will refer to the concept of probability used in mathematical statistics and its scientific applications. However, there is a minority of those who regard a certain nonstatistical concept as the only scientific concept of probability. Since either side holds that its concept is the only correct one, neither seems willing to relinquish the term "probability." Finally, there are a few people—and among them this author—who believe that an unbiased examination must come to the conclusion that both concepts are necessary for science, though in different contexts.

I will now explain both concepts—distinguishing them as "statistical probability" and "inductive probability"—and indicate their different functions in science. We shall see, incidentally, that the inductive concept, now advocated by a heretic minority, is not a new invention of the twentieth century, but was the prevailing one in an earlier period and only forgotten later on.

The *statistical concept of probability* is well known to all those who apply in their scientific work the customary methods of mathematical statistics. In this field, exact methods for calculations employing statistical probability are developed and rules for its application are given. In the simplest cases, probability in this sense means the relative frequency with which a certain kind of event occurs within a given reference class, customarily called the "population." Thus, the statement "The probability that an inhabitant of the United States belongs to blood group A is p" means that a fraction p of the inhabitants belongs to this group. Sometimes a statement of statistical probability refers, not to an

---

Published by The Galois Institute of Mathematics and Art, 1955. Reprinted by permission of the author and the publisher.

actually existing or observed frequency, but to a potential one, i.e., to a frequency that would occur under certain specifiable circumstances. Suppose, for example, a physicist carefully examines a newly made die and finds it is a geometrically perfect and materially homogeneous cube. He may then assert that the probability of obtaining an ace by a throw of this die is 1/6. This means that *if* a sufficiently long series of throws with this die were made, the relative frequency of aces would be 1/6. Thus, the probability statement here refers to a potential frequency rather than to an actual one. Indeed, if the die were destroyed before any throws were made, the assertion would still be valid. Exactly speaking, the statement refers to the physical microstate of the die; without specifying its details (which presumably are not known), it is characterized as being such that certain results would be obtained if the die were subjected to certain experimental procedures. Thus the statistical concept of probability is not essentially different from other disposition concepts which characterize the objective state of a thing by describing reactions to experimental conditions, as, for example, the I.Q. of a person, the elasticity of a material object, etc.

Inductive probability occurs in contexts of another kind; it is ascribed to a hypothesis with respect to a body of evidence. The hypothesis may be any statement concerning unknown facts, say, a prediction of a future event, e.g., tomorrow's weather or the outcome of a planned experiment or of a presidential election, or a presumption concerning the unobserved cause of an observed event. Any set of known or assumed facts may serve as evidence; it consists usually in results of observations which have been made. To say that the hypothesis h has the probability p (say, 3/5) with respect to the evidence e, means that for anyone to whom this evidence but no other relevant knowledge is available, it would be reasonable to believe in h to the degree p or, more exactly, it would be unreasonable for him to bet on h at odds higher than $p:(1 - p)$ (in the example, 3:2). Thus inductive probability measures the strength of support given to h by e or the *degree of confirmation* of h on the basis of e. In most cases in ordinary discourse, even among scientists, inductive probability is not specified by a numerical value but merely as being high or low or, in a comparative judgment, as being higher than another probability. It is important to recognize that every inductive probability judgment is relative to some evidence. In many cases no explicit reference to evidence is made; it is then to be understood that the totality of relevant information available to the speaker is meant as evidence. If a member of a jury says that the defendant is very probably innocent or that, of two witnesses A and B who have made contradictory statements, it is more probable that A lied than that B did, he means it with respect to the evidence that was presented in the trial plus any psychological or other relevant knowledge of a general nature he may possess. Probability as understood in contexts of this kind is not frequency. Thus, in our example, the evidence concerning the defendant, which was presented in the trial, may be such that it cannot be ascribed to any other person; and if it could be ascribed to several people, the juror would not know the relative frequency of innocent persons among them. Thus

97

the probability concept used here cannot be the statistical one. While a statement of statistical probability asserts a matter of fact, a statement of inductive probability is of a purely logical nature. If hypothesis and evidence are given, the probability can be determined by logical analysis and mathematical calculation.

One of the basic principles of the theory of inductive probability is the *principle of indifference*. It says that, if the evidence does not contain anything that would favor either of two or more possible events, in other words, if our knowledge situation is symmetrical with respect to these events, then they have equal probabilities relative to the evidence. For example, if the evidence e_1 available to an observer X_1 contains nothing else about a given die than the information that it is a regular cube, then the symmetry condition is fulfilled and therefore each of the six faces has the same probability 1/6 to appear uppermost at the next throw. This means that it would be unreasonable for X_1 to bet more than one to five on any one face. If X_2 is in possession of the evidence e_2 which, in addition to e_1, contains the knowledge that the die is heavily loaded in favor of one of the faces without specifying which one, the probabilities for X_2 are the same as for X_1. If, on the other hand, X_3 knows e_3 to the effect that the load favors the ace, then the probability of the ace on the basis of e_3 is higher than 1/6. Thus, inductive probability, in contradistinction to statistical probability, cannot be ascribed to a material object by itself, irrespective of an observer. This is obvious in our example; the die is the same for all three observers and hence cannot have different properties for them. Inductive probability characterizes a hypothesis relative to available information; this information may differ from person to person and vary for any person in the course of time.

A brief look at the historical development of the concept of probability will give us a better understanding of the present controversy. The mathematical study of problems of probability began when some mathematicians of the sixteenth and seventeenth centuries were asked by their gambler friends about the odds in various games of chance. They wished to learn about probabilities as a guidance for their betting decisions. In the beginning of its scientific career, the concept of probability appeared in the form of inductive probability. This is clearly reflected in the title of the first major treatise on probability, written by Jacob Bernoulli and published posthumously in 1713; it was called *Ars Conjectandi*, the art of conjecture, in other words, the art of judging hypotheses on the basis of evidence. This book may be regarded as marking the beginning of the so-called classical period of the theory of probability. This period culminated in the great systematic work by Laplace, *Theorie analytique des probabilités* (1812). According to Laplace, the purpose of the theory of probability is to guide our judgments and to protect us from illusions. His explanations show clearly that he is mostly concerned, not with actual frequencies, but with methods for judging the acceptability of assumptions, in other words, with inductive probability.

In the second half of the last century and still more in our century, the application of statistical methods gained more and more ground in science. Thus attention was increasingly focussed on the statistical concept of probability. However, there was no clear awareness of the fact that this development constituted a transition to a fundamentally different meaning of the word "probability." In the 1920's the first probability theories based on the frequency interpretation were proposed by men like the statistician R. A. Fisher, the mathematician R. von Mises, and the physicist-philosopher H. Reichenbach. These authors and their followers did not explicitly suggest to abandon that concept of probability which had prevailed since the classical period and to replace it by a new one. They rather believed that their concept was essentially the same as that of all earlier authors. They merely claimed that they had given a more exact definition for it and had developed more comprehensive theories on this improved foundation. Thus, they interpreted Laplace's word "probability" not in his inductive sense, but in their own statistical sense. Since there is a strong, though by far not complete analogy between the two concepts, many mathematical theorems hold in both interpretations, but others do not. Therefore these authors could accept many of the classical theorems but had to reject others. In particular, they objected strongly to the principle of indifference. In the frequency interpretation, this principle is indeed absurd. In our earlier example with the observer X_1, who knows merely that the die has the form of a cube, it would be rather incautious for him to assert that the six faces will appear with equal frequency. And if the same assertion were made by X_2, who has information that the die is biased, although he does not know the direction of the bias, he would contradict his own knowledge. In the inductive interpretation, on the other hand, the principle is valid even in the case of X_2, since in this sense it does not predict frequencies but merely says, in effect, that it would be arbitrary for X_2 to have more confidence in the appearance of one face than in that of any other face and therefore it would be unreasonable for him to let his betting decisions be guided by such arbitrary expectations. Therefore it seems much more plausible to assume that Laplace meant the principle of indifference in the inductive sense rather than to assume that one of the greatest minds of the eighteenth century in mathematics, theoretical physics, astronomy, and philosophy chose an obvious absurdity as a basic principle.

The great economist John Maynard Keynes made the first attempt in our century to revive the old but almost forgotten inductive concept of probability. In his *Treatise on Probability* (1921) he made clear that the inductive concept is implicitly used in all our thinking on unknown events both in everyday life and in science. He showed that the classical theory of probability in its application to concrete problems was understandable only if it was interpreted in the inductive sense. However, he modified and restricted the classical theory in several important points. He rejected the principle of indifference in its classical form. And he did not share the view of the classical authors that it should be possible in principle to assign a numerical value to the probability of any hypothesis whatsoever. He believed that this could be done only under very special, rarely fulfilled conditions, as in games of chance where there is a well determined number of possible cases, all of them alike in their basic features, e.g., the six possible results of a throw of a die, the possible distributions of cards among the players, the possible final positions of the ball on a roulette table, and the like. He thought that in all other cases at best only comparative judgments of probability could be made, and even these only for hypotheses which belong, so to speak, to the same dimension. Thus one might come to the result that, on the basis of available knowledge, it is more probable that the next child of a specified couple will be male rather than female; but no comparison could be made between the probability of the birth of a male child and the probability of the stocks of General Electric going up tomorrow.

A much more comprehensive theory of inductive probability was constructed by the geophysicist Harold Jeffreys (*Theory of Probability*, 1939). He agreed with the classical view that probability can be expressed numerically in all cases. Furthermore, in view of the fact that science replaces statements in qualitative terms (e.g., "the child to be born will be very heavy") more and more by those in terms of measurable quantities ("the weight of the child will be more than eight pounds"), Jeffreys wished to apply probability also to hypotheses of quantitative form. For this reason, he set up an axiom system for probability much stronger than that of Keynes. In spite of Keynes's warning, he accepted the principle of indifference in a form quite similar to the classical one: "If there is no reason to believe one hypothesis rather than another, the probabilities are equal." However, it can easily be seen that the principle in this strong form leads to contradictions. Suppose, for example, that it is known that every ball in an urn is either blue or red or yellow but that nothing is known either of the color of any particular ball or of the numbers of blue, red, or yellow balls in the urn. Let B be the hypothesis that the first ball to be drawn from the urn will be blue, R, that it will be red, and Y, that it will be yellow. Now consider the hypotheses B and non-B. According to the principle of indifference as used by Laplace and again by Jeffreys, since nothing is known concerning B and non-B, these two hypotheses have equal probabilities, i.e., one half. Non-B means that the first ball is not blue, hence either red or yellow. Thus "R or Y" has probability one half. Since nothing is known concerning R and Y, their probabilities are equal and hence must be one fourth each. On the other hand, if we start with the consideration of R and non-R, we obtain the result that the probability of R is one half and that of B one fourth, which is incompatible with the previous result. Thus Jeffreys's system as it stands is inconsistent. This defect cannot be eliminated by simply omitting the principle of indifference. It plays an essential role in the system; without it, many important results can no longer be derived. In spite of this defect, Jeffreys's book remains valuable for the new light it throws on many statistical problems by discussing them for the first time in terms of inductive probability.

Both Keynes and Jeffreys discussed also the statistical concept of probability, and both rejected it. They believed that all probability statements could be formulated in terms of inductive probability and that therefore there was no need for any probability concept interpreted in terms of frequency. I think that in this point they went too far. Today an increasing number of those who study both sides of the controversy which has been going on for thirty years are coming to the conclusion that here, as often before in the history of scientific thinking, both sides are right in their positive theses, but wrong in their polemic remarks about the other side. The statistical concept, for which a very elaborate mathematical theory exists, and which has been fruitfully applied in many fields in science and industry, need not at all be abandoned in order to make room for the inductive concept. Both concepts are needed for science, but they fulfill quite different functions. Statistical probability characterizes an objective situation, e.g., a state of a physical, biological, or social system. Therefore it is this concept which is used in statements concerning concrete situations or in laws expressing general regularities of such situations. On the other hand, inductive probability, as I see it, does not occur *in* scientific statements, concrete or general, but only in judgments *about* such statements; in particular, in judgments about the strength of support given by one statement, the evidence, to another, the hypothesis, and hence about the acceptability of the latter on the basis of the former. Thus, strictly speaking, inductive probability belongs not to science itself but to the methodology of science, i.e., the analysis of concepts, statements, theories, and methods of science.

The theories of both probability concepts must be further developed. Although a great deal of work has been done on statistical probability, even here some problems of its exact interpretation and its application, e.g., in methods of estimation, are still controversial. On inductive probability, on the other hand, most of the work remains still to be done. Utilizing results of Keynes and Jeffreys and employing the exact tools of modern symbolic logic, I have constructed the fundamental parts of a mathematical theory of inductive probability or inductive logic (*Logical Foundations of Probability*, 1950). The methods developed make it possible to calculate numerical values of inductive probability ("degree of confirmation") for hypotheses concerning either single events or frequencies of properties and to determine estimates of frequencies in a population on the basis of evidence about a sample of the population. A few steps have been made towards extending the theory to hypotheses involving measurable quantities such as mass, temperature, etc.

It is not possible to outline here the mathematical system itself. But I will explain some of the general problems that had to be solved before the system could be constructed and some of the basic conceptions underlying the construction. One of the fundamental questions to be decided by any theory of induction is whether to accept a principle of indifference and, if so, in what form. It should be strong enough to allow the derivation of the desired theorems, but at the same time sufficiently restricted to avoid the contradictions resulting from the classical form.

The problem will become clearer if we use a few elementary concepts of inductive logic. They will now be explained with the help of the first two columns of the accompanying diagram. We consider a set of four individuals, say four balls drawn from an urn. The individuals are described with respect to a given division of mutually exclusive properties; in our example, the two properties black (B) and white (W). An *individual distribution* is specified by ascribing to each individual one property. In our example, there are sixteen individual distributions; they are pictured in the second column (e.g., in the individual distribution No. 3, the first, second, and fourth ball are black, the third is white). A *statistical distribution*, on the other hand, is characterized by merely stating the number of individuals for each property. In the example, we have five statistical distributions, listed in the first column (e.g., the statistical distribution No. 2 is described by saying that there are three B and one W, without specifying *which* individuals are B and which W).

By the *initial probability* of a hypothesis ("probability a priori" in traditional terminology) we understand its probability before any factual knowledge concerning the individuals is available. Now we shall see that, if any initial probabilities which sum up to one are assigned to the individual distributions, all other probability values are thereby fixed. To see how the procedure works, put a slip of paper on the diagram alongside the list of individual distributions and write down opposite each distribution a fraction as its initial probability; the sum of the sixteen fractions must be one, but otherwise you may choose them just as you like. We shall soon consider the question whether some choices might be preferable to others. But for the moment we are only concerned with the fact that any arbitrary choice constitutes one and only one *inductive method* in the sense that it leads to one and only one system of probability values which contain an initial probability for any hypothesis (concerning the given individuals and the given properties) and a relative probability for any hypothesis with respect to any evidence. The procedure is as follows. For any given statement we can, by perusing the list of individual distributions, determine those in which it holds (e.g., the statement "among the first three balls there is exactly one W" holds in distributions Nos. 3, 4, 5, 6, 7, 9). Then we assign to it as initial probability the sum of the initial probabilities of the individual distributions in which it holds. Suppose that an evidence statement e (e.g., "The first ball is B, the second W, the third B") and a hypothesis h (e.g., "The fourth ball is B") are given. We ascertain first the individual distributions in which e holds (in the example, Nos. 4 and 7), and then those among them in which also h holds (only No. 4). The former ones determine the initial probability of e; the latter ones determine that of e and h together. Since the latter are among the former, the latter initial probability is a part (or the whole) of the former. We now divide the latter initial probability by the former and assign the resulting fraction to h as its relative probability with respect to e. (In our example, let us take the values of the initial probabilities of individual distributions given in the diagram for methods I and II, which will soon be explained. In method I the values for

STATISTICAL DISTRIBUTIONS		INDIVIDUAL DISTRIBUTIONS	METOHD I	METHOD II	
Number of Blue	Number of White		Initial Probability of Individual Distributions	Initial Probability of Statistical Distributions	Individual Distributions
1. 4	0	1. ● ● ● ●	1/16	1/5	1/5 = 12/60
2. 3	1	2. ● ● ● ○	1/16	1/5	1/20 = 3/60
		3. ● ● ○ ●	1/16		1/20 = 3/60
		4. ● ○ ● ●	1/16		1/20 = 3/60
		5. ○ ● ● ●	1/16		1/20 = 3/60
3. 2	2	6. ● ● ○ ○	1/16	1/5	1/30 = 2/60
		7. ● ○ ● ○	1/16		1/30 = 2/60
		8. ● ○ ○ ●	1/16		1/30 = 2/60
		9. ○ ● ● ○	1/16		1/30 = 2/60
		10. ○ ● ○ ●	1/16		1/30 = 2/60
		11. ○ ○ ● ●	1/16		1/30 = 2/60
4. 1	3	12. ● ○ ○ ○	1/16	1/5	1/20 = 3/60
		13. ○ ● ○ ○	1/16		1/20 = 3/60
		14. ○ ○ ● ○	1/16		1/20 = 3/60
		15. ○ ○ ○ ●	1/16		1/20 = 3/60
5. 0	4	16. ○ ○ ○ ○	1/16	1/5	1/5 = 12/60

Inductive Probability Methods. (From Rudolf Carnap, "What is Probability?" *Scientific American*, September, 1953.)

Nos. 4 and 7—as for all other individual distributions—are 1/16; hence the initial probability of e is 2/16. That of e and h together is the value of No. 4 alone, hence 1/16. Dividing this by 2/16, we obtain 1/2 as the probability of h with respect to e. In method II, we find for Nos. 4 and 7 in the last column the values 3/60 and 2/60 respectively. Therefore the initial probability of e is here 5/60, that of e and h together 3/60; hence the probability of h with respect to e is 3/5.)

The problem of choosing an inductive method is closely connected with the problem of the principle of indifference. Most authors since the classical period have accepted some form of the principle and have thereby avoided the otherwise unlimited arbitrariness in the choice of a method. On the other hand, practically all authors in our century agree that the principle should be restricted to some well-defined class of hypotheses. But there is no agreement as to the class to be chosen. Many authors advocate either method I or method II, which are exemplified in our diagram. Method I consists in applying the principle of indifference to individual distributions, in other words, in assigning equal initial probabilities to individual distributions. In method II the principle is first applied to the statistical distributions and then, for each statistical distribution, to the corresponding individual distributions. Thus, in our example, equal initial

103

probabilities are assigned in method II to the five statistical distributions, hence 1/5 to each; then this value 1/5 or 12/60 is distributed in equal parts among the corresponding individual distributions, as indicated in the last column.

If we examine more carefully the two ways of using the principle of indifference, we find that either of them leads to contradictions if applied without restriction to all divisions of properties. (The reader can easily check the following results by himself. We consider, as in the diagram, four individuals and a division D_2 into two properties; blue (instead of black) and white. Let h be the statement that all four individuals are white. We consider, on the other hand, a division D_3 into three properties: dark blue, light blue, and white. For division D_2, as used in the diagram, we see that h is an individual distribution (No. 16) and also a statistical distribution (No. 5). The same holds for division D_3. By setting up the complete diagram for the latter division, one finds that there are fifteen statistical distributions, of which h is one, and 81 individual distributions (viz., $3 \times 3 \times 3 \times 3$), of which h is also one. Applying method I to division D_2, we found as the initial probability of h 1/16; if we apply it to D_3, we find 1/81; these two results are incompatible. Method II applied to D_2 led to the value 1/5; but applied to D_3 it yields 1/15. Thus this method likewise furnishes incompatible results.) We therefore restrict the use of either method to one division, viz. The one consisting of all properties which can be distinguished in the given universe of discourse (or which we wish to distinguish within a given context of investigation). If modified in this way, either method is consistent. We may still regard the examples in the diagram as representing the modified methods I and II, if we assume that the difference between black and white is the only difference among the given individuals, or the only difference relevant to a certain investigation.

How shall we decide which of the two methods to choose? Each of them is regarded as *the* reasonable method by prominent scholars. However, in my view, the chief mistake of the earlier authors was their failure to specify explicitly the main characteristic of a reasonable inductive method. It is due to this failure that some of them chose the wrong method. This characteristic is not difficult to find. Inductive thinking is a way of judging hypotheses concerning unknown events. In order to be reasonable, this judging must be guided by our knowledge of observed events. More specifically, other things being equal, a future event is to be regarded as the more probable, the greater the relative frequency of similar events observed so far under similar circumstances. This *principle of learning from experience* guides, or rather ought to guide, all inductive thinking in everyday affairs and in science. Our confidence that a certain drug will help in a present case of a certain disease is the higher the more frequently it has helped in past cases. We would regard a man's behavior as unreasonable if his expectation of a future event were the higher the less frequently he saw it happen in the past, and also if he formed his expectations for the future without any regard to what he had observed in the past. The principle of learning from experience seems indeed so obvious that it might appear superfluous to emphasize it explicitly. In fact, however, even some authors of high rank have advocated an inductive method that violates the principle.

Let us now examine the methods I and II from the point of view of the principle of learning from experience. In our earlier example we considered the evidence e saying that of the four balls drawn the first was B, the second W, the third B; in other words, that two B and one W were so far observed. According to the principle, the prediction h that the fourth ball will be black should be taken as more probable than its negation, non-h. We found, however, that method I assigns probability 1/2 to h, and therefore likewise 1/2 to non-h. And we see easily that it assigns to h this value 1/2 also on any other evidence concerning the first three balls. Thus method I violates the principle. A man following this method sticks to the initial probability value for a prediction, irrespective of all observations he makes. In spite of this character of method I, it was proposed as the valid method of induction by prominent philosophers, among them Charles Sanders Peirce (in 1883) and Ludwig Wittgenstein (in 1921), and even by Keynes in one chapter of his book, although in other chapters he emphasizes eloquently the necessity of learning from experience.

We saw earlier that method II assigns, on the evidence specified, to h the probability 3/5, hence to non-h 2/5. Thus the principle of learning from experience is satisfied in this case, and it can be shown that the same holds in any other case. (The reader can easily verify, for example, that with respect to the evidence that the first three balls are black, the probability of h is 4/5 and therefore that of non-h 1/5.) Method II in its modified, consistent form was proposed by the author in 1945. Although it was often emphasized throughout the historical development that induction must be based on experience, nobody as far as I am aware, succeeded in specifying a consistent inductive method satisfying the principle of learning from experience. (The method proposed by Thomas Bayes (1763) and developed by Laplace—sometimes called "Bayes's rule" or "Laplace's rule of succession"—fulfills the principle. It is essentially method II, but in its unrestricted form; therefore it is inconsistent.) I found later that there are infinitely many consistent inductive methods which satisfy the principle (*The Continuum of Inductive Methods*, 1952). None of them seems to be as simple in its definition as method II, but some of them have other advantages.

Once a consistent and suitable inductive method is developed, it supplies the basis for a *general method of estimation*, i.e., a method for calculating, on the basis of given evidence, an estimate of an unknown value of any magnitude. Suppose that, on the basis of the evidence, there are n possibilities for the value of a certain magnitude at a given time, e.g., the amount of rain tomorrow, the number of persons coming to a meeting, the price of wheat after the next harvest. Let the possible values be $x_1, x_2, \ldots, x_n$, and their inductive probabilities with respect to the given evidence $p_1, p_2, \ldots, p_n$, respectively. Then we take the product p_1x_1 as the expectation value of the first case at the present moment. Thus, if the occurrence of the first case is certain and hence $p_1 = 1$, its expectation value is the full value x_1; if it is just as probable that it will occur as that it will not, and hence $p_1 = 1/2$, its expectation value is half its full value ($p_1x_1 = x_1/2$), etc. We proceed similarly with the other possible values. As estimated or total expectation value of the magnitude on the given evidence we take the sum of the expectation values for the possible cases, that is, $p_1x_1 + p_2x_2 + \ldots + p_nx_n$. (For example, suppose someone considers buying a ticket for a lottery and, on the basis of his knowledge of the lottery procedure, there is a probability of 0.01 that the ticket will win the first prize of $200 and a probability of 0.03 that it will win $50; since there are no other prizes, the probability that it will win nothing is 0.96. Hence the estimate of the gain in dollars is $0.01 \times 200 + 0.03 \times 50 + 0.96 \times 0 = 3.50$. This is the value of the ticket for him and it would be irrational for him to pay more for it.) The same method may be used in order to make a rational decision in a situation where one among various possible actions is to be chosen. For example, a man considers several possible ways for investing a certain amount of money. Then he can—in principle, at least—calculate the estimate of his gain for each possible way. To act rationally, he should then choose that way for which the estimated gain is highest.

Bernoulli and Laplace and many of their followers envisaged the idea of a theory of inductive probability which, when fully developed, would supply the means for evaluating the acceptability of hypothetical assumptions in any field of theoretical research and at the same time methods for determining a rational decision in the affairs of practical life. In the more sober cultural atmosphere of the late nineteenth century and still more in the first half of the twentieth, this idea was usually regarded as a utopian dream. It is certainly true that those audacious thinkers were not as near to their aim as they believed. But a few men dare to think today that the pioneers were not mere dreamers and that it will be possible in the future to make far-reaching progress in essentially that direction in which they saw their vision.

Indifference: The Symmetries of Probability

On estime la probabilité d'un événement par le
nombre des cas favourables divisé par le nombre
des cas possibles. La difficulté ne consiste que
dans l'énumération des cas.

Lagrange, quoted as epitaph to ch. 1 of
J. Bertrand, *Calcul des probabilités*.

SINCE its inception in the seventeenth century, probability theory
has often been guided by the conviction that symmetry can dictate
probability. The conviction is expressed in such slogan formulations
as that equipossibility implies equal probability, and honoured by
such terms as indifference and sufficient reason. As in science
generally we can find here symmetry arguments proper that are
truly a priori, as well as arguments that simply assume contingent
symmetries, and 'arguments' that reflect the thirst for a hidden,
determining reality. The great failure of symmetry thinking was
found here, when indifference disintegrated into paradox; and great
success as well, sometimes real, sometimes apparent. The story is
especially important for philosophy, since it shows the impossibility
of the ideal of logical probability.

1. INTUITIVE PROBABILITY

A traveller approaches a river spanned by bridges that connect its
shores and islands. There has been a great storm the night before,
and each bridge was as likely as not to be washed away. How
probable is it that the traveller can still cross? This puzzle, devised
by Marcus Moore, clearly depends on the pattern of bridges
represented in Figure 12.1.

It also depends on whether the survival of a bridge affects the survival of another. The traveller believes not. Thus for him each bridge had an independent 50 per cent probability of washing away.

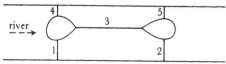

FIG. 12.1. A symmetry argument for probabilities

There is a simple but plodding solution (see *Proofs and illustrations*). But there is a symmetry argument too. Imagine that besides the traveller, there is also a boat moving downstream. The boatman's problem is to get through, which is possible if sufficiently many bridges have been washed away. What is the probability he can get through? Our first observation is that he faces a problem with the same abstract structure. For the traveller, the *entries* are bridges 1 and 2, while for the boatman they are 1 and 4. The *exits* are 4 and 5 for traveller, and are 2 and 5 for boatman. For both there is a *connector*, namely bridge 3. So each sees lying before him the 'maze'

entry   exit
          connector
entry   exit

Good and bad are reversed for traveller and boatman; but suppose that for each, the good state of a bridge has the same independent probability of 50 per cent. Now, by the great Symmetry Requirement, essentially similar problems must have the same solution. Hence:

1. Probability (traveller crosses) = Probability (boat gets through)

But the problems are not only similar; they are also related. For if the traveller has some unbroken path across, the boat cannot get through; and vice versa. Therefore:

2. Probability (boat gets through) = Probability (traveller does not cross)
3. [from 1 and 2] Probability (traveller crosses) = Probability (traveller does not cross)

So it is exactly as likely as not that the traveller will cross—the probability is 50 per cent.

This is a remarkable example, not only as a pure instance of a symmetry argument, but because it introduces all the basic ingredients in the three centuries of controversy over the relation between symmetry and probability. In this problem, the *initial probabilities* are given: 50 per cent for any bridge that it will wash away. We are also given the crucial probability datum about how these eventualities are related: they are *independent*. That means that the collapse of one bridge is neither more nor less probable, on the supposition that some other bridge is washed away. (We are here distinguishing simple probability from *conditional* probability, marked by such terms as 'on the supposition that' or 'given that'.) Then, purely a priori reasoning gives us the probabilities for the events of interest.

The great question for classical probability theory was: can the initial probabilities themselves be deduced too, on the basis of symmetry considerations? If we knew absolutely nothing about storms and bridges, except that one can wash away the other, would rationality not have required us to regard both possible outcomes as equally likely? Once the answer seemed to be obviously *Yes*, and now it seems self-evidently to be *No*, to many of us. But our century also saw the most sophisticated defences of the *yes* answer. And the history of the controversy spun off important and lasting insights.

Proofs and illustrations
In our example, the symmetry transformation used mapped bridge 2 into 4, and vice versa, leaving the others fixed. The *entry-connector-exit* structure is invariant, as is the probability of 'good' (i.e. *whole* for traveller and *broken* for boatman). The reader is invited to consider similar patterns with 1, 3, 4, 5 islands, and to generalize.

The single probability calculus principle that was utilized was— writing '*P*' for 'Probability':

$$P(A) = 1 - P(A)$$

which itself is an immediate corollary to the two axioms

I. $0 = P(\text{contradiction}) \leqslant P(A) \leqslant P(\text{tautology}) = 1$
II. $P(A) + P(B) = P(A \text{ or } B) + P(A \text{ and } B)$

which together exhaust the entire finitary probability theory. For our present purposes, it is not necessary to focus on this calculus (which will be explored further in the next chapter), but the following notions will be relevant (and will be employed intuitively in this chapter):

The *conditional probability* $P(A|B)$ of *A given* that *B* equals $P(A$ and $B)/P(B)$

A and *B* are (*stochastically* or *statistically*) *independent* exactly if $P(A|B) = P(A)$

That conditional probability $P(A|B)$ is defined only if the *antecedent* *B* has probability $P(B) \neq 0$. The independence condition is equivalent to

$$P(B|A) = P(B)$$
$$P(A \text{ and } B) = P(A)P(B)$$

always provided the conditional probabilities are defined. The last equation shows clearly, of course, that the condition is symmetric in *A* and *B*.

2. CELESTIAL PRIOR PROBABILITIES

The modern history of probability began with the Pascal-Fermat correspondence of 1654. The problems they discussed concerned gambling, games of chance. If someone wanted to draw practical advantage from these studies, he would learn from them how to calculate probabilities of winning (or expectation of gain) from initial probabilities in the gambling set-up. But of course he would have to know those initial probabilities already. While we cannot attribute much sophistication here to the gambler, we may plausibly believe that he takes a hard-nosed empirical stance on this. He believes that the dice are fair exactly if all possible numerical combinations come up equally often—and that this assertion is readily testable even in a small number of tosses. Daggers and rapiers will be drawn if a challenged and tested die comes up even three sixes in a row. We know of course from the play *Rosencrantz and Guildenstern Are Dead* how inconclusive such tests must be on a more sophisticated understanding of probability. But the crucial role and status of initial probability hypotheses appears much more clearly in a different sort of problem.

The Academy of Sciences in Paris proposed a prize subject for 1732 and 1734: the configuration of planetary orbits in our solar system. This configuration may be described as follows: each planet orbits in a plane inclined no more than 7.5° to the sun's equator, and the orbits all have the same direction.[2]

The prize was divided between John Bernoulli and his son Daniel. The latter included three arguments that this configuration cannot be attributed to mere chance. Of these the third argument is a typical eighteenth-century 'calculation' of initial probabilities: 7.5° is $\frac{1}{12}$ of 90° (possible maximum inclination of orbit to equator if we ignore direction); there are six (known) planets, so the probability of this configuration happening 'by chance' is $(\frac{1}{12})^6$, which is negligibly small (*circa* 3 in 10 million).

Daniel Bernoulli has here made two assumptions: of a certain *uniformity* (the probability of at most $\frac{1}{12}$ of the maximum, equals $\frac{1}{12}$) and of *independence* (the joint probability of the six statements is the product of their individual probabilities). Before scrutinizing these assumptions, let us look at two more examples.

Buffon, in his *Historie naturelle* gives an argument similar to Daniel Bernoulli's.[3] Buffon says that the mutual inclination of any two planetary orbits is at most 7.5° Taking direction into account, the maximum is 180°, so the chance of this equals $\frac{1}{24}$. Taking now one planet as fixed, we have five others. The joint probability of all five orbits to be inclined no more than 7.5° is therefore $(\frac{1}{24})^5$. This probability (*circa* 1 in 10 million) is approximately three times smaller than the one noted by Bernoulli. Independently Buffon notes that the probability that all six planets should move in the same west to east direction for us, equals $(\frac{1}{2})^6$. It is clear that he is calculating initial probabilities by the same assumptions as Daniel Bernoulli.

In Laplace's writings on celestial mechanics we find another such example.[4] Bernoulli and Buffon argued for a common origin of the planets, that is, a common cause, on the basis of the improbability of mere chance or coincidence. Laplace argues conversely that a certain fact is not initially improbable, and therefore needs no common-cause explanation. The fact in question was that among the many observed comets, not a single hyperbolic trajectory has been reported.[5] Laplace demonstrates that the probability of a comet with hyperbolic orbit is exceedingly low. The demonstration is based on a uniform distribution of probability over the possible directions of motion of comets entering the sun's gravitational field at some large given distance from the sun.

3. INDIFFERENCE AND SUFFICIENT REASON

It is clear that each of these authors is entertaining what we may call a chance hypothesis: that the phenomenon in question arises 'by mere chance', that is, without the presence of causal or other factors constraining the outcome. There is an ambiguity here: are the probabilities assigned the correct ones (*a*) given *no* hypotheses or assumptions about the physical situation, or (*b*) given a substantial, contingent hypothesis about the absence of certain physical features?

If the former is the case, we have here typical symmetry thinking: the fact that certain information is absent in the statement of the problem, is used as a constraint on the solution. If the latter, we are in the presence of a metaphysical assumption, which may have empirical import: that nature, when certain physical constraints are absent, is equally likely to produce any of the unconstrained possibilities, and therefore tends to produce each equally often.

Ian Hacking locates the first theoretical discussion of this topic in Leibniz's memorandum '*De incerti aestimatione*' (1678).[6] In this note Leibniz equates probability with gradations of possibility ('*probabilitas est gradus possibilitas*'). He states the Principle of Indifference, that equipossible cases have the same probability, and asserts that such a principle can be 'proved by metaphysics'.

We can only speculate what metaphysical proof Leibniz envisaged, but it must surely be based on his Principle of Sufficient Reason. Leibniz's programme set out in the *Discourse of Metaphysics* was to deduce the structure of reality from the nature of God. As a first step, this nature entails that God does, or creates, nothing without sufficient reason. In this marriage of metaphysics with divine epistemology, the difference between points (*a*) and (*b*) above vanishes. For Leibniz's God solves the problem of what nature shall do without contributing factors of his own to destroy the symmetries of the problem-as-stated.

This is how Leibniz must have derived symmetry principles governing nature—determining what the real, objective probabilities shall be in a physical situation. We cannot be sure on the basis of this brief note, but he must have given the principle of sufficient reason also this form: that a rational being should assign equal probabilities to distinct possibilities unless there be explicit reason to differentiate them. Since Leibniz clearly appreciated the great value of such an equation for metaphysics, he must have appreciated that strictly speaking, his new beginning for metaphysics effects a collapse of two logically distinct problems.

It was certainly in the terminology of sufficient reasons—perhaps always with a equivocation between (*we have reason*) and (*there is reason*)—that principles of indifference were formulated. There were two; we have seen both at work in the arguments of Bernoulli, Buffon, and Laplace.

The first is the *Principle of Uniform Distribution*. Suppose I shoot bullets at a target and am such a poor marksman that it makes no difference at which point of the target I aim. Then any two equal areas on the target are equally likely to be hit. We call this a uniform distribution. The first indifference principle for assigning probabilities is *to assume a uniform distribution in the absence of reasons to the contrary*.

The second is the *Principle of Stochastic Independence*. I explained independence above; let me illustrate it here. Suppose we are told that 40 per cent of the population smokes and 10 per cent has lung cancer. This gives me the probability that a randomly chosen person is a smoker, or has lung cancer, but does not tell me the joint probability of these two characteristics. There are three cases (see Fig. 12.2). Each of the three lines p, q, r has 10 per cent of the area below it. In the case of the horizontal line q, the joint probability of lung cancer *and* smoking is 10 per cent of 40 per cent, namely 4 per cent. For p it is larger and for r it is smaller. The second indifference principle is *to assume statistical independence, in the absence of reasons to the contrary*.

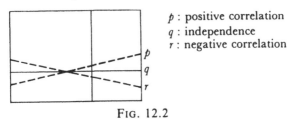

p : positive correlation
q : independence
r : negative correlation

FIG. 12.2

Are these two principles consistent with each other? The joint probability of two events is the same as the ordinary probability of a single complex event. It seems possible therefore that the two principles could be made to apply to the same example, and offer contradictory advice. In the *Proofs and illustrations* we will see that this is not so; the two are consistent with each other.

Proofs and illustrations

Let us consider two variables, say *height h* and *weight w*. Suppose height varies from zero to 10 and weight from zero to 100. Given no other information (hence no reasons to diverge from uniformity or independence), assign probabilities to all possibilities.

The first procedure is to choose uniform distributions for each:

    1. $P(0 \leqslant h \leqslant a) = a/10$ $P(0 \leqslant w \leqslant b) = b/100$

Then calculate the joint probability by assuming independence:

    2. $P(0 \leqslant h \leqslant a \ \ and \ \ 0 \leqslant w \leqslant b) = (a/10)(b/100)$

The other procedure is to look at the complex variable *hw* which has pairs of numbers as values. A person with height 6 and weight 60 has *hw* equal to $<6, 60>$. The big rectangle in Fig. 12.3 encompasses all possibilities ($0 \leqslant h \leqslant 10 \ \ and \ \ 0 \leqslant w \leqslant 100$) while the smaller one describes the possibility of having *hw* fall between $<0, 0>$ and $<a, b>$ in the proper sense of 'between'. Uniformity alone applies now and demands a probability proportional to the area:

Fig. 12.3

    3. $P(<0,0> \leqslant hw \leqslant <a, b>) = ab/1000$

But as we see, 2 and 3 agree. We have proved in effect that if variables *h* and *w* are uniformly distributed and independent, then the complex variable *hw* is uniformly distributed. Hence the two principles are mutually consistent and together constitute the great symmetry principle of classical probability theory—the *Principle of Indifference*.

4. BUFFON'S NEEDLE: EMPIRICAL IMPORT OF INDIFFERENCE

If we must assign initial probabilities, in the absence of relevant information, reason bids us be like Buridan's ass. Do not choose between $P(A) > P(-A)$ and $P(A) < P(-A)$, but set them equal. Similarly in such a case, do not choose between $P(A \text{ and } B) > P(A) \cdot P(B)$ and $P(A \text{ and } B) < P(A) \cdot P(B)$, but set those equal as well. Very well; but will nature oblige us with frequencies to which these initial probabilities have a good fit? Is this dictate of reason one that will let reason unlock the mysteries of nature?

An empiricist will ask these questions with a distinct tinge of mockery to his voice. But here we should report a marvellous example in which calculation by the Principle of Indifference led to beautifully confirmed empirical results. This is Buffon's needle problem. It is much more probative than planetary orbit and comet examples, where one only finds *explanation*—that beautiful but airy creature of the fecund imagination—and not *prediction*.

Buffon's needle problem[7]

Given: a large number of parallel lines are drawn on the floor, and a needle is dropped. What is the probability that the needle cuts one of the lines?

To simplify the problem without loss of essential generality, let the lines be exactly two needle lengths apart. Touching will count as cutting, but clearly at most one line is cut. We may even speak sensibly of the line nearest the needle's point (choose either if the point is exactly halfway between). Then our question is equivalent to: what is the probability that the needle cuts this nearest line? In Fig. 12.4 the needle point is a distance $0 \leqslant d \leqslant 1$ away from line L, and its inclination to L is the angle θ. Thus we have:

favourable cases: the needle cuts L exactly if $d \leqslant y = \sin \theta$

This θ varies from zero to 2π ($= 360$ degrees), and so we can diagram the situation with an area of 1 (needle length) by 2π (radians) as in Fig. 12.5. To distinguish the favourable cases from the unfavourable ones, we draw in the sine curve and shade the area where $y \leqslant d$. Assuming independence and uniform distribution, the probability of the favourable cases must be proportional to the

Symmetry and Logical Probability

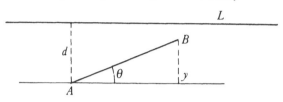

Fig. 12.4. Buffon's needle

shaded area. Since a little calculus quickly demonstrates that this area equals 2, we arrive at the number $2/2\pi$:

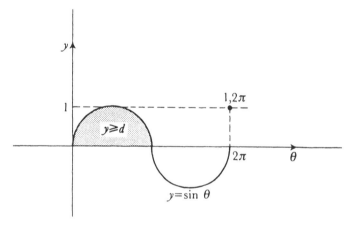

Fig. 12.5. Buffon's probability calculation

The probability of a favourable case equals $1/\pi$, the solution Buffon himself found for his problem.

Since the experiment can be carried out, this is an empirical prediction. It has been carried out a number of times and the outcomes have been in excellent agreement with Buffon's prediction.[8] Now is this not marvellous and a result to make the rationalist metaphysician squeal with delight? For the assumption of symmetry in the probabilities of equipossible cases has here led to a true prediction made a priori.

5. THE CHALLENGE: BERTRAND'S PARADOXES

What I have so far recounted has been very favourable to the Principle of Indifference. Many readers, knowing of its later rejection, but perhaps less familiar with attempts to refine and save it, may already be a little impatient. I will argue for the rejection of its uncritical versions—the empirical phenomena cannot be predicted a priori—but this will be a rejection of naïve symmetry arguments in favour of deeper symmetries, with due respect for the insights that were gained along the way.

We have seen that the Principle has two parts, which are indeed consistent with each other. We have also seen the significant successes of explanations and predictions arrived at in the eighteenth century by means of this Principle. But the challenge to this attempt to calculate initial probabilities on the basis of physical symmetry came exactly from the fundamental principle of symmetry arguments. If two problems are essentially the same, they must receive essentially the same solution. So *a fortiori* if a situation can be equally described in terms of different parameters, we should arrive at the same probabilities if we apply the Principle of Indifference to these other parameters. There will be a logical difficulty—indeed, straightforward inconsistency—if different descriptions of the problem lead via Indifference to distinct solutions.

This logical difficulty with the idea was expounded systematically in a series of paradoxes by Joseph Bertrand at the end of the nineteenth century.[9] Leaving his rather complex geometric examples for *Proofs and Illustrations*, let us turn immediately to a paradigmatic but simple example: the perfect cube factory.[10]

A precision tool factory produces iron cubes with edge length $\leqslant 2$ cm. What is the probability that a cube has length $\leqslant 1$ cm, given that it was produced by that factory?

A naïve application of the Principle of Indifference consists in choosing length l as parameter and assuming a uniform distribution. The answer is then $\frac{1}{2}$. But the problem could have been stated in different words, but logically equivalent form:

Possible cases	Favourable
edge length $\leqslant 2$	length $\leqslant 1$
area of side $\leqslant 4$	area $\leqslant 1$
volume $\leqslant 8$	volume $\leqslant 1$

Treating each statement of the problem naïvely we arrive at answers $\frac{1}{2}, \frac{1}{4}, \frac{1}{8}$. These contradict each other.

The correspondence $l^m \leftrightarrow l^n$, for a parameter l with range $(0, k)$ is one to one, but does not preserve equality of intervals.

Hence uniform distribution on l^m entails non-uniform distribution on l^n. Now sometimes the problem is indeed constrained by symmetries. The cubes example illustrates how these constraints may be so minimal as to leave the set of possible solutions unreduced. More information about the factory could improve the situation. But the Indifference Principle is supposed to fill the gap left by missing information!

Even taken by itself, the example is devastating. But since we shall discuss various attempts to salvage Indifference, it is important to assess two more examples, with somewhat different logical features.

Von Kries posed a problem which is like that of the perfect cube factory, in that several parameters are related by a simple logical transformation. Consider volume and density of a liquid. If mass is set equal to 1, then these parameters are related by:

density = 1/volume;   volume = 1/density.

But a uniform distribution on parameter x is automatically non-uniform on $y = (1/x)$. For example,

x is between 1 and 2 exactly if y is between $\frac{1}{2}$ and 1
x is between 2 and 3 exactly if y is between $\frac{1}{3}$ and $\frac{1}{2}$.

Here the two intervals for x are equal in length, but the corresponding ones for y are not. Thus Indifference appears to give us two conflicting probability assignments again.

Von Mises's example of a Bertrand-type paradox concerned a mixture of two liquids, wine and water. We have a glass container, with a mixture of water and wine. To remove division by zero from every inversion, let the following be data:

the glass contains 10 cc of liquid, of which at least 1 cc is water and at least 1 cc is wine.

What is the probability that at least 5 cc is water? Let the parameters be:

a = proportion of wine to total: $(1/10) \leqslant a \leqslant (9/10)$
b = proportion of water to total: $(1/10) \leqslant b \leqslant (9/10)$
x = proportion of wine to water: $(1/9) \leqslant x \leqslant 9$
y = proportion of water to wine: $(1/9) \leqslant y \leqslant 9$

Obviously $b = (1 - a)$, $x = (a/b)$, $y = 1/x$, and $a = x/(1 + x)$, so descriptions of the situation by means of any parameter can be completely translated into any other parameter. It is easy to see that the same problem recurs. Here are two equal intervals for the proportion of wine to total:

a = Proportion of wine to total	x = Proportion of wine to water
4/10	4/6 = 2/3
5/10	5/5 = 1
6/10	6/4 = 3/2

Since $1 - (2/3)$ is not equal to $(3/2) - 1$, it is clear that a uniform distribution on the proportion a entails a non-uniform proportion on proportion x.

In each case the Principle of Uniformity is applied to one perfectly adequate description of the problem. The statements of the problem, both as to sets of possible cases and set of favourable cases, differ only verbally. But the great underlying principle of symmetry thinking is that essentially similar problems must receive the same solution. Thus the attempt to assign uniform distribution on the basis of symmetries in these *statements* of the problem, is drastically misguided—it violates symmetry in a deeper sense.

Most writers commenting on Bertrand have described the problems set by his paradoxical examples as not well posed. In such a case, the problem as initially stated is really not one problem but many. To solve it we must be told *what* is random; which means, *which* events are equiprobable; which means, *which* parameter should be assumed to be uniformly distributed.

But that response asserts that in the absence of further information we have no way to determine the initial probabilities. In other words, this response rejects the Principle of Indifference altogether. After all, if we were told as part of the problem which parameter should receive a uniform distribution, no such Principle would be needed. It was exactly the function of the Principle to turn an incompletely described physical problem into a definite problem in the probability calculus.

There have been different reactions. We have to list Henri Poincaré, E. T. Jaynes, and Rudolph Carnap among the writers who believed that the Principle of Indifference could be refined and sophisticated, and thus saved from paradox.

Proofs and illustrations

The famous chord problem asks for the problem that a stick, tossed randomly on a circle, will mark out a chord of given length. For a definite standard of comparison we inscribe an equilateral triangle ABC in the circle (see Fig. 12.6). However we draw the triangle, it is clear that the separated arcs, like arc AEB, must each be $\frac{1}{3}$ of the circumference. Thus the length of the side of any such triangle is the same. In fact it is $r\sqrt{3}$, where r is the radius, and the point D is exactly halfway along the radius OE.

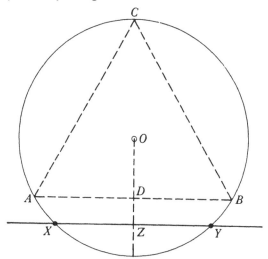

FIG. 12.6. Bertrand's chord problem

What is the probability that chord XY is greater than side AB? If we try to answer this question on the basis of the Principle of Indifference, we actually find three variables which might be asserted to have uniform distribution:

$XY > AB$ exactly if any of the following holds:

(a) $OZ < r/2$
(b) Y is located between $\frac{1}{3}$ and $\frac{2}{3}$ of the circumference away from X, as measured along the circumference
(c) the point Z falls within the 'inner' circle with centre 0 and radius $r/2$.

This gives us three possible applications of the Principle of Uniformity.

Using description (*a*) we reason: *OZ* can be anything from 0 to *r*; the interval [0, *r*/*z*] of favourable cases has length $\frac{1}{2}$ of the interval [0, *r*] of possible cases; hence the probability equals $\frac{1}{2}$ (*Solution A*).

Using (*b*) we reason: each point of contact *X*, *Y* can be any point on the circle. So given the point *X*, we can find point *Y* at any fraction between 0 and 1 of the circumference, measuring counter-clockwise. Of these possible locations, $\frac{1}{3}$ fall in the favourable interval [$\frac{1}{3}$, $\frac{2}{3}$]; hence the probability equals $\frac{1}{3}$ (*Solution B*).

Using (*c*) we note that the centre *Z* of the stick can fall anywhere in the whole circle. In the favourable cases it falls in the 'inner' circle with radius *r*/2—which has an area $\frac{1}{4}$ that of the big circle. Hence the probability equals $\frac{1}{4}$ (*Solution C*).

6. SYMMETRIES TO THE RESCUE?

Henri Poincaré and E. T. Jaynes both argued that if we pay attention to the geometric symmetries in Bertrand's problem, we do arrive at a unique solution.[11] Their general idea applies to all apparent ambiguities in the Principle of Indifference: a careful consideration of the exact symmetries of the problem will remove the inconsistency, provided we focus on the symmetry transformations themselves, rather than on the objects transformed.

In order to show the logical structure very clearly I will concentrate on the simple examples of the perfect cubes, mass versus density, and water mixed with the wine. Let us begin by analysing the intuitive reaction to the cube factory, which led us into paradox. Focusing first on the parameter of length, we used the natural length measure for intervals:

$$m(a, b) = b - a$$

This is the *underlying measure*[12] that gave us our probabilities for cases inside the range [0,2]:

$$P(a, b) = m(a,b)/m(\text{range}) \text{ for } a, b \text{ inside the range.}$$

Now this underlying measure has a very special feature, from the point of view of symmetry:

Translation invariance: if $x' = $ if $x + k$
then $m(a, b) = m(a', b')$.

Up to multiplication by a scalar, m is the unique measure to have this feature. That is easy to see, because one interval can be moved into another by a translation exactly if they have the same length (and are of the same type: open, half-open, etc.).[13] The number K represents the scale, if $m' = Km$, because for example the length in inches is numerically 12 times that in feet. It will not affect the probability at all, because it will cancel out (being present in both numerator and denominator in the equation for P in terms of m). We have therefore the following result:

> *Translation invariant measure.* The probability distribution on a real valued parameter x is uniquely determined, if we are given its range and the requirement that it derive from an underlying measure which is translation invariant.

In what sort of example would the given be exactly as required? Suppose I tell you that Peter is a marksman with no skill whatever, and an unknown target. Now I ask you the probability that his bullet will land between 10 and 20 feet from my heart, given that it lands within 20 feet. Treating this formally, I choose a line that falls on both my heart and the impact point of the bullet, coordinatize this line by choosing a point to call zero, and one foot away from it a point to call $+ 1$. I choose a measure m' on this line, call my heart's coordinate X, and calculate

$$m'(X + 10, X + 20)/m'(X, X + 20)$$

and give you the resulting number as answer. If my procedure was properly in tune with the problem, this answer should better not depend on how I chose the points to call zero and $+ 1$ (which two choices together determined the coordinate X). That entails that m' must be translation invariant, and is therefore now uniquely identified. We note with pleasure that the answer is also not affected by the choice of the foot as unit of measurement—as indeed it should not, because nothing in the problem hinged on its Anglo-Saxon peculiarities.

Now, in what sort of problem is the 'given' so different that this procedure is inappropriate? Obviously, when translation invariance is the wrong symmetry. This happens when the range of the physical quantity in question is not closed under addition and subtraction, for example, if the quantity has an infimum, which acts as natural zero point. For example, no classical object has negative or zero volume, mass, or absolute temperature.

In such a case, the scale or unit may still be irrelevant. For the transformation of the scaling unit consists simply in multiplication by a positive number, which operation does not take us out of this range. Consider now von Kries's problem, which concerns the positive quantities mass and density. With the units of measurement essentially irrelevant we look for an underlying measure

$$M(a, b) = M(ka, kb)$$

for any positive number k (invariance under *dilations*).

There is indeed such a measure, and it is unique in the same sense.[14] That is the *log uniform* distribution:

$$M(a, b) = \log b - \log a$$

where log is the natural logarithm. This function has the nice properties:

$\log(xy) = \log x + \log y$
$\log(x^n) = n\log x$
$\log(1) = 0$

but should be used only for positive quantities, because it moves zero to minus infinity. The first of these equations shows already that M is dilation invariant. The second shows us what is now regarded as equiprobable:

The intervals (b^n, b^{n+k}) all receive the same value $k\log b$, so if within the appropriate range, the following are series of equiprobable cases:

$(0.1, 1), (1, 10), (10, 100), ..., (10^n, 10^{n+l}), ...$
$(0.2, 1), (1, 5), (5, 25), ..., (5^n, 5^{n+l}), ...$

and so forth. A probability measure derived from the log uniform distribution will therefore always give higher probabilities 'closer' to zero, by our usual reckoning.

For example, in the case of temperature we have since Kelvin accepted that this is essentially a positive quantity. Of course we are at liberty to give the name -273 to absolute zero. But this does not remove the infimum; subtraction eventually takes one outside the range. The presence of this infimum creates, or rather is, an asymmetry: it obstructs translation invariance. But it is no obstacle to dilation invariance, so the log uniform distribution is right—it is dictated by the symmetries of the problem.

This reasoning, being rather abstract, may not get us over our initial feeling of surprise. But as Roger Rosencrantz pointed out, we can test all this on the von Kries problem.[15] Our argument implies that von Kries's puzzle is due to focusing on the wrong transformation group. Attention to the right one dictates use of the log uniform distribution. To our delight this removes the conflict:

$$M(1/b \leqslant 1/x \leqslant 1/a) = K(\log a^{-1} - \log b^{-1})$$
$$= K(\log b - \log a)$$
$$= M(a \leqslant x \leqslant b).$$

This is certainly a success for this approach to Indifference.

Consider next the perfect cube factory. Suppose that again we regard the unit of measurement as essentially irrelevant to this problem, conceived in true generality, but observe that length, area, volume are positive quantities. The uniqueness of the log uniform measure for dilation invariance, forces us then to use it as underlying the correct probabilities. This will not help us, unless we ask all our questions about intervals that exclude *zero*; but for them it works wonderfully well:

What is the probability that the length is $\leqslant 2$, given that it is between 1 and 3 inclusive?
What is the probability that the area is $\leqslant 4$, given that it is between 1 and 9 inclusive?
The probability that the length is $\leqslant 2$, given that it is between 1 and 3 inclusive, equals $M(1, 2)/M(1, 3) = \log 2/\log 3 = 0.631$.
The probability that the area is $\leqslant 4$, given that it is between 1 and 9 inclusive, equals $M(1, 4)/M(1, 9) = 2\log 2/2\log 3 = 0.631$.

Thus the two equivalent questions do receive the same answer. The point is perfectly general, because the exponent becomes a multiplier, which appears in both numerator and denominator, and so cancels out. This is again a real success. By showing us how to reformulate the problem, and then using its symmetries to determine a unique solution, this approach has as it were taught us how to understand our puzzled but insistent intuitions.

There is therefore good prima-facie reason to take this approach seriously. In the *Proofs and Illustrations* I shall show how this approach does give us a neat solution for the puzzle of Buffon's needle, construed as a Bertrand problem. But in subsequent sections we'll see that the approach does not generalize sufficiently to save the Principle of Indifference.

Proofs and illustrations
I shall here explain this rescue by geometric symmetries with another illustration. For this purpose I choose Buffon's needle problem again, for properly understood it can itself be described as a rudimentary Bertrand paradox.[16].

Buffon assumes no marksmanship—the location of the lines on the floor does not, as far as we know, affect the location of the fallen needle. So our description of the situation utilizes a frame of reference chosen for convenience, in which we treat as X-axis the line through needle point A which is parallel to the drawn lines, as in Fig. 12.4. Here d is the Y-coordinate of line L, θ the inclination of line AB to the X-axis, and y is Y-coordinate of B, and A is the origin.

Why not assume that y and d are independent and uniformly distributed? We must be careful to describe y so that it does not depend on d. But it is just sine θ, and θ does not depend on d, so that is fine. Thus y ranges from -1 to $+1$ (being measured from the X-axis, chosen so that the line L has equation $Y = 1$). The possible and favourable cases are depicted in Fig. 12.7, and we see that the probability of $y \leqslant d$ equals $\frac{1}{4}$. Hence by applying the Principle of Indifference to Buffon's problem *differently but equivalently described*, we have arrived at a different solution.

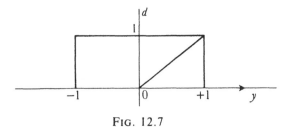

FIG. 12.7

But our description—or rather the solution that utilizes this description in the Principle of Indifference—may itself be faulted for failing to respect geometric symmetry. Consider what happens if the axes are rotated through some angle around point A—that is, the orientation of the lines drawn on the floor is changed. Whatever method of solution we propose, should not make the answer—probability of a cut—depend on this orientation, for the problem remains essentially unchanged. (The aspect varied did not appear at all in the statement of the problem.) How do the two rival solutions vary with respect to this criterion?

Buffon's solution fares very well. For the initial parameter (angle which the needle makes with the X-axis) is changed by adding something (the angle of rotation), modulo 360°. A uniform distribution on that initial parameter induces automatically a uniform distribution also on its transform—equal angular intervals continue to receive equal probability.

But, and here is the rub, if we assume that y is uniformly distributed, it follows that y' (the corresponding coordinate in the rotated frame) is not. The easy way to see this is to look at equal increments in y and notice that they do not correspond to equal increments in y'.

To see this it is necessary to use the formula that transforms coordinates, when the frame is rotated. If the original coordinates of a point are (x, y) they become, upon rotation through angle a around the origin

$$t(x) = x \cos a - y \sin a$$
$$t(y) = x \sin a + y \cos a$$

In our case, point B has coordinates (x, y) but because $AB = 1$ we know that $x^2 + y^2 = 1$. Hence $x = \sqrt{(1 - y^2)}$ and we have

$$t(y) = \sqrt{(1 - y^2)} \sin a + y \cos a$$

Let us now look at two events that have equal probability if y has uniform distribution:

$$[y \,\varepsilon\, (0,\ 1/2)] \qquad [y \,\varepsilon\, (1/2,\ 1)]$$

These are the same events as

$$[t(y) \,\varepsilon\, (t(0),\ t(1/2))] \qquad [t(y) \,\varepsilon\, (t(1/2),\ t(1))]$$

If the variable $t(y)$ has uniform distribution, these events will have equal probabilities only if

$$t(1/2) - t(0) = t(1) - t(1/2)$$

so that is what we need to check. A single counter-example will do, so let us choose the angle of 30° (i.e. $\pi/6$ radians) which has sine $\frac{1}{2}$ and cosine $(\sqrt{3})/2$. Therefore:

$t(0) = \sin\left(\frac{1}{6}\pi\right) = \frac{1}{2}$

$t(\frac{1}{2}) = (\sqrt{3})/2 \sin\frac{1}{6}\pi + \frac{1}{2}\cos\frac{1}{6}\pi$

$\quad = (\sqrt{3})/2\ (\frac{1}{2}) + \frac{1}{2}((\sqrt{3})/2) = (\sqrt{3})/2$

$t(1) = \cos\frac{1}{6}\pi = (\sqrt{3})/2$

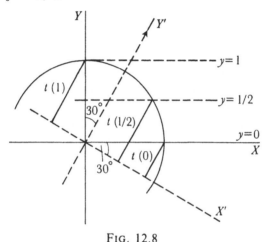

FIG. 12.8

It is very obvious that our desired equation does not hold. Figure 12.8 shows the different cases.

7. PYRRHIC VICTORY AND ULTIMATE DEFEAT

The successes we found in the preceding section, even together with their more sophisticated variants (to be discussed in the *Proofs and Illustrations*), constitute only a Pyrrhic victory. Again we can see this in simple examples, just because of the power of the uniqueness results utilized. Recall that invariance under translations and invariance under dilations each dictate an essentially *unique* answer to all probability questions. What happens when the examples take on more structure?

Peter Milne, writing about Rosencrantz's solution to von Kries's problem, has shown exactly how things go wrong.[17] To show this, he asked how the above results are to be applied to von Mises's water and wine problem. Let us again ask the same question in two different ways, referring back to the notation we had before.

What is the probability that at least 5 cc is water?
b = proportion of water to the total
x = proportion of wine to water.
$P(b \geqslant 0.5 \mid 0.1 \leqslant b \leqslant 0.9) = (\log 0.9 - \log 0.5)/(\log 0.9 - \log 0.1) = 0.267$
$P(x \leqslant 1 \mid 1/9 < x \leqslant 9) = \log 1 - \log(1/9))/(\log 9 - \log(1/9)) = \log 9/2\log 9 = 0.5$

We have received two contradictory answers.

Were we justified in treating the problem in this way? Well, the problem specified cc as unit of measurement but we have just as much warrant to regard this as irrelevant as we had for cm in the cubes problem. If we focus on parameter x here, say, we must treat it in the same way, if we have indeed found the correct form of the Principle of Indifference. Restating the problem then in terms of b, we have not introduced any new information—so we must derive the answer from the probability distribution on x, plus the definition of b in terms of x. Exactly the same would apply if we had started with b, and then moved on to x. But the two end results are not the same, so we have our paradox back.

It is also rather easy to see the pattern that will produce such paradoxes. A translation invariant measure will be well behaved with respect to addition and multiplication, while a dilation invariant measure will be equally good with respect to multiplication and exponentiation. But the relation between b and x uses both sorts of operations:

b = water/(water + wine) = water/10
x = wine/water = (10 − water)/water
hence, water = $10b$ and $x = (100 − b)/b$.

Now neither sort of measure will do. If the required dilation invariance did not dictate an essentially unique measure, we would perhaps have had some leeway to look for something other than logarithms—but we do not.

The history of the Principle of Indifference is instructive. If its mention in scientific sermons serves to remind us to look for symmetries, then it serves well. But a rule to determine initial probabilities a priori it is not. It violates a higher symmetry requirement when it is conceived of in that way.

Even if the Principle were unambiguous, the question whether its results would be probability functions with a good fit to actual frequencies in nature, would anyway be a purely contingent one. To imagine that it would not be—that empirical predictions could be made a priori, by 'pure thought' analysis—is feasible only on the assumption of some metaphysical scheme such as Leibniz's, in which the symmetries of the problems which God selects for attention, determine the structure of reality.[18] But because it is not unambiguous, even that assumption would leave us stranded, unless we knew how God selected his problems.

When E. T. Jaynes[19] discussed Bertrand's chord paradox, although noting that most writers had regarded it as an ill-posed problem, he responded:

But do we really believe that it is beyond our power to predict by 'pure thought' the results of such a simple experiment? The point at issue is far more important than merely resolving a geometric puzzle; for . . . applications of probability theory to physical experiments usually lead to problems of just this type. . . . (p. 478)

Jaynes's analysis of the Bertrand chord problem is along the lines of the preceding section. He shows that there is only one solution which derives from a measure which is invariant under Euclidean transformations.

But when we look more carefully at other parts of Jaynes's paper we see that his more general conclusions nullify the radical tone. Jaynes says of von Mises's water and wine problem, that the fatal ambiguities of the Principle of Indifference remain. More important: the strongest conclusion Jaynes manages to reach is merely one of *advice*, to regard a problem as having a definite solution until the contrary has been proved. The method he advises us to follow is that of symmetry arguments:

To summarize the above results: if we merely specify complete ignorance, we cannot hope to obtain any definite prior distribution, because such a statement is too vague to define any mathematically well-defined problem. We are defining what we mean by complete ignorance far more precisely if we can specify a set of operations which we recognize as transforming the problem into an equivalent one, and the desideratum of consistency then places non-trivial restrictions on the form of the prior.[20]

But as we know, this method always rests on assumptions which may or may not fit the physical situation in reality. Hence it cannot lead to a priori predictions. Success, when achieved, must be attributed to the good fortune that nature fits and continues to fit the general model with which the solution begins.

Proofs and illustrations

Harold Jeffreys introduced the search for invariant priors into the foundations of statistics; there has been much subsequent work along these lines by others.[21] We must conclude with Dawid, however, that the programme 'produces a whole range of choices in some problems, and no prior free from all objections in others' ('Invariant Prior Distributions', 235). I just wish to take up here the elegant logical analysis that Jaynes introduced to generalize the approach which we have been studying in these last two sections.[22]

Here some powerful mathematical theorems come to our aid. For under certain conditions, there exists indeed only one possible probability assignment to a group, so there is no ambiguity.

The general pattern of the approach I have been outlining is as follows. First one selects the correct group of transformations of our sets K which should leave the probability measure invariant. Call the group G. Then one finds the correct probability measure p on this group. Next define

(1) $P(A) = p(\{g \text{ in } G: g(x_0) \varepsilon A\})$

where x_0 is a chosen reference point in the set K on which we want our probability defined. If everything has gone well, P is the probability measure 'demanded' by the group.

What is required at the very least is that (*a*) p is a privileged measure on the group; (*b*) P is invariant under the action of the group; and (*c*) P is independent of the choice of x_0. Mathematics allows these desiderata to be satisfied: if the group G has some 'nice' properties, and we require p to be a left Haar measure (which means that $p(S) = p(\{gg' : g' \varepsilon S\})$) for any part S of G and any member g) then these desired consequences follow, and p, P are essentially unique.[23]

This is a very tight situation, and the required niceties can be expected in geometric models such as are used to define Bertrand's chord problem. But other sorts of models will not be equally nice; and even if they are, different models of the same situation could fairly bring us diverse answers. In any case there is a no a priori reason why all phenomena should fit models with such 'nice' properties only.

8. THE ETHICS OF AMBIGUITY

From the initial example, of a traveller on a treacherous shore, to the partial but impressive successes in the search for invariant priors, I have tried to emphasize how much symmetry considerations tell us. That is the positive side of this definitive dissolution of the idea of unique logical probability. Yet the story is far from complete, and its tactical and strategic suggestions for model construction far from exhausted.

But throughout the history of this subject, there wafts the siren melody of empirical probabilities determined a priori on the basis of pure symmetry considerations. The correct appreciation leads us to exactly the same conclusion as in Chapter 10. Once a problem is modelled, the symmetry requirement may give it a unique, or at least greatly constrained solution. The modelling, however, involves substantive assumptions: an implicit selection of certain parameters as alone relevant, and a tacit assumption of structure in the parameter space. Whenever the consequent limitations are ignored, paradoxes bring us back to our senses—symmetries respected in one modelling of the problem *entail* symmetries broken in another model. As soon as we took the first step, symmetries swept us along in a powerful current—but nature might have demanded a different first step, or embarkation in a different stream.

Facts are ambiguous. It is vain to desire prescience: which resolution of the present ambiguities will later facts vindicate? Our models of the facts, on the other hand, are not ambiguous; they had better not be. To choose one, is therefore a risk. To eliminate the risk is to cease theorizing altogether. That is one message of these paradoxes.

A Subjectivist's Guide to
Objective Chance*

INTRODUCTION

We subjectivists conceive of probability as the measure of reasonable partial belief. But we need not make war against other conceptions of probability, declaring that where subjective credence leaves off, there nonsense begins. Along with subjective credence we should believe also in objective chance. The practice and the analysis of science require both concepts. Neither can replace the other. Among the propositions that deserve our credence we find, for instance, the proposition that (as a matter of contingent fact about our world) any tritium atom that now exists has a certain chance of decaying within a year. Why should we subjectivists be less able than other folk to make sense of that?

Carnap (1945) did well to distinguish two concepts of probability, insisting that both were legitimate and useful and that neither was at fault because it was not the other. I do not think Carnap chose quite the right two concepts, however. In place of his "degree of confirmation" I would put *credence* or *degree of belief*; in place of his "relative

* I am grateful to several people for valuable discussions of this material; especially John Burgess, Nancy Cartwright, Richard Jeffrey, Peter Railton, and Brian Skyrms. I am also much indebted to Mellor (1971), which presents a view very close to mine; exactly how close I am not prepared to say.

frequency in the long run" I would put *chance* or *propensity*, understood as making sense in the single case. The division of labor between
the two concepts will be little changed by these replacements. Credence is well suited to play the role of Carnap's probability$_1$, and
chance to play the role of probability$_2$.

Given two kinds of probability, credence and chance, we can have
hybrid probabilities of probabilities. (Not "second order probabilities", which suggests one kind of probability self-applied.) Chance of
credence need not detain us. It may be partly a matter of chance what
one comes to believe, but what of it? Credence about chance is more
important. To the believer in chance, chance is a proper subject to have
beliefs about. Propositions about chance will enjoy various degrees of
belief, and other propositions will be believed to various degrees conditionally upon them.

As I hope the following questionnaire will show, we have some very
firm and definite opinions concerning reasonable credence about
chance. These opinions seem to me to afford the best grip we have on
the concept of chance. Indeed, I am led to wonder whether anyone *but*
a subjectivist is in a position to understand objective chance!

QUESTIONNAIRE

First question. A certain coin is scheduled to be tossed at noon today.
You are sure that this chosen coin is fair: it has a 50% chance of falling
heads and a 50% chance of falling tails. You have no other relevant
information. Consider the proposition that the coin tossed at noon
today falls heads. To what degree would you now believe that proposition?

Answer. 50%, of course.

(Two comments. (1) It is abbreviation to speak of the coin as fair.
Strictly speaking, what you are sure of is that the entire "chance set-
up" is fair: coin, tosser, landing surface, air, and surroundings together
are such as to make it so that the chance of heads is 50%. (2) Is it
reasonable to think of coin-tossing as a genuine chance process, given
present-day scientific knowledge? I think so: consider, for instance,
that air resistance depends partly on the chance making and breaking
of chemical bonds between the coin and the air molecules it
encounters. What is less clear is that the toss could be designed so that
you could reasonably be sure that the chance of heads is 50% exactly.

If you doubt that such a toss could be designed, you may substitute an example involving radioactive decay.)

Next question. As before, except that you have plenty of seemingly relevant evidence tending to lead you to expect that the coin will fall heads. This coin is known to have a displaced center of mass, it has been tossed 100 times before with 86 heads, and many duplicates of it have been tossed thousands of times with about 90% heads. Yet you remain quite sure, despite all this evidence, that the chance of heads this time is 50%. To what degree should you believe the proposition that the coin falls heads this time?

Answer. Still 50%. Such evidence is relevant to the outcome by way of its relevance to the proposition that the chance of heads is 50%, not in any other way. If the evidence somehow fails to diminish your certainty that the coin is fair, then it should have no effect on the distribution of credence about outcomes that accords with that certainty about chance. To the extent that uncertainty about outcomes is based on certainty about their chances, it is a stable, resilient sort of uncertainty—new evidence won't get rid of it. (The term "resiliency" comes from Skyrms (1977); see also Jeffrey (1965), §12.5.)

Someone might object that you could not reasonably remain sure that the coin was fair, given such evidence as I described and no contrary evidence that I failed to mention. That may be so, but it doesn't matter. Canons of reasonable belief need not be counsels of perfection. A moral code that forbids all robbery may also prescribe that if one nevertheless robs, one should rob only the rich. Likewise it is a sensible question what it is reasonable to believe about outcomes if one is unreasonably stubborn in clinging to one's certainty about chances.

Next question. As before, except that now it is afternoon and you have evidence that became available after the coin was tossed at noon. Maybe you know for certain that it fell heads; maybe some fairly reliable witness has told you that it fell heads; maybe the witness has told you that it fell heads in nine out of ten tosses of which the noon toss was one. You remain as sure as ever that the chance of heads, just before noon, was 50%. To what degree should you believe that the coin tossed at noon fell heads?

Answer. Not 50%, but something not far short of 100%. Resiliency has its limits. If evidence bears in a direct enough way on the outcome—a way which may nevertheless fall short of outright implication—then it may bear on your beliefs about outcomes otherwise than by way of your beliefs about the chances of the outcomes. Resiliency under all evidence whatever would be extremely unreasonable.

We can only say that degrees of belief about outcomes that are based on certainty about chances are resilient under *admissible* evidence. The previous question gave examples of admissible evidence; this question gave examples of inadmissible evidence.

Last question. You have no inadmissible evidence; if you have any relevant admissible evidence, it already has had its proper effect on your credence about the chance of heads. But this time, suppose you are not sure that the coin is fair. You divide your belief among three alternative hypotheses about the chance of heads, as follows.

You believe to degree 27% that the chance of heads is 50%.

You believe to degree 22% that the chance of heads is 35%.

You believe to degree 51% that the chance of heads is 80%.

Then to what degree should you believe that the coin falls heads?

Answer. (27% × 50%) + (22% × 35%) + (51% × 80%); that is, 62%. Your degree of belief that the coin falls heads, conditionally on any one of the hypotheses about the chance of heads, should equal your unconditional degree of belief if you were sure of that hypothesis. That in turn should equal the chance of heads according to the hypothesis: 50% for the first hypothesis, 35% for the second, and 80% for the third. Given your degrees of belief that the coin falls heads, conditionally on the hypotheses, we need only apply the standard multiplicative and additive principles to obtain our answer.

THE PRINCIPAL PRINCIPLE

I have given undefended answers to my four questions. I hope you found them obviously right, so that you will be willing to take them as evidence for what follows. If not, do please reconsider. If so, splendid—now read on.

It is time to formulate a general principle to capture the intuitions that were forthcoming in our questionnaire. It will resemble familiar principles of direct inference except that (1) it will concern chance, not some sort of actual or hypothetical frequency, and (2) it will incorporate the observation that certainty about chances—or conditionality on propositions about chances—makes for resilient degrees of belief about outcomes. Since this principle seems to me to capture all we know about chance, I call it

THE PRINCIPAL PRINCIPLE. Let C be any reasonable initial credence function. Let t be any time. Let x be any real number in the unit interval. Let X be the proposition that the chance, at time t, of A's holding equals x. Let E be any proposition compatible with X that is admissible at time t. Then

$$C(A/XE) = x.$$

That will need a good deal of explaining. But first I shall illustrate the principle by applying it to the cases in our questionnaire.

Suppose your present credence function is $C(-/E)$, the function that comes from some reasonable initial credence function C by conditionalizing on your present total evidence E. Let t be the time of the toss, noon today, and let A be the proposition that the coin tossed today falls heads. Let X be the proposition that the chance at noon (just before the toss) of heads is x. (In our questionnaire, we mostly considered the case that x is 50%). Suppose that nothing in your total evidence E contradicts X; suppose also that it is not yet noon, and you have no foreknowledge of the outcome, so everything that is included in E is entirely admissible. The conditions of the Principal Principle are met. Therefore $C(A/XE)$ equals x. That is to say that x is your present degree of belief that the coin falls heads, conditionally on the proposition that its chance of falling heads is x. If in addition you are sure that the chance of heads is x—that is, if $C(X/E)$ is one—then it follows also that x is your present unconditional degree of belief that the coin falls heads. More generally, whether or not you are sure about the chance of heads, your unconditional degree of belief that the coin falls heads is given by summing over alternative hypotheses about chance:

$$C(A/E) = \Sigma_x C(X_x/E)C(A/X_x E) = \Sigma_x C(X_x/E)x,$$

where X_x, for any value of x, is the proposition that the chance at t of A equals x.

Several parts of the formulation of the Principal Principle call for explanation and comment. Let us take them in turn.

THE INITIAL CREDENCE FUNCTION C

I said: let C be any reasonable initial credence function. By that I meant, in part, that C was to be a probability distribution over (at least) the space whose points are possible worlds and whose regions (sets of worlds) are propositions. C is a non-negative, normalized, finitely additive measure defined on all propositions.

The corresponding conditional credence function is defined simply as a quotient of unconditional credences:

$$C(A/B) =_{df} C(AB)/C(B).$$

I should like to assume that it makes sense to conditionalize on any but the empty proposition. Therefore, I require that C is *regular*: $C(B)$ is zero, and $C(A/B)$ is undefined, only if B is the empty proposition, true at no worlds. You may protest that there are too many alternative possible worlds to permit regularity. But that is so only if we suppose, as I do not, that the values of the function C are restricted to the standard reals. Many propositions must have infinitesimal C-values, and $C(A/B)$ often will be defined as a quotient of infinitesimals, each infinitely close but not equal to zero. (See Bernstein and Wattenberg (1969).) The assumption that C is regular will prove convenient, but it is not justified only as a convenience. Also it is required as a condition of reasonableness: one who started out with an irregular credence function (and who then learned from experience by conditionalizing) would stubbornly refuse to believe some propositions no matter what the evidence in their favor.

In general, C is to be reasonable in the sense that if you started out with it as your initial credence function, and if you always learned from experience by conditionalizing on your total evidence, then no matter what course of experience you might undergo your beliefs would be reasonable for one who had undergone that course of experience. I do not say what distinguishes a reasonable from an unreasonable credence function to arrive at after a given course of experience. We do make the distinction, even if we cannot analyze it; and therefore I may appeal to it in saying what it means to require that C be a reasonable initial credence function.

I have assumed that the method of conditionalizing is *one* reasonable way to learn from experience, given the right initial credence function. I have not assumed something more controversial: that it is the *only* reasonable way. The latter view may also be right (the cases where it seems wrong to conditionalize may all be cases where one departure from ideal rationality is needed to compensate for another) but I shall not need it here.

(I said that C was to be a probability distribution over *at least* the space of worlds; the reason for that qualification is that sometimes one's credence might be divided between different possibilities within a single world. That is the case for someone who is sure what sort of world he lives in, but not at all sure who and when and where in the world he is. In a fully general treatment of credence it would be well to replace the worlds by something like the "centered worlds" of Quine (1969), and the propositions by something corresponding to properties. But I shall ignore these complications here.)

THE REAL NUMBER x

I said: let x be any real number in the unit interval. I must emphasize that "x" is a quantified variable; it is not a schematic letter that may freely be replaced by terms that designate real numbers in the unit interval. For fixed A and t, "the chance, at t, of A's holding" is such a term; suppose we put it in for the variable x. It might seem that for suitable C and E we have the following: if X is the proposition that the chance, at t, of A's holding equals the chance, at t, of A's holding—in other words, if X is the necessary proposition—then

$$C(A/XE) = \text{the chance, at } t, \text{ of } A\text{'s holding.}$$

But that is absurd. It means that if E is your present total evidence and $C(-/E)$ is your present credence function, then if the coin is in fact fair—whether or not you think it is!—then your degree of belief that it falls heads is 50%. Fortunately, that absurdity is not an instance of the Principal Principle. The term "the chance, at t, of A's holding" is a non-rigid designator; chance being a matter of contingent fact, it designates different numbers at different worlds. The context "the proposition that . . . ", within which the variable "x" occurs, is intensional. Universal instantiation into an intensional context with a non-rigid term is a fallacy. It is the fallacy that takes you, for instance, from the true premise "For any number x, the proposition that x is nine is non-contingent" to the false conclusion "The proposition that the number of planets is nine is non-contingent". See Jeffrey (1970) for discussion of this point in connection with a relative of the Principal Principle.

I should note that the values of "x" are not restricted to the standard reals in the unit interval. The Principal Principle may be applied as follows: you are sure that some spinner is fair, hence that it has infinitesimal chance of coming to rest at any particular point; therefore (if your total evidence is admissible) you should believe only to an infinitesimal degree that it will come to rest at any particular point.

THE PROPOSITION *X*

I said: let *X* be the proposition that the chance, at time *t*, of *A*'s holding equals *x*. I emphasize that I am speaking of objective, single-case chance—not credence, not frequency. Like it or not, we have this concept. We think that a coin about to be tossed has a certain chance of falling heads, or that a radioactive atom has a certain chance of decaying within the year, quite regardless of what anyone may believe about it and quite regardless of whether there are any other similar coins or atoms. As philosophers we may well find the concept of objective chance troublesome, but that is no excuse to deny its existence, its legitimacy, or its indispensability. If we can't understand it, so much the worse for us.

Chance and credence are distinct, but I don't say they are unrelated. What is the Principal Principle but a statement of their relation? Neither do I say that chance and frequency are unrelated, but they are distinct. Suppose we have many coin-tosses with the same chance of heads (not zero or one) in each case. Then there is some chance of getting any frequency of heads whatever; and hence some chance that the frequency and the uniform single-case chance of heads may differ, which could not be so if these were one and the same thing. Indeed the chance of difference may be infinitesimal if there are infinitely many tosses, but that is still not zero. Nor do hypothetical frequencies fare any better. There is no such thing as *the* infinite sequence of outcomes, or *the* limiting frequency of heads, that *would* eventuate if some particular coin-toss were somehow repeated forever. Rather there are countless sequences, and countless frequencies, that *might* eventuate and would have some chance (perhaps infinitesimal) of eventuating. (See Jeffrey (1977), Skyrms (1977), and the discussion of "might" counterfactuals in Lewis (1973).)

Chance is not the same thing as credence or frequency; this is not yet to deny that there might be some roundabout way to analyze chance in terms of credence or frequency. I would only ask that no such analysis be accepted unless it is compatible with the Principal Principle. We shall consider how this requirement bears on the prospects for an analysis of chance, but without settling the question of whether such an analysis is possible.

I think of chance as attaching in the first instance to propositions: the chance of an event, an outcome, etc. is the chance of truth of the proposition that holds at just those worlds where that event, outcome, or whatnot occurs. (Here I ignore the special usage of "event" to simply mean "proposition".) I have foremost in mind the chances of truth of propositions about localized matters of particular fact—a certain toss of a coin, the fate of a certain tritium atom on a certain day—but I do not say that those are the only propositions to which chance applies. Not only does it make sense to speak of the chance that a coin will fall heads on a particular occasion; equally it makes sense to speak of the chance of getting exactly seven heads in a particular sequence of eleven tosses. It is only caution, not any definite reason to think otherwise, that stops me from assuming that chance of truth applies to any proposition whatever. I shall assume, however, that the broad class of propositions to which chance of truth applies is closed under the Boolean operations of conjunction (intersection), disjunction (union), and negation (complementation).

We ordinarily think of chance as time-dependent, and I have made that dependence explicit. Suppose you enter a labyrinth at 11:00 a.m., planning to choose your turn whenever you come to a branch point by tossing a coin. When you enter at 11:00, you may have a 42% chance of reaching the center by noon. But in the first half hour you may stray into a region from which it is hard to reach the center, so that by 11:30 your chance of reaching the center by noon has fallen to 26%. But then you turn lucky; by 11:45 you are not far from the center and your chance of reaching it by noon is 78%. At 11:49 you reach the center; then and forevermore your chance of reaching it by noon is 100%.

Sometimes, to be sure, we omit reference to a time. I do not think this means that we have some timeless notion of chance. Rather, we have other ways to fix the time than by specifying it explicitly. In the case of the labyrinth we might well say (before, after, or during your exploration) that your chance of reaching the center by noon is 42%. The understood time of reference is the time when your exploration begins. Likewise we might speak simply of the chance of a certain atom's decaying within a certain year, meaning the chance at the beginning of that year. In general, if A is the proposition that something or other takes place within a certain interval beginning at time t, then we may take a special interest in what I shall call the *endpoint chance* of A's holding: the chance at t, the beginning of the interval in question. If we speak simply of the chance of A's holding, not mentioning a time, it is this endpoint chance—the chance at t of A's holding—that we are likely to mean.

Chance also is world-dependent. Your chance at 11:00 of reaching the center of the labyrinth by noon depends on all sorts of contingent features of the world: the structure of the labyrinth and the speed with which you can walk through it, for instance. Your chance at 11:30 of reaching the center by noon depends on these things, and also on where in the labyrinth you then are. Since these things vary from world to world, so does your chance (at either time) of reaching the center by noon. Your chance at noon of reaching the center by noon is one at the worlds where you have reached the center; zero at all others, including those worlds where you do not explore the labyrinth at all, perhaps because you or it do not exist. (Here I am speaking loosely, as if I believed that you and the labyrinth could inhabit several worlds at once. See Lewis (1968) for the needed correction.)

We have decided this much about chance, at least: it is a function of three arguments. To a proposition, a time, and a world it assigns a real number. Fixing the proposition A, the time t, and the number x, we have our proposition X: it is the proposition that holds at all and only those worlds w such that this function assigns to A, t, and w the value x. This is the proposition that the chance, at t, of A's holding is x.

THE ADMISSIBLE PROPOSITION E

I said: let E be any proposition that is admissible at time t. Admissible propositions are the sort of information whose impact on credence about outcomes comes entirely by way of credence about the chances of those outcomes. Once the chances are given outright, conditionally or unconditionally, evidence bearing on them no longer matters. (Once it is settled that the suspect fired the gun, the discovery of his fingerprint on the trigger adds nothing to the case against him.) The power of the Principal Principle depends entirely on how much is admissible. If nothing is admissible it is vacuous. If everything is admissible it is inconsistent. Our questionnaire suggested that a great deal is admissible, but we saw examples also of inadmissible information. I have no definition of admissibility to offer, but must be content to suggest sufficient (or almost sufficient) conditions for admissibility. I suggest that two different sorts of information are generally admissible.

The first sort is historical information. If a proposition is entirely about matters of particular fact at times no later than t, then as a rule that proposition is admissible at t. Admissible information just before the toss of a coin, for example, includes the outcomes of all previous tosses of that coin and others like it. It also includes every detail—no matter how hard it might be to discover—of the structure of the coin, the tosser, other parts of the set-up, and even anything nearby that might somehow intervene. It also includes a great deal of other information that is completely irrelevant to the outcome of the toss.

A proposition is *about* a subject matter—about history up to a certain time, for instance—if and only if that proposition holds at both or neither of any two worlds that match perfectly with respect to that subject matter. (Or we can go the other way: two worlds match perfectly with respect to a subject matter if and only if every proposition about that subject matter holds at both or neither.) If our world and another are alike point for point, atom for atom, field for field, even spirit for spirit (if such there be) throughout the past and up until noon today, then any proposition that distinguishes the two cannot be entirely about the respects in which there is no difference. It cannot be entirely about what goes on no later than noon today. That is so even if its linguistic expression makes no overt mention of later times; we must beware lest information about the future is hidden in the predicates, as in "Fred was mortally wounded at 11:58". I doubt that any linguistic test of aboutness will work without circular restrictions on the language used. Hence it seems best to take either "about" or "perfect match with respect to" as a primitive.

Time-dependent chance and time-dependent admissibility go together. Suppose the proposition A is about matters of particular fact at some moment or interval t_A, and suppose we are concerned with chance at time t. If t is later than t_A, then A is admissible at t. The Principal Principle applies with A for E. If X is the proposition that the chance at t of A equals x, and if A and X are compatible, then

$$1 = C(A/XA) = x.$$

Put contrapositively, this means that if the chance at t of A, according to X, is anything but one, then A and X are incompatible. A implies that the chance at t of A, unless undefined, equals one. What's past is no longer chancy. The past, unlike the future, has no chance of being any other way than the way it actually is. This temporal asymmetry of chance falls into place as part of our conception of the past as "fixed" and the future as "open"—whatever that may mean. The asymmetry of fixity and of chance may be pictured by a tree. The single trunk is

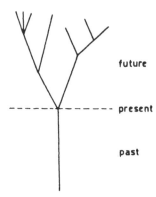

the one possible past that has any present chance of being actual. The many branches are the many possible futures that have some present chance of being actual. I shall not try to say here what features of the world justify our discriminatory attitude toward past and future possibilities, reflected for instance in the judgment that historical information is admissible and similar information about the future is not. But I think they are contingent features, subject to exception and absent altogether from some possible worlds.

That possibility calls into question my thesis that historical information is invariably admissible. What if the commonplace *de facto* asymmetries between past and future break down? If the past lies far in the future, as we are far to the west of ourselves, then it cannot simply be that propositions about the past are admissible and propositions about the future are not. And if the past contains seers with foreknowledge of what chance will bring, or time travelers who have witnessed the outcome of coin-tosses to come, then patches of the past are enough tainted with futurity so that historical information about them may well seem inadmissible. That is why I qualified my claim that historical information is admissible, saying only that it is so "as a rule". Perhaps it is fair to ignore this problem in building a case that the Principal Principle captures our common opinions about chance, since those opinions may rest on a naive faith that past and future cannot possibly get mixed up. Any serious physicist, if he remains at least open-minded both about the shape of the cosmos and about the existence of chance processes, ought to do better. But I shall not; I shall carry on as if historical information is admissible without exception.

Besides historical information, there is at least one other sort of admissible information: hypothetical information about chance itself.

Let us return briefly to our questionnaire and add one further supposition to each case. Suppose you have various opinions about what the chance of heads would be under various hypotheses about the detailed nature and history of the chance set-up under consideration. Suppose further that you have similar hypothetical opinions about other chance set-ups, past, present, and future. (Assume that these opinions are consistent with your admissible historical information and your opinions about chance in the present case.) It seems quite clear to me—and I hope it does to you also—that these added opinions do not change anything. The correct answers to the questionnaire are just as before. The added opinions do not bear in any overly direct way on the future outcomes of chance processes. Therefore they are admissible.

We must take care, though. Some propositions about future chances do reveal inadmissible information about future history, and these are inadmissible. Recall the case of the labyrinth: you enter at 11:00, choosing your turns by chance, and hope to reach the center by noon. Your subsequent chance of success depends on the point you have reached. The proposition that at 11:30 your chance of success has fallen to 26% is not admissible information at 11:00; it is a giveaway about your bad luck in the first half hour. What is admissible at 11:00 is a conditional version: if you were to reach a certain point at 11:30, your chance of success would then be 26%. But even some conditionals are tainted: for instance, any conditional that could yield inadmissible information about future chances by *modus ponens* from admissible historical propositions. Consider also the truth-functional conditional that if history up to 11:30 follows a certain course, then you will have a 98% chance of becoming a monkey's uncle before the year is out. This conditional closely resembles the denial of its antecedent, and is inadmissible at 11:00 for the same reason.

I suggest that conditionals of the following sort, however, are admissible; and indeed admissible at all times. (1) The consequent is a proposition about chance at a certain time. (2) The antecedent is a proposition about history up to that time; and further, it is a complete proposition about history up to that time, so that it either implies or else is incompatible with any other proposition about history up to that time. It fully specifies a segment, up to the given time, of some possible course of history. (3) The conditional is made from its consequent and antecedent not truth-functionally, but rather by means of a strong conditional operation of some sort. This might well be the counterfactual conditional of Lewis (1973); but various rival versions would serve as well, since many differences do not matter for the case at hand. One feature of my treatment will be needed; however: if the antecedent of one of our conditionals holds at a world, then both or neither of the conditional and its consequent hold there.

These admissible conditionals are propositions about how chance depends (or fails to depend) on history. They say nothing, however, about how history chances to go. A set of them is a theory about the way chance works. It may or may not be a complete theory, a consistent theory, a systematic theory, or a credible theory. It might be a miscellany of unrelated propositions about what the chances would be after various fully specified particular courses of events. Or it might be systematic, compressible into generalizations to the effect that after any course of history with property J there would follow a chance distribution with property K. (For instance, it might say that any coin with a certain structure would be fair.) These generalizations are universally quantified conditionals about single-case chance; if lawful, they are probabilistic laws in the sense of Railton (1978). (I shall not consider here what would make them lawful; but see Lewis (1973), §3.3, for a treatment that could cover laws about chance along with other laws.) Systematic theories of chance are the ones we can express in language, think about, and believe to substantial degrees. But a reasonable initial credence function does not reject any possibility out of hand. It assigns some non-zero credence to any consistent theory of chance, no matter how unsystematic and incompressible it is.

Historical propositions are admissible; so are propositions about the dependence of chance on history. Combinations of the two, of course, are also admissible. More generally, we may assume that any Boolean combination of propositions admissible at a time also is admissible at that time. Admissibility consists in keeping out of a forbidden subject matter—how the chance processes turned out—and there is no way to break into a subject matter by making Boolean combinations of propositions that lie outside it.

There may be sorts of admissible propositions besides those I have considered. If so, we shall have no need of them in what follows.

This completes an exposition of the Principal Principle. We turn next to an examination of its consequences. I maintain that they include all that we take ourselves to know about chance.

THE PRINCIPLE REFORMULATED

Given a time t and world w, let us write P_{tw} for the *chance distribution* that obtains at t and w. For any proposition A, $P_{tw}(A)$ is the chance, at time t and world w, of A's holding. (The domain of P_{tw} comprises those propositions for which this chance is defined.)

Let us also write H_{tw} for the *complete history* of world w up to time t: the conjunction of all propositions that hold at w about matters of particular fact no later than t. H_{tw} is the proposition that holds at exactly those worlds that perfectly match w, in matters of particular fact, up to time t.

Let us also write T_w for the *complete theory of chance* for world w: the conjunction of all the conditionals from history to chance, of the sort just considered, that hold at w. Thus T_w is a full specification, for world w, of the way chances at any time depend on history up to that time.

Taking the conjunction $H_{tw}T_w$, we have a proposition that tells us a great deal about the world w. It is nevertheless admissible at time t, being simply a giant conjunction of historical propositions that are admissible at t and conditionals from history to chance that are admissible at any time. Hence the Principal Principle applies:

$$C(A/XH_{tw}T_w) = x$$

when C is a reasonable initial credence function, X is the proposition that the chance at t of A is x, and $H_{tw}T_w$ is compatible with X.

Suppose X holds at w. That is so if and only if x equals $P_{tw}(A)$. Hence we can choose such an X whenever A is in the domain of P_{tw}. $H_{tw}T_w$ and X both hold at w, therefore they are compatible. But further, $H_{tw}T_w$ implies X. The theory T_w and the history H_{tw} together are enough to imply all that is true (and contradict all that is false) at world w about chances at time t. For consider the strong conditional with antecedent H_{tw} and consequent X. This conditional holds at w, since by hypothesis its antecedent and consequent hold there. Hence it is implied by T_w, which is the conjunction of all conditionals of its sort that hold at w; and this conditional and H_{tw} yield X by *modus ponens*. Consequently, the conjunction $XH_{tw}T_w$ simplifies to $H_{tw}T_w$. Provided that A is in the domain of P_{tw} so that we can make a suitable choice of X, we can substitute $P_{tw}(A)$ for x, and $H_{tw}T_w$ for $XH_{tw}T_w$, in our instance of the Principal Principle. Therefore we have

THE PRINCIPAL PRINCIPLE REFORMULATED. Let C be any reasonable initial credence function. Then for any time t, world w, and proposition A in the domain of P_{tw}

$$P_{tw}(A) = C(A/H_{tw}T_w).$$

In words: the chance distribution at a time and a world comes from any reasonable initial credence function by conditionalizing on the complete history of the world up to the time, together with the complete theory of chance for the world.

This reformulation enjoys less direct intuitive support than the original formulation, but it will prove easier to use. It will serve as our point of departure in examining further consequences of the Principal Principle.

CHANCE AND THE PROBABILITY CALCULUS

A reasonable inital credence function is, among other things, a probability distribution: a non-negative, normalized, finitely additive measure. It obeys the laws of mathematical probability theory. There are well-known reasons why that must be so if credence is to rationalize courses of action that would not seem blatantly unreasonable in some circumstances.

Whatever comes by conditionalizing from a probability distribution is itself a probability distribution. Therefore a chance distribution is a probability distribution. For any time t and world w, P_{tw} obeys the laws of mathematical probability theory. These laws carry over from credence to chance via the Principal Principle. We have no need of any independent assumption that chance is a kind of probability.

Observe that although the Principal Principle concerns the relationship between chance and credence, some of its consequences concern chance alone. We have seen two such consequences. (1) The thesis that the past has no present chance of being otherwise than it actually is. (2) The thesis that chance obeys the laws of probability. More such consequences will appear later.

CHANCE AS OBJECTIFIED CREDENCE

Chance is an objectified subjective probability in the sense of Jeffrey (1965), §12.7. Jeffrey's construction (omitting his use of sequences of partitions, which is unnecessary if we allow infinitestimal credences) works as follows. Suppose given a partition of logical space: a set of mutually exclusive and jointly exhaustive propositions. Then we can define the *objectification* of a credence function, with respect to this partition, at a certain world, as the probability distribution that comes from the given credence function by conditionalizing on the member of the given partition that holds at the given world. Objectified credence is credence conditional on the truth—not the whole truth, however, but exactly as much of it as can be captured by a member of the partition without further subdivision of logical space. The member of the partition that holds depends on matters of contingent fact, varying from one world to another; it does not depend on what we think (except insofar as our thoughts are relevant matters of fact) and we may well be ignorant or mistaken about it. The same goes for objectified credence.

Now consider one particular way of partitioning. For any time t, consider the partition consisting of the propositions $H_{tw}T_w$ for all worlds w. Call this the *history-theory partition* for time t. A member of this partition is an equivalence class of worlds with respect to the relation of being exactly alike both in respect of matters of particular fact up to time t and in respect of the dependence of chance on history. The Principal Principle tells us that the chance distribution, at any time t and world w, is the objectification of any reasonable credence function, with respect to the history-theory partition for time t, at world w. Chance is credence conditional on the truth—*if* the truth is subject to censorship along the lines of the history-theory partition, and *if* the credence is reasonable.

Any historical proposition admissible at time t, or any admissible conditional from history to chance, or any admissible Boolean combination of propositions of these two kinds—in short, any sort of admissible proposition we have considered—is a disjunction of members of the history-theory partition for t. Its borders follow the lines of the partition, never cutting between two worlds that the partition does not distinguish. Likewise for any proposition about chances at t. Let X be the proposition that the chance at t of A is x, let Y be any member of the history-theory partition for t, and let C be any reasonable initial credence function. Then, according to our reformulation of the Principal Principle, X holds at all worlds in Y if $C(A/Y)$ equals x, and at no worlds in Y otherwise. Therefore X is the disjunction of all members Y of the partition such that $C(A/Y)$ equals x.

We may picture the situation as follows. The partition divides logical space into countless tiny squares. In each square there is a black region where A holds and a white region where it does not. Now blur the focus, so that divisions within the squares disappear from view. Each square becomes a grey patch in a broad expanse covered with varying shades of grey. Any maximal region of uniform shade is a proposition specifying the chance of A. The darker the shade, the higher is the uniform chance of A at the worlds in the region. The worlds themselves are not grey—they are black or white, worlds where A holds or where it doesn't—but we cannot focus on single worlds, so they all seem to be the shade of grey that covers their region. Admissible propositions, of the sorts we have considered, are regions that may cut across the contours of the shades of grey. The conjunction of one of these admissible propositions and a proposition about the chance of A is a region of uniform shade, but not in general a maximal uniform region. It consists of some, but perhaps not all, the members Y of the partition for which $C(A/Y)$ takes a certain value.

We derived our reformulation of the Principal Principle from the original formulation, but have not given a reverse derivation to show the two formulations equivalent. In fact the reformulation may be weaker, but not in any way that is likely to matter. Let C be a reasonable initial credence function; let X be the proposition that the chance at t of A is x; let E be admissible at t (in one of the ways we have considered) and compatible with X. According to the reformulation, as we have seen, XE is a disjunction of incompatible propositions Y, for each of which $C(A/Y)$ equals x. If there were only finitely many Y's, it would follow that $C(A/XE)$ also equals x. But the implication fails in certain cases with infinitely many Y's (and indeed we would expect the history-theory partition to be infinite) so we cannot quite recover the original formulation in this way. The cases of failure are peculiar, however, so the extra strength of the original formulation in ruling them out seems unimportant.

KINEMATICS OF CHANCE

Chance being a kind of probability, we may define conditional chance in the usual way as a quotient (leaving it undefined if the denominator is zero):

$$P_{tw}(A/B) =_{df} P_{tw}(AB)/P_{tw}(B).$$

To simplify notation, let us fix on a particular world—ours, as it might be—and omit the subscript "w"; let us fix on some particular reasonable initial credence function C, it doesn't matter which; and let us fix on a sequence of times, in order from earlier to later, to be called 1, 2, 3, ... (I do not assume they are equally spaced.) For any time t in our sequence, let the proposition I_t be the complete history of our chosen world in the interval from time t to time $t + 1$ (including $t + 1$ but not t). Thus I_t is the set of worlds that match the chosen world perfectly in matters of particular fact throughout the given interval.

A complete history up to some time may be extended by conjoining complete histories of subsequent intervals. H_2 is $H_1 I_1$, H_3 is $H_1 I_1 I_2$, and so on. Then by the Principal Principal we have:

$$P_1(A) = C(A/H_1 T),$$

$$P_2(A) = C(A/H_2 T) = C(A/H_1 I_1 T) = P_1(A/I_1),$$

$$P_3(A) = C(A/H_3 T) = C(A/H_1 I_1 I_2 T) = P_2(A/I_2)$$

$$= P_1(A/I_1 I_2),$$

and in general

$$P_{t+n+1}(A) = P_t(A/I_t \ldots I_{t+n}).$$

In words: a later chance distribution comes from an earlier one by conditionalizing on the complete history of the interval in between.

The evolution of chance is parallel to the evolution of credence for an agent who learns from experience, as he reasonably might, by conditionalizing. In that case a later credence function comes from an earlier one by conditionalizing on the total increment of evidence gained in the interval in between. For the evolution of chance we simply put the world's chance distribution in place of the agent's credence function, and the totality of particular fact about a time in place of the totality of evidence gained at that time.

In the interval from t to $t + 1$ there is a certain way that the world will in fact develop: namely, the way given by I_t. And at t, the last moment before the interval begins, there is a certain chance that the world will develop in that way: $P_t(I_t)$, the endpoint chance of I_t. Likewise for a longer interval, say from time 1 to time 18. The world will in fact develop in the way given by $I_1 \ldots I_{17}$, and the endpoint chance of its doing so is $P_1(I_1 \ldots I_{17})$. By definition of conditional chance

$$P_1(I_1 \ldots I_{17}) = P_1(I_1) \cdot P_1(I_2/I_1) \cdot P_1(I_3/I_1 I_2) \ldots P_1(I_{17}/I_1 \ldots I_{16}),$$

and by the Principal Principle, applied as above,

$$P_1(I_1 \ldots I_{17}) = P_1(I_1) \cdot P_2(I_2) \cdot P_3(I_3) \ldots P_{17}(I_{17}).$$

In general, if an interval is divided into subintervals, then the endpoint chance of the complete history of the interval is the product of the endpoint chances of the complete histories of the subintervals.

Earlier we drew a tree to represent the temporal asymmetry of chance. Now we can embellish the tree with numbers to represent the kinematics of chance. Take time 1 as the present. Worlds—those of them that are compatible with a certain common past and a certain common theory of chance—lie along paths through the tree. The numbers on each segment give the endpoint chance of the course of history represented by that segment, for any world that passes through that segment. Likewise, for any path consisting of several segments, the product of numbers along the path gives the endpoint chance of the course of history represented by the entire path.

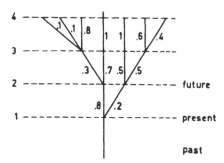

CHANCE OF FREQUENCY

Suppose that there is to be a long sequence of coin tosses under more or less standardized conditions. The first will be in the interval between time 1 and time 2, the second in the interval between 2 and 3, and so on. Our chosen world is such that at time 1 there is no chance, or negligible chance, that the planned sequence of tosses will not take place. And indeed it does take place. The outcomes are given by a sequence of propositions A_1, A_2, ... Each A_t states truly whether the toss between t and $t + 1$ fell heads or tails. A conjunction $A_1 . . . A_n$ then gives the history of outcomes for an initial segment of the sequence.

The endpoint chance $P_1(A_1 . . . A_n)$ of such a sequence of outcomes is given by a product of conditional chances. By definition of conditional chance,

$$P_1(A_1 \ldots A_n) = P_1(A_1) \cdot P_1(A_2/A_1) \cdot P_1(A_3/A_1A_2) \ldots$$
$$\cdot P_1(A_n/A_1 \ldots A_{n-1}).$$

Since we are dealing with propositions that give only incomplete histories of intervals, there is no general guarantee that these factors equal the endpoint chances of the A's. The endpoint chance of A_2, $P_2(A_2)$, is given by $P_1(A_2/I_1)$; this may differ from $P_1(A_2/A_1)$ because the complete history I_1 includes some relevant information that the incomplete history A_1 omits about chance occurrences in the first interval. Likewise for the conditional and endpoint chances pertaining to later intervals.

Even though there is no general guarantee that the endpoint chance of a sequence of outcomes equals the product of the endpoint chances of the individual outcomes, yet it may be so if the world is right. It may be, for instance, that the endpoint chance of A_2 does not depend on those aspects of the history of the first interval that are omitted from A_1—it would be the same regardless. Consider the class of all possible complete histories up to time 2 that are compatible both with the previous history H_1 and with the outcome A_1 of the first toss. These give all the ways the omitted aspects of the first interval might be. For each of these histories, some strong conditional holds at our chosen world that tells what the chance at 2 of A_2 would be if that history were to come about. Suppose all these conditionals have the same consequent: whichever one of the alternative histories were to come about, it would be that X, where X is the proposition that the chance at 2 of A_2 equals x. Then the conditionals taken together tell us that the endpoint chance of A_2 is independent of all aspects of the history of the first interval except the outcome of the first toss.

In that case we can equate the conditional chance $P_1(A_2/A_1)$ and the endpoint chance $P_2(A_2)$. Note that our conditionals are of the sort implied by T, the complete theory of chance for our chosen world. Hence A_1, H_1, and T jointly imply X. It follows that A_1H_1T and XA_1H_1T are the same proposition. It also follows that X holds at our chosen world, and hence that x equals $P_2(A_2)$. Note also that A_1H_1T is admissible at time 2. Now, using the Principal Principle first as reformulated and then in the original formulation, we have

$$P_1(A_2/A_1) = C(A_2/A_1H_1T) = C(A_2/XA_1H_1T) = x = P_2(A_2).$$

If we also have another such battery of conditionals to the effect that the endpoint chance of A_3 is independent of all aspects of the history of the first two intervals except the outcomes A_1 and A_2 of the first two tosses, and another battery for A_4, and so on, then the multiplicative rule for endpoint chances follows:

$$P_1(A_1 \ldots A_n) = P_1(A_1) \cdot P_2(A_2) \cdot P_3(A_3) \ldots P_n(A_n).$$

The conditionals that constitute the independence of endpoint chances mean that the incompleteness of the histories $A_1, A_2, \ldots$ doesn't matter. The missing part wouldn't make any difference.

We might have a stronger form of independence. The endpoint chances might not depend on *any* aspects of history after time 1, not even the outcomes of previous tosses. Then conditionals would hold at our chosen world to the effect that if any complete history up to time 2 which is compatible with H_1 were to come about, it would be that X (where X is again the proposition that the chance at 2 of A_2 equals x). We argue as before, leaving out A_1: T implies the conditionals, H_1 and T jointly imply X, H_1T and XH_1T are the same, X holds, x equals $P_2(A_2)$, H_1T is admissible at 2; so, using the Principal Principle in both formulations, we have

$$P_1(A_2) = C(A_2/H_1T) = C(A_2/XH_1T) = x = P_2(A_2).$$

Our strengthened independence assumption implies the weaker independence assumption of the previous case, wherefore

$$P_1(A_2/A_1) = P_2(A_2) = P_1(A_2).$$

If the later outcomes are likewise independent of history after time 1, then we have a multiplicative rule not only for endpoint chances but also for unconditional chances of outcomes at time 1:

$$P_1(A_1 \ldots A_n) = P_1(A_1) \cdot P_1(A_2) \cdot P_1(A_3) \ldots P_1(A_n).$$

Two conceptions of independence are in play together. One is the familiar probabilistic conception: A_2 is independent of A_1, with respect to the chance distribution P_1, if the conditional chance $P_1(A_2/A_1)$ equals the unconditional chance $P_1(A_2)$; equivalently, if the chance $P_1(A_1A_2)$ of the conjunction equals the product $P_1(A_1) \cdot P_1(A_2)$ of the chances of the conjuncts. The other conception involves batteries of strong conditionals with different antecedents and the same consequent. (I consider this to be *causal* independence, but that's another story.) The conditionals need not have anything to do with probability; for instance, my beard does not depend on my politics since I would have such a beard whether I were Republican, Democrat, Prohibitionist, Libertarian, Socialist Labor, or whatever. But one sort of consequent that can be independent of a range of alternatives, as we have seen, is a consequent about single-case chance. What I have done is to use the Principal Principle to parlay battery-of-conditionals independence into ordinary probabilistic independence.

If the world is right, the situation might be still simpler; and this is the case we hope to achieve in a well-conducted sequence of chance trials. Suppose the history-to-chance conditionals and the previous history of our chosen world give us not only independence (of the stronger sort) but also uniformity of chances: for any toss in our sequence, the endpoint chance of heads on that toss would be h (and the endpoint chance of tails would be $1 - h$) no matter which of the possible previous histories compatible with H_1 might have come to pass. Then each of the A_i's has an endpoint chance of h if it specifies an outcome of heads, $1 - h$ if it specifies an outcome of tails. By the multiplicative rule for endpoint chances,

$$P_1(A_1 \ldots A_n) = h^{fn} \cdot (1 - h)^{(n-fn)}$$

where f is the frequency of heads in the first n tosses according to $A_1 \ldots A_n$.

Now consider any other world that matches our chosen world in its history up to time 1 and in its complete theory of chance, but not in its sequence of outcomes. By the Principal Principle, the chance distribution at time 1 is the same for both worlds. Our assumptions of independence and uniformity apply to both worlds, being built into the shared history and theory. So all goes through for this other world as it did for our chosen world. Our calculation of the chance at time 1 of a sequence of outcomes, as a function of the uniform single-case chance of heads and the length and frequency of heads in the sequence, goes for any sequence, not only for the sequence $A_1, A_2, \ldots$ that comes about at our chosen world.

Let F be the proposition that the frequency of heads in the first n tosses is f. F is a disjunction of propositions each specifying a sequence of n outcomes with frequency f of heads; each disjunct has the same chance at time 1, under our assumptions of independence and uniformity; and the disjuncts are incompatible. Multiplying the number of these propositions by the uniform chance of each, we get the chance of obtaining some or other sequence of outcomes with frequency f of heads:

$$P_1(F) = \frac{n! \cdot h^{fn} \cdot (1 - h)^{(n-fn)}}{(fn)! \cdot (n - fn)!}.$$

The rest is well known. For fixed h and n, the right hand side of the

155

equation peaks for *f* close to *h*; the greater is *n*, the sharper is the peak. If there are many tosses, then the chance is close to one that the frequency of heads is close to the uniform single-case chance of heads. The more tosses, the more stringent we can be about what counts as "close". That much of frequentism is true; and that much is a consequence of the Principal Principle, which relates chance not only to credence but also to frequency.

On the other hand, unless *h* is zero or one, the right hand side of the equation is non-zero. So, as already noted, there is always some chance that the frequency and the single-case chance may differ as badly as you please. That objection to frequentist analyses also turns out to be a consequence of the Principal Principle.

EVIDENCE ABOUT CHANCES

To the subjectivist who believes in objective chance, particular or general propositions about chances are nothing special. We believe them to varying degrees. As new evidence arrives, our credence in them should wax and wane in accordance with Bayesian confirmation theory. It is reasonable to believe such a proposition, like any other, to the degree given by a reasonable initial credence function conditionalized on one's present total evidence.

If we look at the matter in closer detail, we find that the calculations of changing reasonable credence involve *likelihoods*: credences of bits of evidence conditionally upon hypotheses. Here the Principal Principle may act as a useful constraint. Sometimes when the hypothesis concerns chance and the bit of evidence concerns the outcome, the reasonable likelihood is fixed, independently of the vagaries of initial credence and previous evidence. What is more, the likelihoods are fixed in such a way that observed frequencies tend to confirm hypotheses according to which these frequencies differ not too much from uniform chances.

To illustrate, let us return to our example of the sequence of coin tosses. Think of it as an experiment, designed to provide evidence bearing on various hypotheses about the single-case chances of heads. The sequence begins at time 1 and goes on for at least *n* tosses. The evidence gained by the end of the experiment is a proposition *F* to the effect that the frequency of heads in the first *n* tosses was *f*. (I assume that we use a mechanical counter that keeps no record of individual tosses. The case in which there is a full record, however, is little different. I also assume, in an unrealistic simplification, that no other evidence whatever arrives during the experiment.) Suppose that at time 1 your credence function is $C(-/E)$, the function that comes from our chosen reasonable initial credence function C by conditionalizing on your total evidence E up to that time. Then if you learn from experience by conditionalizing, your credence function after the experiment is $C(-/FE)$. The impact of your experimental evidence F on your beliefs, about chances or anything else, is given by the difference between these two functions.

Suppose that before the experiment your credence is distributed over a range of alternative hypotheses about the endpoint chances of heads in the experimental tosses. (Your degree of belief that none of these hypotheses is correct may not be zero, but I am supposing it to be negligible and shall accordingly neglect it.) The hypotheses agree that these chances are uniform, and each independent of the previous course of history after time 1; but they disagree about what the uniform chance of heads is. Let us write G_h for the hypothesis that the endpoint chances of heads are uniformly h. Then the credences $C(G_h/E)$, for various h's, comprise the *prior distribution* of credence over the hypotheses; the credences $C(G_h/FE)$ comprise the *posterior distribution*; and the credences $C(F/G_hE)$ are the likelihoods. Bayes' Theorem gives the posterior distribution in terms of the prior distribution and the likelihoods:

$$C(G_h/FE) = \frac{C(G_h/E) \cdot C(F/G_hE)}{\Sigma_h \left[C(G_h/E) \cdot C(F/G_hE) \right]}.$$

(Note that "h" is a bound variable of summation in the denominator of the right hand side, but a free variable elsewhere.) In words: to get the posterior distribution, multiply the prior distribution by the likelihood function and renormalize.

In talking only about a single experiment, there is little to say about the prior distribution. That does indeed depend on the vagaries of initial credence and previous evidence.

Not so for the likelihoods. As we saw in the last section, each G_h implies a proposition X_h to the effect that the chance at 1 of F equals x_h, where x_h is given by a certain function of h, n, and f. Hence G_hE and X_hG_hE are the same proposition. Further, G_hE and X are compatible (unless G_hE is itself impossible, in which case G_h might as well be omitted from the range of hypotheses). E is admissible at 1, being about matters of particular fact—your evidence—at times no later than 1. G_h also is admissible at 1. Recall from the last section that what makes such a proposition hold at a world is a certain relationship between that world's complete history up to time 1 and that world's history-to-chance conditionals about the chances that would follow various complete extensions of that history. Hence any member of the history-theory partition for time 1 either implies or contradicts G_h; G_h is therefore a disjunction of conjunctions of admissible historical propositions and admissible history-to-chance conditionals. Finally, we supposed that C is reasonable. So the Principal Principle applies:

$$C(F/G_hE) = C(F/X_hG_hE) = x_h.$$

The likelihoods are the endpoint chances, according to the various hypotheses, of obtaining the frequency of heads that was in fact obtained.

When we carry the calculation through, putting these implied chances for the likelihoods in Bayes' theorem, the results are as we would expect. An observed frequency of f raises the credences of the hypotheses G_h with h close to f at the expense of the others; the more sharply so, the greater is the number of tosses. Unless the prior distribution is irremediably biased, the result after enough tosses is that the lion's share of the posterior credence will go to hypotheses putting the single-case chance of heads close to the observed frequency.

CHANCE AS A GUIDE TO LIFE

It is reasonable to let one's choices be guided in part by one's firm opinions about objective chances or, when firm opinions are lacking, by one's degrees of belief about chances. *Ceteris paribus*, the greater chance you think a lottery ticket has of winning, the more that ticket should be worth to you and the more you should be disposed to choose it over other desirable things. Why so?

There is no great puzzle about why credence should be a guide to life. Roughly speaking, what makes it be so that a certain credence function is *your* credence function is the very fact that you are disposed to act in more or less the ways that it rationalizes. (Better: what makes it be so that a certain reasonable initial credence function and a certain reasonable system of basic intrinsic values are both yours is that you are disposed to act in more or less the ways that are rationalized by the pair of them together, taking into account the modification of credence by conditionalizing on total evidence; and further, you would have been likewise disposed if your life history of experience, and consequent modification of credence, had been different; and further, no other such pair would fit your dispositions more closely.) No wonder your credence function tends to guide your life. If its doing so did not accord to some considerable extent with your dispositions to act, then it would not be your credence function. You would have some other credence function, or none.

If your present degrees of belief are reasonable—or at least if they come from some reasonable initial credence function by conditionalizing on your total evidence—then the Principal Principle applies. Your credences about outcomes conform to your firm beliefs and your partial beliefs about chances. Then the latter guide your life because the former do. The greater chance you think the ticket has of winning, the greater should be your degree of belief that it will win; and the greater is your degree of belief that it will win, the more, *ceteris paribus*, it should be worth to you and the more you should be disposed to choose it over other desirable things.

PROSPECTS FOR AN ANALYSIS OF CHANCE

Consider once more the Principal Principle as reformulated:

$$P_{tw}(A) = C(A/H_{tw}T_w).$$

Or in words: the chance distribution at a time and a world comes from any reasonable initial credence function by conditionalizing on the complete history of the world up to the time, together with the complete theory of chance for the world.

Doubtless it has crossed your mind that this has at least the form of an analysis of chance. But you may well doubt that it is informative as an analysis; that depends on the distance between the analysandum and the concepts employed in the analysans.

Not that it has to be informative *as an analysis* to be informative. I hope I have convinced you that the Principal Principle is indeed informative, being rich in consequences that are central to our ordinary ways of thinking about chance.

There are two different reasons to doubt that the Principal Principle qualifies as an analysis. The first concerns the allusion in the analysans to reasonable initial credence functions. The second concerns the allusion to complete theories of chance. In both cases the challenge is the same: could we possibly get any independent grasp on this concept, otherwise than by way of the concept of chance itself? In both cases my provisional answer is: most likely not, but it would be worth trying. Let us consider the two problems in turn.

It would be natural to think that the Principal Principle tells us nothing at all about chance, but rather tells us something about what makes an initial credence function be a reasonable one. To be reasonable is to conform to objective chances in the way described. Put this strongly, the response is wrong: the Principle has consequences, as we noted, that are about chance and not at all about its relationship to credence. (They would be acceptable, I trust, to a believer in objective single-case chance who rejects the very idea of degree of belief.) It tells us more than nothing about chance. But perhaps it is divisible into two parts: one part that tells us something about chance, another that takes the concept of chance for granted and goes on to lay down a criterion of reasonableness for initial credence.

Is there any hope that we might leave the Principal Principle in abeyance, lay down other criteria of reasonableness that do not mention chance, and get a good enough grip on the concept that way? It's a lot to ask. For note that just as the Principal Principle yields some consequences that are entirely about chance, so also it yields some that are entirely about reasonable initial credence. One such consequence is as follows. There is a large class of propositions such that if Y is any one of these, and C_1 and C_2 are any two reasonable initial credence functions, then the functions that come from C_1 and C_2 by conditionalizing on Y are exactly the same. (The large class is, of course, the class of members of history-theory partitions for all times.) That severely limits the ways that reasonable initial credence functions may differ, and so shows that criteria adequate to pick them out must be quite strong. What might we try? A reasonable initial credence function ought to (1) obey the laws of mathematical probability theory; (2) avoid dogmatism, at least by never assigning zero credence to possible propositions and pehaps also by never assigning infinitesimal credence to certain kinds of possible propositions; (3) make it possible to learn from experience by having a built-in bias in favor of worlds where the future in some sense resembles the past; and perhaps (4) obey certain carefully restricted principles of indifference, thereby respecting certain symmetries. Of these, critera (1)–(3) are all very well, but surely not yet strong enough. Given C_1 satisfying (1)–(3), and given any proposition Y that holds at more than one world, it will be possible to distort C_1 very slightly to produce C_2, such that $C_1(\text{—}/Y)$ and $C_2(\text{—}/Y)$ differ but C_2 also satisfies (1)–(3). It is less clear what (4) might be able to do for us. Mostly that is because (4) is less clear *sim-*

pliciter, in view of the fact that it is not possible to obey too many different restricted principles of indifference at once and it is hard to give good reasons to prefer some over their competitors. It also remains possible, of course, that some criterion of reasonableness along different lines than any I have mentioned would do the trick.

I turn now to our second problem: the concept of a complete theory of chance. In saying what makes a certain proposition be the complete theory of chance for a world (and for any world where it holds), I gave an explanation in terms of chance. Could these same propositions possibly be picked out in some other way, without mentioning chance?

The question turns on an underlying metaphysical issue. A broadly Humean doctrine (something I would very much like to believe if at all possible) holds that all the facts there are about the world are particular facts, or combinations thereof. This need not be taken as a doctrine of analyzability, since some combinations of particular facts cannot be captured in any finite way. It might be better taken as a doctrine of supervenience: if two worlds match perfectly in all matters of particular fact, they match perfectly in all other ways too—in modal properties, laws, causal connections, chances, ... It seems that if this broadly Humean doctrine is false, then chances are a likely candidate to be the fatal counter-instance. And if chances are not supervenient on particular fact, then neither are complete theories of chance. For the chances at a world are jointly determined by its complete theory of chance together with propositions about its history, which latter plainly are supervenient on particular fact.

If chances are not supervenient on particular fact, then neither chance itself nor the concept of a complete theory of chance could possibly be analyzed in terms of particular fact, or of anything supervenient thereon. The only hope for an analysis would be to use something in the analysans which is itself not supervenient on particular fact. I cannot say what that something might be.

How might chance, and complete theories of chance, be supervenient on particular fact? Could something like this be right: the complete theory of chance for a world is that one of all possible complete theories of chance that somehow best fits the global pattern of outcomes and frequencies of outcomes? It could not. For consider any such global pattern, and consider a time long before the pattern is complete. At that time, the pattern surely has some chance of coming about and some chance of not coming about. There is surely some chance of a very different global pattern coming about; one which, according to the proposal under consideration, would make true some different complete theory of chance. But a complete theory of chance is not something that could have some chance of coming about or not coming about. By the Principal Principle,

$$P_{tw}(T_w) = C(T_w/H_{tw}T_w) = 1.$$

If T_w is something that holds in virtue of some global pattern of particular fact that obtains at world w, this pattern must be one that has no chance at any time (at w) of not obtaining. If w is a world where many matters of particular fact are the outcomes of chance processes, then I fail to see what kind of global pattern this could possibly be.

But there is one more alternative. I have spoken as if I took it for granted that different worlds have different history-to-chance conditionals, and hence different complete theories of chance. Perhaps this is not so: perhaps all worlds are exactly alike in the dependence of chance on history. Then the complete theory of chance for every world, and all the conditionals that comprise it, are necessary. They are supervenient on particular fact in the trivial way that what is non-contingent is supervenient on anything—no two worlds differ with respect to it. Chances are still contingent, but only because they depend on contingent historical propositions (information about the details of the coin and tosser, as it might be) and not also because they depend on a contingent theory of chance. Our theory is much simplified if this is true. Admissible information is simply historical information; the history-theory partition at t is simply the partition of alternative complete histories up to t; for any reasonable initial credence function C

$$P_{tw}(A) = C(A/H_{tw}),$$

so that the chance distribution at t and w comes from C by conditionalizing on the complete history of w up to t. Chance is reasonable credence conditional on the whole truth about history up to a time. The broadly Humean doctrine is upheld, so far as chances are concerned: what makes it true at a time and a world that something has a certain chance of happening is something about matters of particular fact at that time and (perhaps) before.

What's the catch? For one thing, we are no longer safely exploring the consequences of the Principal Principle, but rather engaging in speculation. For another, our broadly Humean speculation that history-to-chance conditionals are necessary solves our second problem by making the first one worse. Reasonable initial credence functions are constrained more narrowly than ever. Any two of them, C_1 and C_2, are now required to yield the same function by conditionalizing on the complete history of any world up to any time. Put it this way: according to our broadly Humean speculation (and the Principal Principle) if I were perfectly reasonable and knew all about the course of history up to now (no matter what that course of history actually is, and no matter what time is now) then there would be only one credence function I could have. Any other would be unreasonable.

It is not very easy to believe that the requirements of reason leave so little leeway as that. Neither is it very easy to believe in features of the world that are not supervenient on particular fact. But if I am right, that seems to be the choice. I shall not attempt to decide between the Humean and the anti-Humean variants of my approach to credence and chance. The Principal Principle doesn't.

REFERENCES

Bernstein, Allen R. and Wattenberg, Frank, "Non-Standard Measure Theory", in *Applications of Model Theory of Algebra, Analysis, and Probability*, ed. by W. Luxemburg, Holt, Reinhart, and Winston, 1969.

Carnap, Rudolf, "The Two Concepts of Probability", *Philosophy and Phenomenological Research* 5 (1945), 513–32.

Jeffrey, Richard C., *The Logic of Decision*, McGraw-Hill, 1965.

Jeffrey, Richard C., review of articles by David Miller *et al.*, *Journal of Symbolic Logic* 35 (1970), 124–27.

Jeffrey, Richard C., "Mises Redux", in *Basic Problems in Methodology and Linguistics: Proceedings of the Fifth International Congress of Logic, Methodology and Philosophy of Science*, Part III, ed. by R. Butts and J. Hintikka, D. Reidel, Dordrecht, Holland, 1977.

Lewis, David, "Counterpart Theory and Quantified Modal Logic", *Journal of Philosophy* 65 (1968), 113–26.

Lewis, David, *Counterfactuals*, Blackwell, 1973.

Mellor, D. H., *The Matter of Chance*, Cambridge University Press, 1971.

Quine, W. V., "Propositional Objects", in *Ontological Relativity and Other Essays*, Columbia University Press, 1969.

Railton, Peter, "A Deductive-Nomological Model of Probabilistic Explanation", *Philosophy of Science* 45 (1978), 206–26.

Skyrms, Brian, "Resiliency, Propensities, and Causal Necessity", *Journal of Philosophy* 74 (1977), 704–13.

Postscripts to

"A Subjectivist's Guide to Objective Chance"

A. NO ASSISTANCE NEEDED[1]

Henry Kyburg doubts that the Principal Principle has as much scope as my praise of it would suggest. He offers a continuation of my questionnaire, says that his added questions fall outside the scope of the Principal Principle, and suggests that we need some Assistant Principle to deal with them. His first added question is as follows.[2]

> *Question.* You are sure that a certain coin is fair. It was tossed this morning, but you have no information about the outcome of the toss. To what degree should you believe the proposition that it landed heads?

> *Answer.* 50 per cent, of course.

That's the right answer (provided the question is suitably interpreted). But the Principal Principle, unassisted, does suffice to yield that answer. What we must bear in mind is that the Principle relates time-dependent chance to time-dependent admissibility of evidence; and that it applies to any time, not only the present.

Kyburg thinks the Principle falls silent "since there is *no* chance that the coin fell other than the way it did," and quotes me to the effect that "what's past is no longer chancy." Right. We won't get anywhere if we apply the Principle to *present* chances. But what's past *was* chancy, if indeed the coin was fair; so let's see what we get by applying the Principle to a past time, and working back to present credences. Notation:

---

[1] In writing this postscript, I have benefited from a discussion by W. N. Reinhardt (personal communication, 1982). Reinhardt's treatment and mine agree on most but not all points.

[2] Henry E. Kyburg, Jr., "Principle Investigation," *Journal of Philosophy* 78 (1981): 772–78.

t: a time just before the toss,

C: a reasonable initial credence function that will yield my later credences by conditionalizing on total evidence,

C_0: my present credence function,

A: the proposition that the coin fell heads,

X: the proposition that the coin was fair, that is that its chance at t of falling heads was 50%,

E: the part of my present total evidence that is admissible at t,

F: the rest of my present total evidence.

Since *ex hypothesi* I'm certain of X, we have

(1) $C_0 = C_0(-/X)$.

By definition of C, we have

(2) $C_0 = C(-/EF)$.

Assuming that F is irrelevant to the tosses, we have

(3) $C(A/XEF) = C(A/XE)$.

By the Principal Principle, applied not to the present but to t, we have

(4) $C(A/XE) = 50\%$.

Now, by routine calculation from (1)–(4) we have

(5) $C_0(A) = 50\%$.

which answers Kyburg's question.

Step (3) deserves further examination, lest you suspect it of concealing an Assistant Principle. Recall that F is the part of my present total evidence that was not admissible already at time t. Presumably it consists of historical information about the interval between t and the present. For historical information about earlier times would be already admissible at t; and historical information about later times, or nonhistorical information, could scarely be part of my present total evidence. (Here, as in the paper, I set aside strange possibilities in which the normal asymmetries of time break down. So far as I can tell, Kyburg is content to join me in so doing.) Thus if I had watched the toss, or otherwise received information about its outcome, that information would be included in F.

However, Kyburg stipulated in his question that "you have no information about the outcome of the toss". We might reasonably construe that to mean that no information received between t and the present is evidentially relevant to whether the coin fell heads, with evidential relevance construed in the usual way in terms of credence. Then (3) comes out as a stipulated condition of the problem, not some extra principle.

There is a different, stricter way that Kyburg's stipulation might perhaps be construed. It might only exclude information that settles the outcome decisively, leaving it open that I have information that bears evidentially on the outcome without settling it. For instance, it might be that the tosser promised to phone me if the toss fell heads, I got no phone call, but that is far from decisive because my phone is not reliable. On that construal, we are not entitled to assume (3). But on that construal Kyburg's answer is wrong; or anyway it isn't right as a matter of course on the basis of what he tells us; so we don't want any principle that delivers that answer.

Kyburg has a second added question to challenge the Principal Principle.

> *Question.* As above, but you know that the coin was tossed 100 times, and landed heads 86 times. To what degree should you believe the proposition that it landed heads on the first toss?

> *Answer.* 86 per cent.

The strategy for getting the Principal Principle to yield an answer is the same as before, but the calculation is more complicated. Notation as before, except for

A: the proposition that the coin fell heads *on the first toss*,

B: the proposition that the coin fell heads 86 times out of 100,

X: the proposition that the coin was fair, that is that its chance at t of falling heads was 50% *on each toss*,

F: the rest of my present total evidence, besides the part that was admissible at t, *and also besides the part B*,

x: the fraction of heads-tails sequences of length 100 in which there are 86 heads.

Our equations this time are as follows. They are justified in much the same way as the like-numbered equations above. But this time, to get the new (2) we split the present total evidence into three parts B, E, and F. And to get the new (4), we use the Principal Principle repeatedly to multiply endpoint chances, as was explained in the section of the paper dealing with chance of frequency.

(1) As before;

(2) $C_0 = C(-/BEF)$;

(3) $C(A/XBEF) = C(A/XBE)$;

(4) $C(AB/XE) = x \cdot 86\%$, $C(B/XE) = x$;

(5) $C_0(A) = 86\%$.

Kyburg also thinks I need an extra "Principle of Integration" which I neglected to state. But this principle, it turns out, has nothing especially to do with chance! It is just a special case of a principle of infinite additivity for credences. Indeed it could be replaced, at the point where he claims I tacitly used it, by *finite* additivity of credences. (And finite additivity goes without saying, though I nevertheless did say it.) To be sure, if we want to treat credences in the setting of nonstandard analysis, we are going to want some kind of infinite additivity. And some kind of infinite additivity comes automatically when we start with finite additivity and then treat some infinite sets as if they were finite. It is an interesting question what kind of infinite addivity of credences we can reasonably assume in the nonstandard setting. But this question belongs entirely to the theory of credence—not to the connection between chance and credence that was the subject of my paper.

B. CHANCE WITHOUT CHANCE?

Isaac Levi thinks that I have avoided confronting "the most important problem about chance"; which problem, it seems, is the reconciliation of chances with determinism, or of chances with different chances.[3] Consider a toss of coin. Levi writes that

> . . . in typical cases, the agent will and should be convinced that information exists (though inaccessible to him) which is highly relevant [to the outcome]. Thus, the agent may well be convinced that a complete history through [the onset of the toss] will include a specification of the initial mechanical state of the coin upon being tossed and boundary conditions which, taken together, determine the outcome to be heads up or tails up according to physical laws.

> . . . given the available knowledge of physics, we cannot [deny that the mechanical state of the coin at the onset of the toss determines the out-

[3] Isaac Levi, review of *Studies in Inductive Logic and Probability*, ed. by R. C. Jeffrey, *Philosophical Review* 92 (1983): 120–21.

come] provided we can assume the motion of the coin . . . to be sealed off from substantial external influences. But even if we allow for fluctuations in the boundary conditions, we would not suppose them so dramatic as to permit large deviations from 0 or 1 to be values of the chances of heads. . . .

And yet

> Lewis, however, appears ready to assign .5 to the chance of [the] coin landing heads up. . . .

So how do I square the supposition that the chance of heads is 50% with the fact that it is zero or one, or anyway it does not deviate much from zero or one?

I don't. If the chance is zero or one, or close to zero or one, then it cannot also be 50%. To the question how chance can be reconciled with determinism, or to the question how disparate chances can be reconciled with one another, my answer is: *it can't be done.*

It was not I, but the hypothetical "you" in my example, who appeared ready to assign a 50% chance of heads. If my example concerned the beliefs of an ignoramus, it is none the worse for that.

I myself am in a more complicated position than the character in this example. (That is why I made an example of him, not me.) I would not give much credence to the proposition that the coin has a chance of heads of 50% exactly. I would give a small share of credence to the proposition that it is zero exactly, and an equal small share to the proposition that it is one exactly. I would divide most of the rest of my credence between the vicinity of 50%, the vicinity of zero, and the vicinity of one.

The small credence I give to the extremes, zero and one exactly, reflects my slight uncertainty about whether the world is chancy at all. Accepted theory says it is, of course; but accepted theory is not in the best of foundational health, and the sick spot (reduction of the wave function brought on by measurement) is the very spot where the theory goes indeterministic. But most of my credence goes to the orthodox view that there are plenty of chance processes in microphysics. And not just the microphysics of extraordinary goings-on in particle accelerators! No; for instance the making and breaking of chemical bonds is chancy, so is the coherence of solids that stick together by means of chemical bonding, so is the elasticity of collisions between things that might bond briefly before they rebound,. . . . So is any process whatever that could be disrupted by chance happenings nearby—and infallible "sealing off" is not to be found.

In Levi's physics, a coin coming loose from fingers and tumbling in air until it falls flat on a table is a classical system, an oasis of determinism in a chancy microworld. I do not see how that can be. The coin, and the fingers and the air and the table, are too much a part of that microworld. There are also the external influences, which cannot be dismissed either by requiring them to be substantial or by invoking fictitious seals; but never mind, let us concentrate on the toss itself. There is chance enough in the processes by which the coin leaves the fingers; in the processes whereby it bounces off air molecules and sends them recoiling off, perhaps to knock other molecules into its path; in the process whereby the coin does or doesn't stretch a bit as it spins, thereby affecting its moment of inertia; and in the processes whereby it settles down after first touching the table. In ever so many minute ways, what happens to the coin is a matter of chance.

But all those chance effects are *so* minute.—But a tossed coin is *so* sensitive to minute differences. Which dominates—minuteness or sensitivity? That is a question to be settled not by asking what a philosopher would find it reasonable to suppose, but by calculation. The calculations would be difficult. We may not make them easier by approximations in which expected values replace chance distributions. I have not heard of anyone who has attempted these calculations, and of course they are far beyond my own power. Maybe they are beyond the state of the art altogether. Without them, I haven't a clue whether the minuteness of the chance effects dominates, in which case the chance of heads is indeed close to zero or one; or whether instead the sensitivity dominates, in which case the chance of heads is close to 50%. Hence my own distribution of credence.

The hypothetical "you" in my example has a different, simpler distribution. Why? He might be someone who has done the calculations and found that the sensitivity dominates. Or he might have been so foolish as to intuit that the sensitivity would dominate. Or he might be altogether misinformed.

Well-informed people often say that ordinary gambling devices are deterministic systems. Why? Perhaps it is a hangover of instrumentalism. If we spoke as instrumentalists, we would be right to say so— meaning thereby not that they really *are* deterministic, but rather that it is sometimes instrumentally useful to pretend that they are. To the extent that it is feasible to predict gambling devices at all—we can't predict heads or tails, but we can predict, for instance, that the coin won't tumble in mid-air until next year, and won't end up sticking to the wall—deterministic theories are as good predictive instruments as can be had. Perhaps when the instrumentalist expert says that tossed coins are deterministic, the philosopher misunderstands him, and thinks he means that tossed coins are deterministic.

Can it be that Levi himself was speaking as an instrumentalist in the passages I cited? If so, then the problem of reconciling chance and determinism is not very hard. It is just the problem of reconciling truth *simpliciter* with truth in fiction. In truth, nobody lived at 221B Baker Street; in fiction, Holmes lived there. In truth, most likely, the coin is chancy; in fiction, it is deterministic. No worries. The character in my example, of course, was meant to be someone who believed that the chance of heads was 50% in truth—not in fiction, however instrumentally useful such fiction might be.

There is no chance without chance. If our world is deterministic there are no chances in it, save chances of zero and one. Likewise if our world somehow contains deterministic enclaves, there are no chances in those enclaves. If a determinist says that a tossed coin is fair, and has an equal chance of falling heads or tails, he does not mean what I mean when he speaks of chance. Then what *does* he mean? This, I suppose, is the question Levi would like to see addressed. It is, of course, a more urgent question for determinists than it is for me.

That question has been sufficiently answered in the writings of Richard Jeffrey and Brian Skyrms on objectified and resilient credence.[4] Without commiting themselves one way or the other on the question of determinism, they have offered a kind of counterfeit chance to meet the needs of the determinist. It is a relative affair, and apt to go indeterminate, hence quite unlike genuine chance. But what better could a determinist expect?

According to my second formulation of the Principal Principle, we have the history-theory partition (for any given time); and the chance distribution (for any given time and world) comes from any reasonable initial credence function by conditionalizing on the true cell of this partition. That is, it is objectified in the sense of Jeffrey. Let us note three things about the history-theory partition.

(1) It seems to be a natural partition, not gerrymandered. It is what we get by dividing possibilities as finely as possible in certain straightforward respects.

[4] Richard C. Jeffrey, *The Logic of Decision* (New York: McGraw-Hill, 1965; second edition, Chicago: University of Chicago Press, 1983) Section 12.7; Brian Skyrms, "Resiliency, Propensities, and Causal Necessity," *Journal of Philosophy* 74 (1977): 704–13; Brian Skyrms, *Causal Necessity* (New Haven: Yale University Press, 1980).

(2) It is to some extent feasible to investigate (before the time in question) which cell of this partition is the true cell; but

(3) it is unfeasible (before the time in question, and without peculiarities of time whereby we could get news from the future) to investigate the truth of propositions that divide the cells.

Hence if we start with a reasonable initial credence function and do enough feasible investigation, we may expect our credences to converge to the chances; and no amount more feasible investigation (before the time) will undo that convergence. That is, after enough investigation, our credences become resilient in the sense of Skyrms. And our credences conditional on cells of the partition are resilient from the outset.

Conditions (1)–(3) characterize the history-theory partition; but not uniquely. Doubtless there are other, coarser partitions, that also satisfy the conditions. How feasible is feasible? Some investigations are more feasible than others, depending on the resources and techniques available, and there must be plenty of boundaries to be drawn between the feasible and the unfeasible before we get to the ultimate boundary whereby investigations that divide the history-theory cells are the most unfeasible of all. Any coarser partition, if it satisfies conditions (1)–(3) according to some appropriate standards of feasible investigation and of natural partitioning, gives us a kind of counterfeit chance suitable for use by determinists: namely, reasonable credence conditional on the true cell of that partition. Counterfeit chances will be relative to partitions; and relative, therefore, to standards of feasibility and naturalness; and therefore indeterminate unless the standards are somehow settled, or at least settled well enough that all remaining candidates for the partition will yield the same answers. Counterfeit chances are therefore not the sort of thing we would want to find in our fundamental physical theories, or even in our theories of radioactive decay and the like. But they will do to serve the conversational needs of determinist gamblers.

C. LAWS OF CHANCE

Despite the foundational problems of quantum mechanics, it remains a good guess that many processes are governed by probabilistic laws of nature. These laws of chance, like other laws of nature, have the form of universal generalizations. Just as some laws concern forces, which are magnitudes pertaining to particulars, so some laws concern single-case chances, which likewise are magnitudes pertaining to particulars.

For instance, a law of chance might say that for any tritium atom and any time when it exists, there is such-and-such chance of that atom decaying within one second after that time.[5] What makes it at least a regularity—a true generalization—is that for each tritium atom and time, the chance of decay is as the law says it is. What makes it a law, I suggest, is the same thing that gives some others regularities the status of laws: it fits into some integrated system of truths that combines simplicity with strength in the best way possible.[6]

This is a kind of regularity theory of lawhood; but it is a collective and selective regularity theory. Collective, since regularities earn their lawhood not by themselves, but by the joint efforts of a system in which they figure either as axioms or as theorems. Selective, because not just any regularity qualifies as a law. If it would complicate the otherwise best system to include it as an axiom, or to include premises that would imply it, and if it would not add sufficient strength to pay its way, then it is left as a merely accidental regularity.

Five remarks about the best-system theory of lawhood may be useful before we return to our topic of how this theory works in the presence of chance.

---

[5] Peter Railton employs laws of chance of just this sort to bring probabilistic explanation under the deductive-nomological model. The outcome itself cannot be deduced, of course; but the single-case chance of it can be. See Railton, "A Deductive- Nomological Model of Probabilistic Explanation," *Philosophy of Science* 45 (1978): 206–26; and the final section of my "Causal Explanation" in this volume.

[6] I advocate a best-system theory of lawhood in *Counterfactuals* (Oxford: Blackwell, 1973), pp. 73–75. Similar theories of lawhood were held by Mill and, briefly, by Ramsey. See John Stuart Mill, *A System of Logic* (London: Parker, 1843), Book III, Chapter IV, Section 1; and F. P. Ramsey, "Universals of Law and of Fact," in his *Foundations* (London: Routledge & Kegan Paul, 1978). For further discussion, see John Earman, "Laws of Nature: The Empiricist Challenge," in *D. M. Armstrong*, ed. by Radu J. Bogdan (Dordrecht: Reidel, 1984).

Mill's version is not quite the same as mine. He says that the question what are the laws of nature could be restated thus: "What are the fewest general propositions from which all the uniformities which exist in the universe might be deductively inferred?"; so it seems that the ideal system is supposed to be complete as regards uniformities, that it may contain only general propositions as axioms, and that its theorems do not qualify as laws.

It is not clear to me from his brief statement whether Ramsey's version was quite the same as mine. His summary statement (after changing his mind) that he had taken laws to be "consequences of those propositions we should take as axioms if we knew everything and organized it as simply as possible into a deductive system" (*Foundations*, p. 138) is puzzling. Besides Ramsey's needless mention of knowledge, his "it" with antecedent "everything" suggests that the ideal system is supposed to imply everything true. Unless Ramsey made a stupid mistake, which is impossible, that cannot have been his intent; it would make all regularities come out as laws.

(1) The standards of simplicity, of strength, and of balance between them are to be those that guide us in assessing the credibility of rival hypotheses as to what the laws are. In a way, that makes lawhood depend on us—a feature of the approach that I do not at all welcome! But at least it does not follow that lawhood depends on us in the most straightforward way: namely, that if our standards were suitably different, then the laws would be different. For we can take our actual standards as fixed, and apply them in asking what the laws would be in various counterfactual situations, including counterfactual situations in which people have different standards—or in which there are no people at all. Likewise, it fortunately does not follow that the laws are different at other times and places where there live people with other standards.

(2) On this approach, it is not to be said that certain generalizations are *lawlike* whether or not they are true, and the laws are exactly those of the lawlikes that are true. There will normally be three possibilities for any given generalization: that it be false, that it be true but accidental, and that it be true as a law. Whether it is true accidentally or as a law depends on what else is true along with it, thus on what integrated systems of truths are available for it to enter into. To illustrate the point: it may be true accidentally that every gold sphere is less than one mile in diameter; but if gold were unstable in such a way that there was no chance whatever that a large amount of gold could last long enough to be formed into a one-mile sphere, then this same generalization would be true as a law.

(3) I do not say that the competing integrated systems of truths are to consist entirely of regularities; however, only the regularities in the best system are to be laws. It is open that the best system might include truths about particular places or things, in which case there might be laws about these particulars. As an empirical matter, I do not suppose there are laws that essentially mention Smith's garden, the center of the earth or of the universe, or even the Big Bang. But such laws ought not to be excluded *a priori*.[7]

(4) It will trivialize our comparisons of simplicity if we allow our competing systems to be formulated with just any hoked-up primi-

---

[7] In defense of the possibility that there might be a special law about the fruit in Smith's garden, see Michael Tooley, "The Nature of Laws," *Canadian Journal of Philosophy* 7 (1977): 667–98, especially p. 687; and D. M. Armstrong, *What is a Law of Nature?* (Cambridge: Cambridge University Press, 1983), Sections 3.I, 3.II, and 6.VII. In "The Universality of Laws," *Philosophy of Science* 45 (1978): 173–81, John Earman observes that the best-system theory of lawhood avoids any *a priori* guarantee that the laws will satisfy strong requirements of universality.

tives. So I take it that this kind of regularity theory of lawhood requires some sort of inegalitarian theory of properties: simple systems are those that come out formally simple when formulated in terms of perfectly natural properties. Then, sad to say, it's useless (though true) to say that the natural properties are the ones that figure in laws.[8]

(5) If two or more systems are tied for best, then certainly any regularity that appears in all the tied systems should count as a law. But what of a regularity that appears in some but not all of the tied systems? We have three choices: it is not a law (take the intersection of the tied systems); it is a law (take the union); it is indeterminate whether it is law (apply a general treatment for failed presuppositions of uniqueness). If required to choose, I suppose I would favor the first choice; but it seems a reasonable hope that nature might be kind to us, and put some one system so far out front that the problem will not arise. Likewise, we may hope that some system will be so far out front that it will win no matter what the standards of simplicity, strength, and balance are, within reason. If so, it will also not matter if these standards themselves are unsettled. To simplify, let me ignore the possibility of ties, or of systems so close to tied that indeterminacy of the standards matters; if need be, the reader may restore the needed complications.

To return to laws of chance: if indeed there are chances, they can be part of the subject matter of a system of truths; then regularities about them can appear as axioms or theorems of the best system; then such regularities are laws. Other regularities about chances might fail to earn a place in the best system; those ones are accidental. All this is just as it would be for laws about other magnitudes. So far, so good.

But there is a problem nearby; not especially a problem about laws of chance, but about laws generally in a chancy world. We have said that a regularity is accidental if it cannot earn a place in the best system: if it is too weak to enter as an axiom, and also cannot be made to follow as a theorem unless by overloading the system with particular information. That is one way to be accidental; but it seems that a regularity might be accidental also for a different and simpler reason. It might hold merely by chance. It might be simple and powerful and well deserve a place in the ideal system and yet be no law. For it might have, or it might once have had, some chance of failing to hold; whereas it seems very clear, *contra* the best-system theory as so far stated, that no genuine law ever could have had any chance of not holding. A world of

---

[8] See my "New Work for a Theory of Universals," *Australasian Journal of Philosophy* 61 (1983): 343–77, especially pp. 366–68.

lawful chance might have both sorts of accidental regularitites, some disqualified by their inadequate contribution to simplicity and strength and others by their chanciness.

Suppose that radioactive decay is chancy in the way we mostly believe it to be. Then for each unstable nucleus there is an expected lifetime, given by the constant chance of decay for a nucleus of that species. It might happen—there is some chance of it, infinitesimal but not zero—that each nucleus lasted for precisely its expected lifetime, no more and no less. Suppose that were so. The regularity governing lifetimes might well qualify to join the best system, just as the corresponding regularity governing *expected* lifetimes does. Still, it is not a law. For if it were a law, it would be a law with some chance—in fact, an overwhelming chance—of being broken. That cannot be so.[9]

(Admittedly, we do speak of defeasible laws, laws with exceptions, and so forth. But these, I take it, are rough-and-ready approximations to the real laws. There real laws have no exceptions, and never had any chance of having any.)

Understand that I am not supposing that the constant chances of decay are *replaced* by a law of constant lifetimes. That is of course possible. What is not possible, unfortunately for the best-system theory, is for the constant chances to remain and to coexist with a law of constant lifetimes.

If the lifetimes chanced to be constant, and if the matter were well investigated, doubtless the investigators would come to believe in a law of constant lifetimes. But they would be mistaken, fooled by a deceptive coincidence. It is one thing for a regularity to be a law; another thing for it to be so regarded, however reasonably. Indeed, there are philosphers who seem oblivious to the distinction; but I think these philosophers misrepresent their own view. They are sceptics; they do not believe in laws of nature at all, they resort to regarded-as-law regularities as a substitute, and they call their substitute by the name of the real thing.

---

[9] At this point I am indebted to correspondence and discussion with Frank Jackson, arising out of his discussion of "Hume worlds" in "A Causal Theory of Counterfactuals," *Australasian Journal of Philosophy* 55 (1977): 3–21, especially pp. 5–6. A Hume world, as Jackson describes it, is "a possible world where every particular fact is as it is in our world, but there are no causes or effects at all. Every regular conjunction is an accidental one, not a causal one." I am not sure whether Jackson's Hume world is one with chances — lawless chances, of course — or without. In the former case, the bogus laws of the Hume world would be like our bogus law of constant lifetimes, but on a grander scale.

So the best-system theory of lawhood, as it stands, is in trouble. I propose this correction. Previously, we held a competition between all true systems. Instead, let us admit to the competition only those systems that are true not by chance; that is, those that not only are true, but also have never had any chance of being false. The field of eligible competitors is thus cut down. But then the competition works as before. The best system is the one that achieves as much simplicity as is possible without excessive loss of strength, and as much strength as is possible without excessive loss of simplicity. A law is a regularity that is included, as an axiom or as a theorem, in the best system.

Then a chance regularity, such as our regularity of constant lifetimes, cannot even be included in any of the competing systems. A *fortiori*, it cannot be included in the best of them. Then it cannot count as a law. It will be an accidental regularity, and for the right reason: because it had a chance of being false. Other regularities may still be accidental for our original reason. These would be regularities that never had any chance of being false, but that don't earn their way into the best system because they don't contribute enough to simplicity and strength. For instance suppose that (according to regularities that do earn a place in the best system) a certain quantity is strictly conserved, and suppose that the universe is finite in extent. Then we have a regularity to the effect that the total of this quantity, over the entire universe, always equals a certain fixed value. This regularity never had any chance of being false. But it is not likely to earn a place in the best system and qualify as a law.

In the paper, I made much use of the history-to-chance conditionals giving hypothetical information about the chance distribution that would follow a given (fully specified) initial segment of history. Indeed, my reformulation of the Principal Principle involves a "complete theory of chance" which is the conjunction of all such history-to-chance conditionals that hold at a given world, and which therefore fully specifies the way chances at any time depend on history up to that time.

It is to be hoped that the history-to-chance conditionals will follow, entirely or for the most part, from the laws of nature; and, in particular, from the laws of chance. We might indeed impose a requirement to that effect on our competing systems. I have chosen not to. While the thesis that chances might be entirely governed by law has some plausibility, I am not sure whether it deserves to be built into the analysis of lawhood. Perhaps rather it is an empirical thesis: a virtue that we may hope distinguishes our world from more chaotic worlds.

At any rate, we can be sure that the history-to-chance conditionals will not conflict with the system of laws of chance. Not, at any rate, in what they say about the outcomes and chances that would follow any initial segment of history that ever had any chance of coming about. Let *H* be a proposition fully specifying such a segment. Let *t* be a time at which there was some chance that *H* would come about. Let *L* be the conjunction of the laws. There was no chance, at *t*, of *L* being false. Suppose for *reductio* first that we have a history-to-chance conditional "if *H*, then *A*" (where *A* might, for instance, specify chances at the end-time of the segment); and second that *H* and *L* jointly imply not-*A*, so that the conditional conflicts with the laws. The conditional had no chance at *t* of being false—this is an immediate consequence of the reformulated Principal Principle. Since we had some chance at *t* of *H*, we had some chance of *H* holding along with the conditional, hence some chance of *H* and *A*. And since there was no chance that *L* would be false, there was some chance that all of *H*, *A*, and *L* would hold together, so some chance at *t* of a contradiction. Which is impossible: there never can be any chance of a contradiction.

A more subtle sort of conflict also is ruled out. Let *t*, *L*, and *H* be as before. Suppose for *reductio* first that we have a history-to-chance conditional "if *H*, then there would be a certain positive chance of *A*"; and second that *H* and *L* jointly imply not-*A*. This is not the same supposition as before: after all, it would be no contradiction if something had a positive chance and still did not happen. But it is still a kind of conflict: the definiteness of the law disagrees with the chanciness of the conditional. To rule it out, recall that we had at *t* some chance of *H*, but no chance of the conditional being false; so at *t* there was a chance of *H* holding along with the conditional; so at *t* there was a chance that, later, there would be a chance of *A* following the history *H*; but chanciness does not increase with time (assuming, as always, the normal asymmetries); an earlier chance of a later chance of something implies an earlier chance of it; so already at *t* there was some chance of *H* and *A* holding together. Now we can go on as before: we have that at *t* there was no chance that *L* would be false, so some chance that all of *H*, *A*, and *L* would hold together, so some chance at *t* of a contradiction; which is impossible.

The best-system theory of lawhood in its original form served the cause of Humean supervenience. History, the pattern of particular fact throughout the universe, chooses the candidate systems, and the standards of selection do the rest. So no two worlds could differ in laws without differing also in their history. But our correction spoils that. The laws—laws of chance, and other laws besides—supervene now on the pattern of particular chances. If the chances in turn somehow supervene on history, then we have Humean supervenience of the laws as well; if not, not. The corrected theory of lawhood starts with the chances. It does nothing to explain them.

Once, *circa* 1975, I hoped to do better: to extend the best-system approach in such a way that it would provide for the Humean supervenience of chances and laws together, in one package deal. This was my plan. We hold a competition of deductive systems, as before; but we impose less stringent requirements of eligibility to enter the competition, and we change the terms on which candidate systems compete. We no longer require a candidate system to be entirely true, still less do we require that it never had any chance of being false. Instead, we only require that a candidate system be true in what it says about history; we leave it open, for now, whether it also is true in what it says about chances. We also impose a requirement of coherence: each candidate system must imply that the chances are such as to give that very system no chance at any time of being false. Once we have our competing systems, they vary in simplicity and in strength, as before. But also they vary in what I shall call *fit*: a system fits a world to the extent that the history of that world is a comparatively probable history according to that system. (No history will be very probable; in fact, any history for a world like ours will be very improbable according to any system that deserves in the end to be accepted as correct; but still, some are more probable than others.) If the histories permitted by a system formed a tree with finitely many branch points and finitely many alternatives at each point, and the system specified chances for each alternative at each branch point, then the fit between the system and a branch would be the product of these chances along that branch; and likewise, somehow, for the general, infinite case. (Never mind the details if, as I think, the plan won't work anyway.) The best system will be the winner, now, in a three-way balance between simplicity, strength, and fit. As before, the laws are the generalizations that appear as axioms or theorems in the best system; further, the true chances are the chances as they are according to the best system. So it turns out that the best system is true in its entirety— true in what it says about chances, as well as in what it says about history. So the laws of chance, as well as other laws, turn out to be true; and further, to have had no chance at any time of being false. We have our Humean supervenience of chances and of laws; because history selects the candidate systems, history determines how well each one fits, and our standards of selection do the rest. We will tend, *ceteris paribus*, to get the proper agreement between frequencies and uniform chances, because that agreement is conducive to fit. But we leave it open that frequencies may chance to differ from the uniform chances, since *ceteris* may not be *paribus* and the chances are under pressure not only to fit the frequencies but also to fit into a simple and strong system. All this seems very nice.

But it doesn't work. Along with simpler analyses of chance in terms of actual frequency, it falls victim to the main argument in the last section of the paper. Present chances are determined by history up to now, together with history-to-chance conditionals. These conditionals are supposed to supervene, via the laws of chance of the best system, on a global pattern of particular fact. This global pattern includes future history. But there are various different futures which have some present chance of coming about, and which would make the best system different, and thus make the conditionals different, and thus make the present chances different. We have the actual present chance distribution over alternative futures, determined by the one future which will actually come about. Using it, we have the expected values of the present chances: the average of the present chances that would be made true by the various futures, weighted by the chances of those futures. But these presently expected values of present chances may differ from the actual present chances. A peculiar situation, to say the least.

And worse than peculiar. Enter the Principal Principle: it says first that if we knew the present chances, we should conform our credences about the future to them. But it says also that we should conform our credences to the expected values of the present chances.[10] If the two

---

[10] Let A be any proposition; let $P_1, P_2, \ldots$ be a partition of propositions to the effect that the present chance of A is $x_1, x_2, \ldots$, respectively; let these propositions have positive present chances of $y_1, y_2, \ldots$, respectively; let C be a reasonable initial credence function; let E be someone's present total evidence, which we may suppose to be presently admissible. Suppose that $C(-/E)$ assigns probability 1 to the propositions that the present chance of P_1 is y_1, the present chance of P_2 is $y_2, \ldots$. By additivity,

(1) $C(A/E) = C(A/P_1E)C(P_1/E) + C(A/P_2E)C(P_2/E) + \ldots$

By the Principal Principle,

(2) $C(P_1/E) = y_1,$

$C(P_2/E) = y_2,$

$\ldots$

and

differ, we cannot do both. So if the Principle is right (and if it is possible to conform our credences as we ought to), the two cannot differ. So a theory that says they can is wrong.

That was the strategy behind my argument in the paper. But I streamlined the argument by considering one credence in particular. Let T be a full specification of history up to the present and of present chances; and suppose for *reductio* that F is a nonactual future, with some positive present chance of coming about, that would give a different present distribution of chances. What is a reasonable credence for F conditionally on T? Zero, because F contradicts T. But not zero, by the Principal Principle, because it should equal the positive chance of F according to T. This completes the *reductio*.

This streamlining might hide the way the argument exploits a predicament that arises already when we consider chance alone. Even one who rejects the very idea of credence, and with it the Principal Principle, ought to be suspicious of a theory that permits discrepancies between the chances and their expected values.

If anyone wants to defend the best-system theory of laws and chances both (as opposed to the best-system theory of laws, given chances), I suppose the right move would be to cripple the Principal Principle by declaring that information about the chances at a time is *not*, in general, admissible at that time; and hence that hypothetical information about chances, which can join with admissible historical information to imply chances at a time, is likewise inadmissible. The reason would be that, under the proposed analysis of chances, information about present chances is a disguised form of inadmissible information about future history—to some extent, it reveals the outcomes of matters that are presently chancy. That crippling stops all versions of our *reductio* against positive present chances of futures that would

(3) $C(A/P_1E) = x_1$,

$C(A/P_2E) = x_2$,

. . .

(Since the $C(P_i/E)$'s are positive, the $C(A/P_iE)$'s are well defined.) So we have the prescription

(4) $C(A/E) = y_1x_1 + y_2x_2 + \ldots$

that the credence is to be equal to the expected value of chance.

yield different present chances.[11] I think the cost is excessive; in ordinary calculations with chances, it seems intuitively right to reply on this hypothetical information. So, much as I would like to use the best-system approach in defense of Humean supervenience, I cannot support this way out of our difficulty.

I stand by my view, in the paper, that if there is any hope for Humean supervenience of chances, it lies in a different direction: the history-to-chance conditionals must supervene trivially, by not being contingent at all. As noted, that would impose remarkably stringent standards on reasonable belief. To illustrate: on this hypothesis, enough purely historical information would suffice to tell a reasonable believer whether the half-life of radon is 3.825 days or 3.852. What is more: enough purely historical information *about any initial segment of the universe*, however short, would settle the half-life! (It might even be a segment before the time when radon first appeared.) For presumably the half-life of radon is settled by the laws of chance; any initial segment of history, aided by enough noncontingent history-to-chance conditionals, suffices to settle any feature of the world that never had a chance to be otherwise; and the laws are such a feature. But just how is the believer, however reasonable, supposed to figure out the half-life given his scrap of ancient history? We can hope, I suppose, that some appropriate symmetries in the space of possibilities would do the trick. But it seems hard to connect these hoped-for symmetries with anything we now know about the workings of radioactive decay!

D. RESTRICTED DOMAINS

In reformulating the Principal Principle, I took care not to presuppose that the domain of a chance distribution would include all propositions. Elsewhere I was less cautious. I am grateful to Zeno Swijtink for

[11] As to the version in the paper: declaring hypothetical information about chances inadmissible blocks my reformulation of the Principal Principle, and it was this reformulation that I used in the *reductio*.

As to the version in the previous footnote: if information about present chances is inadmissible, then it becomes very questionable whether the total evidence E can indeed be admissible, given that $C(-/E)$ assigns probability 1 to propositions about present chance.

As to the streamlined version in this postscript: T includes information about present chances, and its partial inadmissibility would block the use of the Principal Principle to prescribe positive credence for F conditionally on T.

pointing out (personal communication, 1984) that if I am to be uniformly noncommital on this point, two passages in my final section need correction.

I say that if C_1 and C_2 are any two reasonable initial credence functions, and Y is any member of the history-theory partition for any time, then $C_1(-/Y)$ and $C_2(-/Y)$ are "exactly the same." Not so. The most I can say is that they agree exactly in the values they assign to the propositions in a certain (presumably large) set; namely, the domain of the chance distribution implied by Y. My point stands: I have a consequence of the Principal Principle that is entirely about credence, and that limits the ways in which reasonable initial credence functions can differ.

Later I say that these differences are—implausibly—even more limited on the hypothesis that the complete theory of chance is the same for all worlds. The same correction is required, this time with complete histories in place of history-theory conjunctions. Again my point stands. The limitation of difference is less than I said, but still implausibly stringent. Unless, of course, there are very few propositions which fall in the domains of chance distributions; but that hypothesis also is very implausible, and so would not save the day for a noncontingent theory of chance and for Humean supervenience.

My reason for caution was not that I had in mind some interesting class of special propositions—as it might be, about free choices—that would somehow fail to have well-defined chances. Rather, I thought it might lead to mathematical difficulties to assume that a probability measure is defined on all propositions without exception. In the usual setting for probability theory—values in the standard reals, sigma-additivity—that assumption is indeed unsafe: by no means just any measure on a restricted domain of subsets of a given set can be extended to a measure on all the subsets. I did not know whether there would be any parallel difficulty in the nonstandard setting; it probably depends on what sort of infinite additivity we wish to assume, just as the difficulty in the standard setting arises only when we require more than finite additivity.

Plainly this reason for caution is no reason at all to think that the domains of chance distributions will be notably sparser than the domains of idealized credence functions.

Theories of Probability
Colin Howson

ABSTRACT

My title is intended to recall Terence Fine's excellent survey, *Theories of Probability* [1973]. I shall consider some developments that have occurred in the intervening years, and try to place some of the theories he discussed in what is now a slightly longer perspective. Completeness is not something one can reasonably hope to achieve in a journal article, and any selection is bound to reflect a view of what is salient. In a subject as prone to dispute as this, there will inevitably be many who will disagree with any author's views, and I take the opportunity to apologize in advance to all such people for what they will see as the narrowness and distortion of mine.

1 Introduction

Carnap [1950] noted that the mathematical theory of probability seems to be a syntax with not one but two interpretations, one epistemic and the other objective, one relating to our knowledge of the world and the other to the world independently of our knowledge. This dual nature of probability has been a persistent feature of its history. It is there already in Laplace's distinction between 'probabilité' (which he related 'in part to our ignorance, in part to our knowledge') and 'possibilité' (Laplace [1820]), and was also noted by Poisson, who retained the word 'probabilité' for the epistemic concept but employed the word 'chance' for the other. What has happened since then by way of further elucidation has been the registration

Colin Howson

of a number of candidate explicatory theories. As explications of the epistemic concept we have been offered, at various times and by various people: degree of rational belief, degree of partial entailment, degree of partial truth, fair betting quotient. Explications of the objective notion include: degree of possibility, limiting relative frequency, 'almost sure' long-run relative frequency in independent trials, chance, and propensity.

Over the years the critical process has narrowed the field down to the point where today's student of philosophical probability confronts the following principal accounts: the *Bayesian* theory of epistemic probability, and the *limiting relative frequency, propensity, 'prequentialist'*, and *chance* theories of objective probability. There are also a number of *nonstandard* epistemic probability theories, i.e. theories in which one or more axioms of the probability calculus (usually the additivity axiom) is infringed. In what follows I shall give an account of these various approaches to probability, identifying what seem to me their strengths and weaknesses.

2 Epistemic probability

2.1 The Bayesian theory[1]

2.1.1 Betting quotients and degrees of belief

The fundamental principles of the Bayesian theory are that (a) between the certainty that a proposition A is true and the certainty that it is false, one may have intermediate degrees of belief in A, (b) these can be represented numerically, and (c) if they are rational in a minimal sense then, measured in the scale of the closed unit interval, they satisfy the finitely additive probability axioms. Many but not all Bayesians believe that degrees of belief are not merely finitely additive but also countably additive.

The probability axioms have been approached from two directions. One, indicated first by Ramsey [1926] and in more detail by de Finetti [1937], argues that the axioms are rationality principles constraining the *betting quotients* you would be willing to offer in certain types of bet, and identifies these quotients with your degrees of belief in the propositions bet on. The other (where Ramsey [1926] is the seminal work) arrives at the axioms via utility theory. We shall examine the two approaches in turn.

2.1.2 Bets and degrees of belief

A bet on A is a contract whereby one party (the bettor-on) agrees to give

---
[1] My discussion of the Bayesian theory owes a great debt to Armendt's [1994], which I shall state at the outset rather than by too-numerous interruptions to the text.

another (the bettor-against) R units of value, in exchange for Q units if A turns out to be true. The ratio R/Q is the odds on A, and the *betting quotient* on A is the normalized odds p = R/(R + Q). S = R + Q is called the stake. Advocates of the betting approach identify your degree of belief in A with the betting quotient on A you would be willing to nominate in a bet in which the stake, and whether you are to be bettor-on or against, is decided by your opponent.

The probability axioms can now be represented as rationality constraints on your degrees of belief, since by the so-called Dutch Book argument any infringement of the axioms can be exploited to give your opponent a sure gain at your expense (a clear statement, as well as the first explicit use of this argument, is in de Finetti [1937]). Conversely, it is easily shown that no Dutch Book can be made against a system of betting quotients satisfying the finitely additive probability axioms.

The argument can be generalized to partial beliefs as elicited not only in certain kinds of bets, but by any in a wide class of what are called *scoring rules*. A scoring rule for a proposition A assigns a penalty (positive or negative) depending on a number p, representing the penalisee's uncertainty about A, and the truth value of A. A bet is a scoring rule of the form $S(I(A) - p)$, where $I(A)$, the indicator of A, is 1 if A is true and 0 if A is false. A simple scoring rule yielding only positive penalties is the quadratic rule $(I(A) - p)^2$. Lindley has shown [1982] that, for any positive scoring rule, and any set $p_1, \ldots, n_n$ of degrees of belief in corresponding propositions, then, if the p_i jointly minimize the penalty over all possible states of affairs, either they directly satisfy the finitely additive probability axioms or some transform of them does. For the quadratic rule the p_i directly satisfy the axioms.

But as critics have not been slow to point out, the assumption that you will agree to have your beliefs elicited by a bet of the type above, or in general by a scoring rule, is unrealistic, even if the penalties are small. Some people (like Mellor [1971]) have tried to avoid the objection by assuming a scenario in which you are compelled to participate, but this proposal is vulnerable to the objection that the loss-avoidance constraints represented by the probability axioms cannot be assumed to apply outside such coercive circumstances.

A less problematic strategy is to identify your degree of belief in a proposition with the betting quotient you merely believe fair, whether you bet or not (Howson and Urbach [1993], Ch. 5). Of course, in most circumstances it is unlikely that you will be able to identify a unique such number. One response is to admit interval-valued belief functions, and the implications of doing so have been extensively studied and will be discussed later. Even if the idealization of point-valued degrees of belief is granted (as most Bayesians in practice do), there remains the problem of justifying the probability axioms. While the non-negativity axiom, and the axiom which states that necessary propositions have probability 1, can quite easily be justified as consistency constraints (Howson and Urbach [1993], ibid.), the additivity axiom is less straightforward. The standard argument, stripped of unnecessary frills, is this. Let A and B be two mutually exclusive propositions. Define the sum X + Y of the two options X = '1 if A, 0 if not', and Y = '1 if B, 0 if not' to be the option whose payoffs for each distribution of truth values over A and B is the sum of the payoffs from X and Y. So defined, X + Y is equivalent to the option '1 if AvB, 0 if not', as is simple to check. Let your fair price for X be p and for Y be q (i.e. p is your fair betting quotient on A, and q your fair betting quotient on B). Assume (value-additivity assumption) that the fair price for the sum option is the sum of the fair prices of each. It follows that p + q is the fair price for the option '1 if AvB, 0 if not'; i.e. p + q is your fair betting quotient on AvB, and we obtain the finite additivity axiom for probabilities (there is a similar argument for countable additivity).

The flaw in this argument, as Schick [1986] has pointed out, is that the value-additivity assumption seems to be an independent assumption which can be infringed without inconsistency. Some (like Skyrms [1987], Ch. V.7) have argued that utility theory provides its justification. For if we define the fair price for any bet $B(C)$ on a proposition C to be its expected value relative to your belief distribution, then because expected value is a linear functional we can indeed infer that the fair price for $B(AvB)$ is the sum of the fair prices of $B(A)$ and $B(B)$. Utility seems to have been quietly assumed anyway in the definition of your fair betting quotient. For, as we have seen, another way of representing a bet with stake S and betting quotient p is as a contract in which a price, pS, is paid for the option of receiving S if A is true and nothing if not. This means that you believe the fair price for the option is proportional to S, and, given what we know about diminishing marginal utility of money, it is false unless S is measured in utility units.

But the expected utility principle is usually itself taken to be based on the assumption that your belief distribution obeys the probability calculus, so it might seem that we are arguing in a circle. A way out of the circle would be to show that the expected utility principle and the representation of partial belief by a probability function are *both* consequences of a more fundamental set of rationality postulates. This is just what the seminal work of Ramsey, Savage, and others is widely believed to have done, and to a discussion of that work we now turn.

2.1.3 Utilities

Ramsey [1926] seems to have been the first to show how a utility scale can be calibrated. His basic tool was the notion of an 'ethically neutral proposition believed to degree 1/2'. An ethically neutral proposition is one such that, considering it in itself, you are indifferent between its being true and being false. Such a proposition, call it A, is believed to degree 1/2 if you are indifferent between the option 'receive X if A is true, Y if not', and 'receive Y if A is true, X if not', yet you prefer X to Y.

The calibration of the utility scale now proceeds as follows. Take any two things X and Y between which you have a preference; suppose you prefer X to Y. Assign X and Y any two real numbers x and y respectively such that y < x. Suppose C is an ethically neutral proposition believed to degree 1/2. Consider the option 'receive X if C is true, Y if not'. According to an axiom Ramsey lays down, there is a Z in the domain of your preference-relation which is as desirable to you as any option of the form 'receive X if C, Y if not'. Assign Z the real number $z = (x + y)/2$. Further utility points midway between x and z and z and y can be interpolated by the same method, and so on indefinitely. Additional axioms postulate the existence of other appropriate points (like continuity points) on the utility scale, and using these it is possible to show that your preferences can be mapped in an order-preserving way into the real numbers. Furthermore, all and only the affine functions of this mapping are order-preserving correspondences. Each can be regarded as a (real-valued) utility assignment to your preferences, differing only in scale unit and origin.

This implies that ratios of utility differences do not depend on the representative utility function chosen, and Ramsey defined degrees of belief explicitly as ratios of utility-differences, as follows. Suppose you are indifferent between X and the option 'Y if A, Z if not'. Then your degree of belief p(A) in A is defined to be

$$\frac{u(X) - u(Z)}{u(Y) - u(Z)}$$

which, as Ramsey observes, is equivalent to defining your degree of belief in terms of generalized, utility-based odds. Rather arbitrarily it is postulated that the ratio is independent of the particular quantities X, Y, and Z. At any rate, it is now possible to prove that (i) degrees of belief are finitely-additive probabilities, and (ii) the utility of the option 'Y if A, Z if not' is equal to its expected utility p(A)u(Y) + (1 − p(A))u(Z).

Ramsey's work for a long time was not widely known, and for this and other reasons most decision-theorists have tended to adopt the alternative axiomatic development of utility leading to (i) and (ii) due to L. J. Savage

[1954], which itself draws on earlier work of de Finetti. In Savage's system the probability function is implicitly rather than explicitly defined as the unique finitely additive probability function representing an ordering of propositions (which Savage, in keeping with the mathematical tradition stemming from the work of Kolmogorov, calls events). The ordering, a so-called *qualitative probability ordering*, is determined by preferences between options like the ones Ramsey considered. The preferences are constrained by certain consistency principles, which to some extent match corresponding axioms of Ramsey's. As do Ramsey's, Savage's axioms also generate a class of utility functions each of which is an affine transformation of another in the class, and such that preferences between options is represented by the ordering by magnitude of their expected utilities, relative to any one such utility function.

Savage's axioms are conceptually more distant from the expected utility principle than are Ramsey's, which virtually postulate that preferences have a structure isomorphic with the reals. Savage's axioms describe the structure of consistent preferences, initially over acts and derivatively over outcomes, in a way which appeals more explicitly to rationality considerations. Thus his first axiom asserts that, among other things, preferences should be transitive and that if one act is preferred to another, then the second is not preferred to the first. Another axiom, the 'sure thing principle', says that if you would prefer doing f to doing g if a proposition B is true, and you would not otherwise prefer doing g to doing f, then you should prefer doing f to g in any case.

Innocuous though these principles seem, they generate some surprising and often strongly counterintuitive conclusions, like the well-known Allais and Ellsberg paradoxes (for a recent discussion see Gärdenfors and Sahlin [1988]). Even properties as transparent as transitivity itself turn out to be rather hard to justify in a non-circular way. One popular attempt is the so-called money-pump argument. Suppose you prefer X to Y; then presumably your evaluation of X is that of Y plus some value premium p. Now suppose you prefer X to Y and Y to Z but you also prefer Z to X. Then the value of X is that of Y + p, the value of Y is that of Z + q, and the value of Z is that of X + r. And off we start again. In forever cycling round the loop, anyone prepared to implement your evaluations would be a money-pump.

Potential money-pumping can arise even from transitive preferences, if there are infinitely many of them: for example, if you deem that X_2 is preferred to X_1, X_3 to X_2, X_4 to X_3, etc., all by equal amounts. The salient feature of the money-pump argument, presumably, is that intransitivity amounts to evaluating a good at a higher value than itself. But is this true? Admittedly (a) the combination $E(X, Y, Z) = E(X, Y) + E(Y, Z) + E(Z, X)$

of the pairwise exchanges of X for Y, of Y for Z, and of Z for X, is an indirect way of making the pairwise exchange $E(X, X)$, and (b) the price you think worth paying for $E(X, X)$ is, presumably, 0, but even if you think the pairwise exchanges are worth separately p, q, and r units respectively, why should you equate the value of $E(X, Y, Z)$, i.e. $p + q + r$, with that of $E(X, X)$, i.e. 0?

To say that you should makes implicit appeal to a principle of path-independence, or *extensionality*, already introduced into decision theory by Arrow [1982]. Deductive logic has for long been concerned only with extensional properties; this means, for example, that $\exists x(x = x)$ is logically equivalent to 'Mary is happy or Mary is not happy', though the intensions, or meanings, of the two sentences are quite different. If we follow Ramsey [1926] and regard the Bayesian theory as *logic*, we might then prescribe that the values of extensionally equivalent exchanges be identical. Then indeed the option '1 if A∨B, 0 if not', should be valued as the sum of values of the options '1 if A, 0 if not' and '1 if B, 0 if not', where A and B are exclusive, and the exchange $E(X, Y, Z)$ should have the same value as the exchange $E(X, X)$. If someone asks 'that is all very well, but what difference does observing such rules make in practice?', there is a well-known answer:

> If anyone's mental condition violated these laws, his choice would depend on the precise form in which the options ,were offered him, which would be absurd. He could have a book made against him by a cunning better and would then stand to lose in any event (Ramsey [1926], p. 78).

These observations are unlikely to bring to a sudden stop a dispute that has been so prolonged. Nevertheless, the proposal to treat the Bayesian theory as an extensional logic of partial belief does not seem an unreasonable one. Perhaps that is all that one should venture. At any rate, it is now time to discuss another controversial area in the Bayesian theory, the rule of conditionalization.

2.1.4 Bayesian conditionalization

Your degrees of belief in some propositions may be *conditional* on some other propositions being true. In terms of fair bets a degree of belief in A conditional on B is glossed as a fair betting-quotient in a corresponding conditional bet, i.e. a bet on A whose activation is conditional on B's truth. It is easy to show (Howson and Urbach [1993] pp. 82–4, for example) that suitable bets on A&B and B determine a conditional bet on A given B, with betting-quotient p/q where p is the betting-quotient on A&B and q is that on B, $q > 0$; hence if $p(A|B)$ is your degree of belief in A conditional on B's being true, and $p(B) > 0$, then, granted the validity of this sort of argument, consistency requires that $p(A|B) = p(A\&B)/p(B)$, which is sometimes called the *multiplication axiom*.

The multiplication axiom says nothing, however, about how your beliefs should be updated, or *changed* on the receipt of new information; for this reason it and the other probability axioms are often referred to as *synchronic* principles. But Bayesians do famously have a rule for updating belief, called the Rule of Conditionalization, which is as follows. Suppose that your degrees of belief before learning the truth of a proposition E are represented by a probability function p_1, with a corresponding conditional probability functions $p_1(. |E)$, and that your degrees of belief consequent on learning E are represented by p_2. The rule of conditionalization is that, for all considered propositions A,

$$p_2(A) = p_1(A|E) \qquad (1)$$

when you come to learn E and no more. On (1) is based the whole of Bayesian methodology (Horwich [1982], Howson and Urbach [1993], Earman [1992]). Since the theory of consistent degrees of belief was more or less invented with the methodological application in mind, (1) is clearly of fundamental importance in that theory. Nevertheless, as Hacking [1967] pointed out to the considerable embarrassment of contemporary Bayesians, the standard arguments, i.e. the ones we have reviewed above, for consistent degrees of belief being formally probabilities, do not appear to extend to justifying (1).

2.1.5 Diachronic Dutch Books

In 1973, however, Paul Teller, attributing its original authorship to David Lewis, published a Dutch Book argument for (1) (Teller [1973]). Lewis–Teller showed that if you followed any other rule than (1) for determining p_2 on receipt of information E, then a person who knew this rule could in principle exploit the difference between $p_2(A)$ and $p_1(A|E)$ to guarantee a loss come what may to anybody willing to bet at those rates. The strategy is to induce them to make a conditional bet before the truth of E is known. If E is true, they will then bet at a different rate on A, i.e. at $p_2(A)$, which will net them a loss overall (the directions of the bets will be determined by whether $p_1(A|E)$ is greater or less than $p_2(A)$), while if E is false a suitable side-bet will also generate a loss. It appears from the Lewis–Teller argument that anyone who adopts a rule for updating belief different from (1) is infringing a rule no different in kind from the probability axioms themselves.

The appearance is deceptive. Your conditional probability $p_1(A|E)$

represents your current view of how likely A is to be true in an E-world, i.e. a world in which E is true. $p_2(A)$ represents your view *at some later date* of how likely A is to be true in the E-world you are then in. As Christensen [1991] pointed out, there is no inconsistency revealed by these being different, any more than there is by your believing A for certain today and its negation ¬A for certain tomorrow. Only if both A and ¬A were in your set of currently accepted propositions would there be inconsistency. *Mutatis mutandis* the same is true in the present case. Moreover, it remains true *even if you know now what $p_2(A)$ will be in the event of E's being true, and know that it differs from $p_1(A|E)$*, as the following example, due to Richmond Thomason, shows. A husband announces 'if my wife is unfaithful, I shall never know'—the wife being known to be expert in deception. The corresponding conditional probability he ascribes to his not knowing that his wife is unfaithful [A], given his wife's infidelity [E], is presumably 1 or near 1. Yet learning that his wife was unfaithful he could scarcely consistently assign probability close to 1 to not knowing what he has just learnt, contrary to the precept laid down in (1).[2]

Some interesting conclusions follow from these observations. One is that vulnerability to certain loss is not always an indicator of inconsistency. In Thomason's example your fair betting-quotients are eminently sensible; yet were you to be willing to bet at those rates, a Lewis–Teller Dutch Book could be made against you. There are Dutch Books and Dutch Books. Only one that can be made against a set of betting-quotients all of which are synchronously believed fair plausibly indicates incoherence; the Lewis–Teller diachronic Dutch Book shows only that it can be irrational to bet honestly against an opponent who knows as much as you.

Another conclusion to be drawn is that (1), far from being a general consistency constraint on a par with the probability axioms, will actually generate inconsistency if followed universally, and in the unfaithful wife and credulous husband we have an example of where this happens (another example, involving the taking of a mind-altering drug, has generated a considerable literature and a spurious principle, the so-called Reflection Principle; see van Fraassen [1984], Christensen [1991], and Howson [1993]). But does this mean that Bayesians must abandon (1)? As Hacking remarked, Bayesianism's losing Conditionalization would be like salt losing its savour (Hacking [1967]). Nevertheless, we have to admit that (1) is not universally valid. The question is, what are the conditions under which it is valid? I shall postpone answering this question, to the

---

[2] The situation is slightly complicated by the fact that if he does believe that A is false, then he should conditionalise on the stronger statement E & ¬A. Since $P(A|E\&\neg A) = 0$ if $P(E\&\neg A) > 0$, for an unambiguous counterexample we need to make $P(A)$ actually equal to 1, so that the conditional probability is undefined.

Colin Howson

extent it can be answered, until we have considered a generalization of (1) due to Jeffrey.

2.1.6 Probability kinematics

(1) enjoys doubtful status not only because it seems to have no justification of the sort the probability axioms themselves are claimed to have by Bayesians, but also because it seems to represent an ideal of acquiring evidence that is too far short of the reality to be comfortable. This is for two reasons. (a) E in (1) must be learned with *certainty*; indeed, (1) implies just this, for $p_2(E) = p_1(E|E) = 1$. This also implies that $p_2(E)$ cannot itself be revised by any further evidence according to (a): probabilities of unity are absorbing barriers; once there, you stay there. (b) Much of the evidence we acquire 'from without', as it were, and which causes us to change our beliefs, is not even propositional in character. Richard Jeffrey [1983] suggested a method, which has since become known as the *rule of Jeffrey Conditionalization*, or alternatively *probability kinematics*, which promises to meet both these objections simultaneously. Jeffrey supposed that as a result of some experience, an agent's degree of belief in a proposition E changes exogenously from $p_1(E)$ to $p_2(E)$, where $p_2(E)$ is less than 1 (in his example, seeing a piece of fabric in dim light causes us to revise our degree of belief that it is green, but not to certainty that it is or certainty that it is not). He then proposed that our belief function be updated on all other propositions A in its domain according to the rule

$$p_2(A) = p_1(A|E)p_2(E) + p_1(A|\neg E)p_2(\neg E) \qquad (2)$$

More generally, if the members of a finite partition E_i, and not just the pair $\{E, \neg E\}$, change their probabilities exogenously to $p_2(E_i)$, the Jeffrey rule becomes

$$p_2(A) = \Sigma p_1(A|E_i)p_2(E_i) \qquad (2')$$

Jeffrey's rule has some important properties. (i) It reduces continuously to the rule of (Bayesian) conditionalization when $p_2(E) = 1$, i.e. when E is learned for certain (or in (2′) when one member of the partition has its probability increased to 1). (ii) Probability distributions updated according to it are revisable, so long as $p_2(E)$ does not become 1 or 0. This means that further experiential inputs may convince me that my first function p_1 was after all correct; i.e. the progression $p_1 \rightarrow p_2 \rightarrow p_3 = p_1$ is possible, where p_2 and p_3 are updates of their predecessors by the Jeffrey rule. (iii) A necessary and sufficient condition for it is that $p_1(A| \pm E) = p_2(A| \pm E)$ for every A, where $+E$ is E and $-E$ is $\neg E$; for (2′) the condition is that $p_1(A|E_i) = p_2(A|E_i)$, $1 \leq i \leq n$. (iv) (2′) generates the same function p_2 which minimizes the *information* or *cross-entropy* functional $I(p_1, p_2) = \Sigma p_2(x_i) \log [p_2(x_i)/p_1(x_i)]$, where x_i are a finite or denumerably infinite set of elementary possibilities, subject to the constraint that the E_i take the values $p_2(E_i)$. (v) A Dutch Book argument, analogous to the Lewis–Teller Dutch Book argument for (1), can be constructed against anyone who announces their intention of violating (iii), and hence (2) (Armendt [1980]).

We shall consider the significance of these properties in reverse order. The Dutch Book argument for (2′) is subject to the same objection as was the Lewis–Teller one for (1): it demonstrates nothing more than the inadvisability of betting honestly against someone who knows your updating strategy. This leaves the question of justifying (2′) open. The fact that (2′) minimizes $I(p_1, p_2)$ has been considered (for example, by Williams [1980]) as a justification of (2′) on the ground that $I(p_1, p_2)$ is a measure of the 'closeness' in function space of the two distributions p_1 and p_2. However, since $I(p_1, p_2)$ is not the only possible such measure, and since it is technically not a distance metric (it is not symmetric), there must be some doubt about this claim. Even were it true, some further argument would be necessary to show why the 'closest' distribution to p_1 is the appropriate updated belief function (for some technical details and a good general discussion see Diaconis and Zabell [1982]). Similar remarks apply to van Fraassen's symmetry argument for (2) ([1989], pp. 334–7). Symmetry principles can be invoked in all sorts of ways. A symmetry principle is that every bijection of the possibility space into itself should be measure-preserving for 'ignorance' measures, but this is impossible (we get the 'paradoxes of geometrical probability'). The fact that (2) and therefore (1) are not of universal validity is equally good reason for rejecting the symmetry principle assumed in van Fraassen's argument.

Property (iii), which says that (2′) is valid just so long as all probabilities conditional on the E_i remain unchanged by the change from $p_1(E_i)$ to $p_2(E_i)$. And this, at least in one way, answers the question asked at the end of the previous subsection: under what conditions is (1) valid? The answer is that, when $p_2(E)$ goes to 1, (1) is valid if and only if $p_2(A|E) = p_1(A|E)$ for every A, since this is the form that (iii) takes in this case (that this is a complete answer to the question seems to be a view which Jeffrey himself endorses ([1992], p. 81)). Despite its apparent facility, the answer has the considerable advantage over 'deeper' ones that it invokes in its support nothing more than the probability calculus itself.

To accept the (conditional) validity of the Jeffrey rule has an important consequence for the applicability of (1). We observed that (1) implies the assignment $p_2(E) = 1$, and that this, once made, cannot be revised by conditionalization however much subsequently gathered evidence suggests that it is wrong. From the point of view of its methodological application, this would be a serious defect of the Bayesian theory incorporating just the rule (1). Adding Jeffrey's rule allows the difficulty to be circumvented in a natural way: (i) allows us to regard (1) as a convenient and simple approximation of (2) when $p_2(E)$ is sufficiently large; and (ii) tells us that this evaluation of E can then consistently be revised downward if so desired.

2.1.7 Objective Bayesianism

This review of the contemporary Bayesian theory has so far not discussed the so-called Objective Bayesians, who believe that the Bayesian theory should incorporate further principles about the appropriate form of prior distribution to adopt for characteristic classes of problem. The leading modern exponents of this view, which has a long pedigree which includes Bayes and Laplace, are Harold Jeffreys [1961], E. T. Jaynes [1968, 1973, 1983], and Roger Rosenkrantz [1977, 1981]. One of the historically if not practically most important of these classes of problem is that where there is held to be no relevant prior information, except the limits of the range of the random variable(s) defining the possibility space. The demand here is that the priors are 'informationless', i.e. that they contain no information beyond the constraints imposed by background information. Where the possibility space is represented by a bounded interval in Euclidean n-space the informationless prior is just the uniform distribution over the interval. Other cases are less straightforward; where the interval is the positive or negative real line, for example, the informationless prior density is usually held to be of the form $|x|^{-1}$, which is 'improper', i.e. its integral diverges; where the interval is the whole real line, the informationless prior is an improper uniform density.

Improper distributions generate paradoxes, however (Dawid, Stone, and Zidek [1973]), a fact which might seem as symptomatic of their essentially arbitrary character. A standard criticism of the Bayes–Laplace theory was that the space of possibilities can usually be represented in more than one way, and in the continuous case the density may transform into a quite different type of function from one representation (random variable) to the next. While Jaynes has argued [1973] that the background constraints in any problem can sometimes be represented by a characteristic group of transformations which uniquely determines the prior density, this does not remove the arbitrariness: it merely pushes it one stage farther back. Problems by themselves do not in general uniquely determine prior distributions, so if some group of transformations does so determine one, that group cannot be uniquely determined merely by the problem but has at some point invoked additional information to determine it. This tells us that no prior distribution represents *only* background information:

inevitably it represents a view of how likely the various alternatives are to be true. This is not to say that for some problems there will not be a degree of consensus about the appropriate prior; the point is merely that the search for informationless priors is misconceived, and it is one that the majority of Bayesians now seems to accept (Jaynes's theory of invariant priors is critically discussed in Milne [1983]).

2.2 Non-standard epistemic probabilities

The last thirty years have witnessed the development of several types of non-standard probability theory, often called non-additive probability theory. The best known of these are the Dempster–Shafer theory, and the closely related theory of upper and lower probabilities. The characteristic items in the Dempster–Shafer theory are *frames of discernment*, so-called, and *belief functions*, whose domains are the set of all subsets of a frame. A frame of discernment is a space of hypotheses which are jointly taken to be exhaustive, and the belief functions assign real numbers in the closed unit interval to each subset of a frame Ω, with 1 assigned to Ω and 0 to $\emptyset$. Where Bel is a belief function, Bel(A) is supposed to represent an agent's degree of belief, in the light of the evidence, that the true hypothesis is in A. Two features make this theory radically unlike a probability theory. One is that the belief assigned to a particular subset and that assigned to its complement are functionally independent of each other, and the other is Dempster's method of combining belief functions based on independent bodies of evidence (for a fuller discussion see any or all of Shafer [1976], Williams [1978], Howson and Urbach [1993], Ch. 15, Kyburg [1987]).

Belief functions share many of the formal properties of lower probabilities. Axioms for upper and lower probabilities were first produced by Koopman [1940], and their properties were later derived independently as constraints on an indeterminate belief measure by Good [1962], and Smith [1961]. Standard Bayesian probabilities are based on the assumption that the supremum of the betting-quotients which give an advantage to the bettor-on is equal to the infimum of the quotients which give an advantage to the bettor-against. Relaxing this assumption and employing similar consistency considerations to those described in Section 2.1, it can be shown that the suprema and infima obey the axioms of upper and lower probabilities respectively.

These axioms are mainly in the form of inequalities, which include the principles of superadditivity of lower probabilities and of subadditivity of upper probabilities: that is to say, $P_*(A) + P_*(B) \leq P_*(AvB) \leq P^*(AvB) \leq P^*(A) + P^*(B)$, where P_* is the lower and P^* the upper

195

probability function, from which it immediately follows that $P_*(A) \leq P^*(A)$ (a thorough exposition is to be found in Walley [1991]). Milne has shown [1993] that the calculus of upper and lower probabilities has, *mutatis mutandis*, also the same dual interpretation as the standard probability calculus, both as a system of consistency constraints on belief, and as limits of long-run relative frequencies, in the latter case the limits supremum and infimum respectively of the relative frequencies of a particular character in an infinite sequence of characters.

Other non-additive theories are Shortliffe and Buchanan's [1975] MYCIN calculus of certainty factors, the theory of fuzzy probabilities (Dubois and Prade [1989]), L. J. Cohen's theory of what he calls non-Pascalian probability [1989], and Spohn's theory of non-probabilistic belief functions (Spohn [1990]). All these theories were developed with the aim of mathematically modelling rational belief, and belief-change in the light of evidence, and their divergence from the formal probability model is defended on the ground of what are claimed to be deficiencies in the latter—primarily the alleged inability of probabilistic theories to model ignorance, and their making belief in a hypothesis functionally dependent on belief in its negation. This is not the place either to investigate the justice of these claims or the structure of those theories; for this the reader is invited to consult the references given.

3 Objective probability

3.1 Limiting relative frequency: von Mises' theory

The idea that the calculus of probabilities could be interpreted as a calculus of relative frequencies within suitable infinite sequences seems to have been due originally to John Venn, the author of Venn diagrams. At a time when the Bayes–Laplace theory was becoming regarded as irremediably subjective and arbitrary, the notion of an objective, frequency-based theory was attractive, and proved increasingly so. Hans Reichenbach [1949], Richard von Mises [1939, 1964], and recently Bas van Fraassen [1980], have produced versions of such a theory, though van Fraassen's is a development of Reichenbach's. Because von Mises' theory is the most promising as a foundation of statistics (since it alone incorporates a principle of randomness), I shall confine the discussion to that theory.

Von Mises's theory is based on his notion of a *collective*, which is an infinite sequence W of attributes from some finite or denumerably infinite set A of attributes, satisfying two conditions:

(i) for each attribute A in A, the relative frequency of A in W tends to be a finite limit;

(ii) there is no infinite subsequence W' of W determined by a place-selection, in which any attribute in A has a limiting relative frequency different from that in W. A place-selection is an effectively computable function f whose domain is the set of all initial segments $(w_1, w_2, \ldots, w_n)$, $n \geq 1$, such that $f(w_1, w_2, \ldots, w_n)$ takes the values 0 or $n + 1$, and such that if $\{m_1, m_2, m_3, \ldots\}$ is the set of all nonzero values of f, $m_1 < m_2 < m_3 < \ldots$, then $W' = (w_{m1}, w_{m2}, w_{m3}, \ldots)$.

A collective W satisfying (i) and (ii) is an idealized mathematical model of the successive outcomes, classified by the attributes in A, of a data source S repeatedly observed under the same conditions. The probability of the attribute A relative to S is defined to be the limiting relative frequency of A in W, which exists by (i). (ii) expresses the condition that there is no successful gambling system for S, i.e. no effective way of selecting outcomes on which to bet, which uses only information about earlier outcomes, for which the odds differ from those in W. Church [1940], using the theory of recursive functions, gave a precise characterization of the notion of an effectively computable function, and Wald [1938] showed that there existed an uncountable infinity of collectives with any finite or denumerably infinite set of attributes.

Previous proposers of probability-as-limiting-relative-frequency, like Venn himself, had laid down only the condition (i) that relative frequencies converge. The addition of (ii), which von Mises called the *Axiom of Randomness*, was a considerable innovation. Implicitly, it makes von Mises' theory a theory of random sampling, for it can be shown that successive members of W are probabilistically independent and identically distributed. Precisely, if W is partitioned into consecutive n-tuples of attributes, the result can be shown to be a collective $W_{(n)}$, with attribute set A^n, i.e. the set of all n-tuples of members of A, such that in the derived collective $W_{(n)}$, (a) the probability of an n-tuple in A^n having A at its ith place is the same for all $i \leq n$, and equal to the probability of A relative to W, and (b) the probability in $W_{(n)}$ of an n-tuple having any k attributes at specified points, $k \leq n$, is equal to the product of the probabilities of each occurring at that point (von Mises [1964]). It follows that W determines in an intuitively natural way the probabilistic properties of any n-fold random sample from the same data source S, for the set Ω of all denumerably infinite sequences of members of A can, by partitioning them in the manner above, equally represent all possible sequences of n outcomes generated by S, $n = 1, 2, 3, \ldots$; Ω is *universal* for such representations, in other words.

3.1.1 The role of limit theorems

Von Mises' frequency theory is conceptually a sophisticated one. It has also come in for a good deal of criticism, and was for long regarded as fatally discredited. One focus of criticism is the explicit definition of probability as limiting relative frequency. Many if not most statisticians and mathematical probabilists, including Kolmogorov, Neyman, Doob, Cramér, and Feller, among others, regard the relation between probability and long-run frequency as less determinate, and defined by the limit theorems of the mathematical theory, like the weak and strong laws of large numbers and the central limit theorem. The weak and strong laws are proved for a possibility space Ω consisting of either n-fold or infinite sequences of outcomes, which are classified as either 0 or 1 by a corresponding sequence $X_1, X_2, \ldots$ of random variables defined on Ω. The weak law of large numbers says that if the X_i are independent and identically distributed (i.i.d. in statisticians' parlance), taking the value 1 with common probability p and 0 with probability $1 - p$, then for every $\epsilon > 0$ the probability of the event $|n^{-1}\Sigma X_i - p| < \epsilon$ tends to 1 as n tends to infinity (the sum is from $i = 1$ to n). The strong laws says that if Ω is infinite and the probability function countably additive, then $\lim n^{-1}\Sigma X_i$ exists and is equal to p except for a set of probability 0.

One thing to be clear about is that there is no conflict, as has sometimes been claimed, between the various laws of large numbers and von Mises' theory. Since his theory, as we noted, is a theory of independent identically distributed random variables both the weak law and an approximate form of the strong law for finite sequences are actually logical consequences of it. To resolve the issue it must therefore be decided whether the laws of large numbers in their full measure-theoretic context represent merely a better theory of the long-run stability of frequencies in repeated outcomes of chance mechanisms than von Mises'. There seems room for doubt. Without some independent gloss of the phrases 'independent, identically distributed random variables' and 'with probability close to [equal to] 1', the laws of large numbers tell us nothing at all about the relation between such probabilities and relative frequencies in successive outcomes of a chance process.

A century and a half ago A. A. Cournot tried to answer this objection by proposing that we treat small probabilities as impossibilities, so that the assumptions of independence etc. become in effect testable hypotheses which can be falsified by observing whether the observed relative frequencies lie within the predicted bands. But a simple consequence of Cournot's rule is that almost every hypothesis of use to statistics is a priori declared false by it. For example, consider the hypothesis which says that a sequence X_i of random variables is i.i.d. with a specified probability p strictly between 0 and 1. Then whatever small number ϵ is declared to be the lower limit of physical probability, for some value of n each sequence of outcomes will be assigned a probability smaller than ϵ. Probabilities even of 0 cannot be consistently regarded as impossibilities so long as one countenances the possible truth of hypotheses ascribing continuous distributions to any variate. Gillies [1973] has proposed a more elaborate version of Cournot's rule, but this turns out to be equally unsound (Redhead [1974]). Nor can the apparently more sophisticated Neyman–Pearson theory of statistical tests help out here, for the desirability of minimizing type 1 and type 2 errors explicitly assumes that probabilities approximate long-run relative frequencies.

The conclusion seems to be that long-run frequencies cannot be shown to approximate probabilities without explicitly building that condition into the meaning of the probability function, in which case the limiting relative frequency definition becomes virtually obligatory. Such a definition also has the merit that it is demonstrable that the probability axioms are satisfied, with however two exceptions: countable additivity is not guaranteed (Giere [1976], p. 326), and the domain of the probability function is not always closed under the finite Boolean operations (a fact discovered independently by de Finetti and Rubin, and reported in van Fraassen [1980], p. 184). Von Mises did later add countable additivity over a countable set **A** of attributes as a defining condition on a collective, from which it follows that the probability function is countably additive on the power set of **A** (von Mises [1964], pp. 18–20).

3.1.2 Randomness

Randomness in von Mises' theory is a property of infinite sequences only. Yet our experience is restricted necessarily to finite sequences, and it is on these presumably that our informal notion of randomness is based. Accordingly, several people, including Popper [1959a], Kolmogorov [1965], and Martin-Löf [1966], have developed theories of random finite sequences. Of these, Kolmogorov's, in which a finite sequence is random if (roughly) the shortest computer program which reproduces it is as long as the sequence itself, has generated a flourishing mathematical theory. More detailed discussions of Kolmogorov's theory can be found in Earman [1986] and Fine [1973], in which Fine also proves the very interesting result that randomness in Kolmogorov's sense *by itself* implies apparent convergence of relative frequencies ([1965], p. 93, Theorem 2).

While the fruitfulness of theories like Kolmogorov's vindicates a view of randomness as a property of finite sequences, it does not necessarily foreclose others. Indeed, there is a very well-known theory of randomness which has proved of central importance to science, for which von Mises' theory of random collectives is nothing less than a concrete model. This is the statisticians' theory of randomness, expressed in the definition of a random sample, i.e. a sequence of i.i.d. random variables. As we have seen, a consequence of von Mises' definition is that a collective is in effect just a sequence of n-fold random samples in this sense, for all $n \geq 1$. To this extent von Mises' own axiom of randomness is a virtue, not a vice of his theory, enabling it to deliver the weak and other laws of large numbers without additional special hypotheses of independence and constant probability.

3.1.3 Other objections

Ville [1939] showed that there are collectives W, with attribute space $\{0, 1\}$, in which the relative frequency of 1s exceeds 1/2 in every initial segment of W, yet such that the limiting relative frequency of 1 is 1/2. He saw this as conflicting with von Mises' declared objective of so defining a collective that it is demonstrably immune to a gambling system. Yet in the collective above anyone consistently betting on 1 will always remain ahead. So far all this shows, however, is that in some respects von Mises' theory has unintended counterintuitive consequences. So have most formal theories which attempt to formalize a preformal intuitive domain. The important question is whether such consequences are seriously prejudicial to the overall enterprise; if, as it seems to be, this enterprise is to construct a model of random sampling the answer is, provisionally, no.

The infinitistic nature of a von Mises collective has raised worries of a more philosophical character. Those attached to a positivistic epistemology, like de Finetti and Jeffrey, regard the modelling of finite repetitions of an experiment by an infinite collective as wrong in principle, if not meaningless (de Finetti [1974], Jeffrey [1992], p. 11). Against this, however, is the opinion of a leading statistician that *only* 'asymptotic' objective theories can be adequate to the explanatory task (Dawid [1985b], p. 1528). Furthermore, convergence-of-opinion theorems proved within the Bayesian theory show that, despite their infinitary character, von Mises collectives satisfy a criterion of empirical significance, even if it is a fairly minimal one. For example, suppose that P is a countably additive Bayesian probability function defined on a σ-field $\mathbf{F}$ of subsets of the set Ω of all infinite sequences of 0s and 1s. Let H be the subset of Ω consisting of all those sequences which are von Mises collectives with probability p of the attribute 1; H is the extension, in other words, of the hypothesis that the data source generates von Mises collectives in which the attribute 1 has probability p. Let $\mathbf{F}_n$ be the sub σ-field of $\mathbf{F}$ generated by observations of the first n outcomes of the source. Let $I_H(s)$, $s \in \Omega$, be the indicator of H, i.e. the measurable function which takes the value 1 if H is true for s, and 0 if not. According to a well-known theorem (Halmos [1950], p. 215, Theorem B),

$$P(H|F_n)(s) \to I_H(s)$$

for all sequences s in Ω except a set of P-probability 0.

It follows that on pain of inconsistency you must be certain that—i.e. assign probability 1 to the proposition that—your identifications will in the limit prove correct. However, since the set of sequences to which you assign probability 0 can be so large that it includes all but one, the result just stated allows for the possibility of your being mistaken quite a lot of the time (a recent thorough discussion of the significance of such convergence of opinion results can be found in Earman [1992]). A more regulated way of obtaining a posterior probability distribution for H with variable p is to multiply the likelihood of H by the prior probability of H and normalize, according to Bayes' theorem. If the random variable Z_n counts the number of 1s in a sequence s in Ω, and your data are that $Z_n = r$, then there is a coherence argument that the likelihood $P(Z_n = r|H)$ should be set equal to $^nC_r p^r (1 - p)^{n-r}$ (Howson and Urbach [1993], pp. 344-7); this is an instance of what used to be called the Principle of Direct Probability, which Lewis has more recently called the Principal Principle (Lewis [1980]). As n grows large the function of p, $^nC_r p^r (1 - p)^{n-r}$, becomes increasingly peaked at the point $p = r/n$, a feature which, so long as the prior is not extreme, will eventually become reflected in the posterior distribution.

3.2 Dawid's prequential theory

In a series of papers [1982, 1984, 1985a, 1985b], Dawid has attempted to develop a theory of objective probability, which he calls the prequential theory, not vulnerable to objections he regards as especially damaging to existing frequency accounts, particularly von Mises'. Dawid's principal objection to von Mises' theory is that its fundamental datum— the collective—is, he claims, merely the typical data generated by one particular probabilistic model, that of Bernoulli trials with parameter p. Dawid's goal is to construct a criterion that will determine, for *any* probabilistic model M, the class of data sequences which can be said to be explained by M. First I shall give a brief account of Dawid's theory, and then return to consider his charge against von Mises'.

Dawid's criterion is developed for what he calls a probability forecasting system, a concept of great generality which subsumes that of a probabilistic model. Briefly, a probability forecasting system assigns to each in an infinite sequence of two-valued random variables A_i a probability

whose value at each index i may depend not only on i but also on the
observed values of the A_j, $j < i$. These probability forecasts might be made
on the basis of a particular model (in which case the forecasting system is
just the model probability distribution for A_i conditional on the values of
the previous A_j), but equally they might not. The criterion Dawid proposes
is *calibration*. A forecasting system F is calibrated with respect to an
outcome sequence e if in every subsequence s of e selected by a place
selection, the average probability of the first k events in s, as given by F, is
asymptotically equal to the relative frequency of their occurrence.

Dawid had earlier proved [1982] that if Π is a distribution over the
σ-field generated by all finite sequences $(A_1, \ldots, A_n)$, $n \geq 1$, then the
forecasting system F based on Π is calibrated with respect to a set of
outcome sequences having Π probability 1. In other words, any model
explains 'almost all' the outcome sequences it generates; also, a link is
forged, in the limit, between relative frequency and probability, 'almost
always'. A simple example: the generation by the data source of a von
Mises collective with two attributes and probabilities p, 1 − p respectively
is explained by the model of Bernoulli trials with parameter p, since such
collectives are exactly the outcome sequences for which the forecasting
system based on that model is calibrated (and they are a set of probability 1
given that model).

To sum up: models give rise to forecasting systems via which they
explain just those outcomes sequences for which they are calibrated.
This account is very attractive, but it is vulnerable to an important
objection with which we are already familiar. Why are probability 1
characteristics relevant to a discussion of empirical adequacy? What
empirical difference would it make to construct a criterion which is
satisfied only for sets of outcomes of probability .5, or even 0? 'A occurs
with probability 1' sounds as if it ought to be (very) significant, but, as we
know from the earlier discussion of the laws of large numbers (of which
Dawid's theorem is another), by itself it has no empirical significance.
Explicit frequency theories, like von Mises', were constructed precisely to
endow the mathematical theory with empirical significance, and von
Mises' arguably does so when combined with a Bayesian methodology.
Dawid himself advocates a Bayesian methodology ([1985b], p. 127), but
the link between model probabilities and Bayesian probabilities remains to
be established, and the earlier discussion of von Mises' theory suggests that
only a probability defined explicitly in terms of frequencies seems able to
accomplish this.

The charge of restricted applicability, which Dawid brings against von
Mises' theory, must be conceded in some part. But the restriction is no
disadvantage if it establishes von Mises' theory as a model, in the

set-theoretic sense, of the theory of random sampling, i.e. of sequences of independent, identically distributed random variables. Since such sequences are traditionally the theoretical locus for evaluating arbitrary statistical models, a legitimate role for von Mises's theory is that, combined with the Bayesian apparatus for constructing posterior distributions, it provides the final link between model and reality.

3.3 Popper's propensity theory

Popper has raised another interesting objection to von Mises' theory, which is that it does not seem to provide any account of what are called *objective single-case probabilities*. Single-case probabilities are probabilities which can be ascribed to a given event's occurring at a particular spatio-temporal location. Clearly, Bayesians do this as a matter of course, but their probability function is explicitly subjectivistic. Popper's intention was to construct a theory in which an *objective* single-case probability function is defined, and equal to a limiting relative frequency in an infinite sequence.

For this to be possible, Popper argued, certain conditions must be placed on the sequence, conditions which, he claimed, are not met in von Mises' theory. Popper illustrates this claim by means of an example, in which a finite number of throws of a fair die are interpolated into an infinite sequence of outcomes of throws of a biased die [1959b]. While the augmented infinite sequence of outcomes is formally a collective if the sequence of throws of the biased die is, the limiting behaviour in the hybrid sequence will reflect only the throws of the biased die. Hence, Popper concluded, those limiting relative frequencies will not be equal to the probability of any outcome at any given throw of the fair die. Popper proposed to eliminate this defect, as he saw it, of von Mises' theory by defining probabilities as equal to limiting relative frequencies only in infinite sequences of repetitions *of the same experimental conditions*, and he called the resulting theory the propensity theory, because in it probabilities reveal a propensity or disposition of the experimental conditions to generate on repetition its own characteristic limiting relative frequencies (Popper [1959b, 1967]). The hybrid sequence in his example mixes two different set-ups, each generating a quite different relative frequency.

Has Popper identified a genuine defect in von Mises' theory? Von Mises' own writings strongly suggest that the intended model of his theory *was* repeatable trials endowed with dispositions to generate characteristic limiting relative frequencies (see, for example, von Mises [1939], p. 18). Moreover, it is not clear that single-case probabilities *can* be consistently defined in the way Popper suggests. Von Mises actually denied that it made any sense to talk about objective single-case probabilities, and there is a good ground for his scepticism. For a single trial, like the throw of a die, or a woman who has smoked heavily all her adult life and will or will not contract lung cancer in her sixtieth year, or the drawing of a lottery ticket, will not in general instantiate a unique experimental protocol, or, in the older parlance, fall into a unique reference class. On the contrary, such 'chance events' will in general fall into a number of reference classes characterized by quite different statistics. In Popper's own example, the first throw of the fair die will also be a throw of that die in which various other relevant causal parameters take specific values, and there is no reason to suppose that the experimental protocol obtained by fixing all these values will yield a limiting relative frequency of a six equal to one-sixth; it is much more likely to be nearer one or zero.

The problem of the ambiguous reference class indicates that as it stands Popper's theory is incoherent—one and the same event will be assigned different 'single case' probabilities. In contexts where further specification will in general alter the statistics (i.e. long-run frequencies), it seems that either one must either adopt a conventionalistic approach, in which the 'single-case' probabilities are defined relative to the most specific reference class for which sufficiently extensive statistics can be obtained (this seems to be an approach favoured by Fisher [1973], p. 35, and later Giere [1973], p. 481), or else restrict the theory to contexts where there is reason to believe that maximally specific reference classes exist.

Because quantum mechanics appears to be irreducibly indeterministic, its domain is often claimed to provide examples of such maximally specific classes, and Popper tells us that he designed his theory with quantum mechanics as its primary field of application [1967]. He claims that his single-case propensity theory allows a unified interpretation of objective probabilities as measures of the degree of possibility of outcomes relative to the way the set-up is prepared. This view, he believes, dissolves one of the principal philosophical problems of quantum mechanics: explaining what happens in the reduction of superpositions on measurement.

Popper's thesis has been strongly challenged. For example, Milne points out that predicating probabilities of sets of experimental conditions successfully *obstructs* Popper's purpose of explaining how such probabilities interfere in the two-slit experiment (Milne [1985]). The current consensus seems to be that Popper's theory does not throw any light on the Measurement Problem (a dissenting voice is Maxwell's [1988]). Grossman [1972] denies that the conceptual problems of quantum mechanics are symptomatic of a mistaken theory of probability, claiming that they are physical problems which analyses of probability can do nothing to elucidate. As is well known, some people believe that quantum mechanics, if it is interpreted realistically, requires a nonBoolean propositional logic, and hence a nonclassical probability theory (a recent advocate of this view is Hughes [1992], Ch. 8). Other people take the line that the standard mathematical formulation of quantum mechanics is embedded in the theory of Hilbert spaces, which can be and is articulated using classical logic; hence, they infer, quantum-mechanical probabilities, defined in the usual Born manner, can be regarded as ordinary, classical probabilities. A recent persuasive defence of the classical view is van Fraassen ([1991], Ch. 5). This is too extensive and difficult a debate even to attempt to enter into here, and I shall simply refer the reader to the sources cited.

3.4 Chance

While Popper's claims on behalf of his propensity theory seem at the least overambitious, his idea of providing a theory of objective single-case probabilities was taken up by Mellor, Lewis, Skyrms, and others, who have proposed theories of what they call chance (though some of the leading ideas are found in Ramsey [1928] and Jeffreys [1961]). While there are differences between these authors' theories, they share the following principles: (i) chance distributions reflect objective properties of the particular situation at the particular time;' (ii) the chances are chances of outcomes occurring at particular times; (iii) the numerical measure of chance is the degree of belief which it is reasonable to invest in the relevant outcome given the objective constraints; and (iv) where these remain constant from trial to trial there is a high (personal) probability that the chances are approximately equal to the relative frequencies which would be observed in a long sequence of trials.

Granted the usual consistency arguments for fair-betting quotients satisfying the probability axioms, (iii) entails that chances satisfy those axioms (Mellor believes that only the unconditional probability axioms are justified, however; but his argument ([1971], p. 48) is not accepted by more orthodox Bayesians). The authors mentioned differ in the way they justify the other claims listed above. I shall discuss their ideas in turn, starting with Skyrms, postponing a critical evaluation to afterwards.

Skyrms draws on earlier ideas of de Finetti, in particular on de Finetti's theory of exchangeable and partially exchangeable events, and on some results of ergodic theory. De Finetti believed that it was possible to reduce what people thought they were referring to, when they talked about objective probabilities, to personal probability distributions which take into account known or believed causal factors present in the set-up, like a coin's mass distribution, a smoker's physical constitution and age, and so on (de Finetti [1931], discussed in Skyrms [1991]). Personal probability distributions which make the component outcomes *exchangeable* within the possible outcome sequences (i.e. such that if one sequence arises from another by a finite permutation, it has the same probability as the other) reflect a belief that the same relevant causal factors are present at each trial.

De Finetti's well-known representation theorem for infinite sequences of exchangeable events implies that with probability 1 limits of relative frequencies exist (though the implication requires countable additivity which de Finetti always resisted). Exchangeable sequences are stationary sequences, or measure-preserving with respect to shifts, i.e. to transformations that send a sequence s in any set of Ω of singly or doubly infinite sequences to the sequence whose ith member is the i + 1th member of s. If in addition it is supposed that those sets of sequences which are invariant under a shift, for example the sets defined in terms of limit properties, have probability of 1 or 0 only, then the ergodic theorem implies the strong law of large numbers, i.e. that with probability 1 the relative frequency of an outcome is equal to its probability (Billingsley [1965], p. 13). De Finetti's representation theorem exhibits the agent's probability as a mixture of ergodic measures (in fact, Bernoullian measures), which, conditioned on increasing sample data, tends to a particular one in the limit (von Plato [1982] discusses the connection between ergodic theory and foundational theories of probability).

Skyrms believes the ergodic theorem shows how to connect chances with frequencies. He takes an agent's assessment of the chance of an outcome A to be their degree of belief in A conditional on (what is believed to be) the true member of a partition representing the different values of all the relevant causal parameters—the chance-determining parameters. These become the invariant sets under a temporal shift representing a subsequent performance of the experiment under the same relevant conditions, and hence preserving the chance measure. If the relevant chance-determining parameters are completely specified by the partition then there is no invariant set B such that conditioning on B will change the chance measure; in the well-known terminology of Skyrms himself, the chance distribution must be *resilient* to all relevant conditions. Resiliency implies that the invariant sets have (chance) measure 0 or 1 (Skyrms [1984], Ch. 3, [1991]), and the ergodic theorem implies that with probability 1 the limiting relative frequency of an outcome is equal to its chance.

In Lewis's theory as presented in Lewis [1980], chance is a function which assigns a real number in the closed unit interval to a triple consisting of a world, a time, and a proposition. The principal constraint imposed on this function is called by Lewis, appropriately enough, the Principal

Principle. According to this, the probability of A, conditional on the chance of A being r, and any other currently true admissible information about chances, is r, where the probability function is what Lewis calls a 'reasonable' one. We have seen how an analogous principle connects probabilities in von Mises Collectives with personal probabilities.

Granted the validity of the consistency arguments for personal probabilities obeying the probability axioms, the Principal Principle guarantees that chances satisfy them also (Lewis, op. cit., p. 277). The chances of successive events may also be independent, according to Lewis, in which case a posterior distribution for a chance, first derived for the special case of a uniform prior by Bayes [1763], can be constructed using Bayes's Theorem, where the likelihood is evaluated by means of the Principal Principle (for a clear account of how this is done see Earman [1992], Ch. 1). It was pointed out earlier, in the discussion of von Mises' theory, that the likelihood becomes sharply peaked in the neighbourhood of $P = r/n$, where r/n is the relative frequency of the outcome in the sample, and P is the chance (in the von Mises case, the collective's probability), and that as long as the prior density is not extreme, the posterior probability of P lying in an arbitrarily small neighbourhood of the relative frequency tends to 1.

In these theories chances are numerically identical with 'objectified' subjective probabilities (the terminology is Jeffrey's ([1965], Ch. 12.7), which is to say that they are measured by degrees of belief allegedly warranted by appropriate facts about the world. Mellor's theory differs from both Lewis's and Skyrms's in distinctive ways, however. Chances are not identified now with conditional probabilities (conditional degrees of belief), as in the other theories, but with unconditional ones, though the identification of chance with 'reasonable' partial belief enables Mellor to argue in the same sort of way that chances are formally probabilities. He invokes the strong law of large numbers to identity chances with long-run frequencies: according to Mellor, the degree of belief warranted by the circumstances currently prevailing is equal to p if there is reason to believe that were the agent to bet often enough at that betting-quotient they would eventually break even. A form of the strong law of large numbers then allegedly shows that the reasonable degree of belief in an outcome is approximately equal to its frequency in a long sequence of such bets (Mellor [1971], p. 162).

Now to criticism, and we shall proceed in reverse, starting with Mellor. His use of the strong law assumes independence, but he gives no indication of how he would justify that assumption. Indeed, it is difficult to see how it could be justified at all within his theory. He tells us that chance is reasonable personal probability, but what could be more reasonable than adjusting your belief about the 'chance' of heads at the next throw to the number observed so far? Yet if the chance is that degree of belief then this is precisely to deny independence.

For this reason independence is problematic for all these theories of chance, based as they are on personal probabilities. A more serious problem is that *indeterminism* is implicit in the strong resiliency condition all these writers lay on the chance measure, namely that it be the degree of belief warranted by the values of all the relevant structural parameters in the set-up at that time, and is unchanged by further conditioning. If the environment of which that set-up is a part is deterministic, then the only resilient probabilities are clearly 1 and 0. But even if the quantum domain is indeterministic (and there is no logical necessity to this), the non-quantum domain is not obviously so. Mellor concedes that there might be causal chains determining the velocity and position of each molecule of a gas, but denies that this precludes treating the classical distributions of the kinetic theory of gases as chance distributions in his sense, since the distribution of the causal chains is not itself necessarily deterministic (op. cit., p. 154). But this argument seems to be fallacious: even were the distribution indeterministic, the chance that the gas will be at a particular phase point at a particular time, given the position and momentum of each molecule (considered as a classical particle) at any previous time, is still 1 or 0 (even if the equations of motion are not humanly integrable).

The problem at bottom is that chance is really just a 'folk' concept, as Lewis admits in as many words ([1980], p. 269): we talk so often about the chance of this or that happening that unless we are careful we come to believe that our attributions of chances express some objective truth of the matter. Lewis even appears to claim (ibid.) that the belief alone justifies endowing the notion of chance with an objective referent! The lack of an explicit argument in Lewis's paper for the existence of chance seems to leave the Principal Principle with an undetermined parameter, 'the chance of A', as the quantity that is supposed to determine our degrees of belief; but it also suggests that Lewis regards 'the chance of A' as *implicitly defined* by the Principal Principle: the chance of A is that quantity which makes our warranted degree of belief in A constant in just the factual circumstances cited.

But implicit definitions do not always uniquely determine an object satisfying them. A *proof* of existence and uniqueness is generally required, and not only is this lacking here, but there are excellent reasons for thinking that none can be given. For all attempts to show that factual evidence determines a uniquely warranted degree of belief in any uncertain proposition have come to nothing. Carnap's systems of inductive logic

[1950, 1952, 1971, 1980] bear eloquent witness to the hopelessness of the task (see Howson and Urbach [1993], Ch. 4, for an extended discussion of Carnap's programme). The conclusion seems inescapable that statements about chances reduce to being merely statements of personal probability.

But we clearly need a theory of objective probability, and science positively demands one. Chance theorists spurn the suggestion that a frequency theory can provide one. Yet their criticisms are not compelling. Lewis objects on the ground that there is some chance that the chance (of heads on a toss of a coin, say) and the relative frequency in a large number of trials will differ. He infers that chance and frequency cannot be the same. This is fallacious; it certainly can be true that there is a positive chance that the chance of an event and the relative frequency of its occurrence in any finite number of trials differ, where the chance is equal to the long-run relative frequency. It is easily shown to be true in von Mises' theory. Also, as we observed, von Mises' theory can be combined with the personalist Bayesian theory to deliver a promising surrogate for objective single-case probabilities. The other stock objection of chance theorists to frequency theories is that the invocation of infinite sequences of trials is neither necessary nor desirable to underpin the use of objective probabilities in science and elsewhere. I think we see now that, desirable or not, it is probably necessary.

4 Conclusion

It would be foolhardy to predict that philosophical probability has entered a final stable phase; surveys of the field tend to have useful lifetimes of a decade or so, at most two. It would also probably be incorrect to pretend that there is likely in the near future to be any settled consensus as to which interpretations of probability make viable and useful theories, and which are dead ends. Indeed, its very success in being incorporated at a fundamental level into the natural and social sciences, which I take to include the theory of rational behaviour, means that probability is going to be constrained by the way these sciences develop. And as we have learned from Popper, that is highly unpredictable. Perhaps in another twenty years time things will look very different.

Department of Philosophy, Logic and Scientific Method
London School of Economics
Houghton Street
London WC2A 2AE
UK

References

Armendt, B. [1980]: 'Is There a Dutch Book Argument for Probability Kinematics?' *Philosophy of Science*, **47**, pp. 583-9.

Armendt, B. [1994]: 'Dutch Books, Additivity, and Utility Theory', *Philosophical Topics*, forthcoming.

Arrow, K. J. [1982]: 'Risk Perception in Psychology and Economics', *Economic Inquiry*, **20** pp. 1-9.

Bayes, T. [1763]: 'An Essay Towards Solving a Problem in the Doctrine of Chances', *Philosophical Transactions of the Royal Society*, **53**, pp. 370-418.

Billingsley, P. [1965]: *Ergodic Theory and Information*, New York, John Wiley.

Carnap, R. [1950]: *Logical Foundations of Probability*, Chicago, University of Chicago Press.

Carnap, R. [1952]: *The Continuum of Inductive Methods*, Chicago, University of Chicago Press.

Carnap, R. [1971, 1980]: 'A Basic System of Inductive Logic', Parts I and II, *Studies in Inductive Logic and Probability*, Vols. 1 and 2, Vol. 1 edited by R. Carnap and R. C. Jeffrey, Vol. 2 edited by R. C. Jeffrey, Berkeley, University of California Press.

Christensen, D. [1991]: 'Clever Bookies and Coherent Beliefs', *The Philosophical Review*, **100**, pp. 229-47.

Church, A. [1940]: 'On the Concept of a Random Sequence', *Bulletin of the American Mathematical Society*, **46**, pp. 130-5.

Cohen, L. J. [1989]: *An Introduction to the Philosophy of Induction and Probability*, Oxford, The Clarendon Press.

Dawid, A. P. [1982]: 'The Well-Calibrated Bayesian', *Journal of the American Statistical Association*, **77**, pp. 605-10.

Dawid, A. P. [1984]: 'Statistical Theory: The Prequential Approach', *Journal of the Royal Statistical Society*, A, **147**, pp. 278-92.

Dawid, A. P. [1985a]: 'Calibration-Based Empirical Probability', *Annals of Statistics*, **13**, pp. 1251-73.

Dawid, A. P. [1985b]: 'Probability, Symmetry and Frequency', *British Journal for the Philosophy of Science*, **36**, pp. 107-28.

Dawid, A. P., Stone, M., and Zidek, J. V. [1973]: 'Marginalisation Paradoxes in Bayesian and Structural Inference', *Journal of the Royal Statistical Society*, B, pp. 189-223.

de Finetti, B. [1931]: 'Probabilism' (English translation of 'Probabilismo'), *Erkenntnis* [1989], pp. 1-55.

de Finetti, B. [1937]: 'Foresight, Its Logical Laws, Its Subjective Sources', in H. Kyburg and H. Smokler (*eds*), *Studies in Subjective Probability*, New York, John Wiley.

de Finetti, B. [1974]: *Theory of Probability*, New York, John Wiley.

Diaconis, P. and Zabell, S. L. [1982]: 'Updating Subjective Probability', *Journal of the American Statistical Association*, **77**, pp. 822-30.

Dubois, D. and Prade, H. [1989]: 'Fuzzy Sets, Probability and Measurement', *European Journal of Operational Research*, **50**, pp. 135–54.

Earman, J. [1980]: *A Primer on Determinism*, Dordrecht, Reidel.

Earman, J. [1992]: *Bayes or Bust? A Critical Examination of Bayesian Confirmation Theory*, Cambridge, MA, MIT Press.

Fine, T. L. [1973]: *Theories of Probability*, New York, Academic Press.

Fisher, R. A. [1973]: *Statistical Methods and Scientific Inference*, third edition, New York, Hafner Press.

Gärdenfors, P. and Sahlin, N.-E. [1988]: *Decision, Probability, and Utility*, Cambridge, Cambridge University Press.

Giere, R. N. [1973]: 'Objective Single-Case Probabilities and the Foundations of Statistics', in P. Suppes *et al.* (*eds*), *Logic, Methodology and Philosophy of Science IV*, North Holland, Amsterdam, pp. 467–83.

Giere, R. N. [1976]: 'A Laplacean Formal Semantics for Single-Case Propensities', *Journal of Philosophical Logic*, **5**, pp. 321–53.

Gillies, D. A. [1973]: *An Objective Theory of Probability*, London, Methuen.

Good, I. J. [1962]: 'Subjective Probability as the Measure of a Non-Measurable Set', in E. Nagel, P. Suppes, and A. Tarski (*eds*), *Logic, Methodology and Philosophy of Science, Proceedings of the 1960 International Conference*, Stanford, Stanford University Press.

Grossman, N. [1972]: 'Quantum Mechanics and Interpretations of Probability Theory', *Philosophy of Science*, **39**, pp. 451–60.

Hacking, I. [1967]: 'Slightly More Realistic Personal Probability', *Philosophy of Science*, **34**, pp. 311–25.

Halmos, P. [1950]: *Measure Theory*, New York, van Nostrand Reinhold.

Horwich, P. [1982]: *Probability and Evidence*, Cambridge: Cambridge University Press.

Howson, C. [1993]: 'Dutch Books and Consistency', in D. Hull, M. Forbes, and K. Okruhlik (*eds*), *PSA 1992*, Vol. 2, East Lansing MI, Philosophy of Science Association, pp. 161–8.

Howson, C. and Urbach, P. [1993]: *Scientific Reasoning: The Bayesian Approach*, second edition, Chicago, Open Court.

Hughes, R. I. G. [1992]: *The Structure and Interpretation of Quantum Mechanics* (paperback edition), Cambridge, MA, Harvard University Press.

Jaynes, E. T. [1968]: 'Prior Probabilities', *Institute of Electrical and Electronic Engineers Transactions on Systems Science and Cybernetics*, SSC-4, pp. 227–41.

Jaynes, E. T. [1973]: 'The Well-Posed Problem', *Foundations of Physics*, **3**, pp. 477–93.

Jaynes, E. T. [1983]: *Papers on Probability, Statistics and Statistical Physics*, edited by R. Rosenkrantz, Dordrecht, Reidel.

Jeffrey, R. C. [1965]: *The Logic of Decision* (second edition 1983) Chicago, University of Chicago Press.

Jeffrey, R. C. [1992]: *Probability and the Art of Judgment*, Cambridge, Cambridge University Press. .

Jeffreys, H. [1961]: *Theory of Probability*, third edition, Oxford, The Clarendon Press.

Kolmogorov, A. N. [1965]: 'Three Approaches to the Quantitative Definition of Information', *Problems in Information Transmission*, 1, pp. 1–7.

Koopman, B. O. [1940]: 'The Bases of Probability', *Bulletin of the American Mathematical Society*, 46, pp. 763–74.

Kyburg, H. E. [1987]: 'Bayesian and Non-Bayesian Evidential Updating', *Artificial Intelligence*, 31, pp. 271–293.

Laplace, P. S. de [1820]: *Philosophical Essay on Probabilities* (translation of *Essai Philosophique sur les Probabilités*), New York, Dover [1951].

Lewis, D. [1980]: 'A Subjectivist's Guide to Objective Chance', in R. C. Jeffrey (*ed.*), *Studies in Inductive Logic and Probability*, Vol. 2, Berkeley, University of California Press, pp. 263–93.

Lindley, D. V. [1982]: 'Scoring Rules and the Inevitability of Probability', *International Statistical Review*, 50, pp. 1–26.

Martin-Löf, P. [1966]: 'The Definition of Random Sequences', *Information and Control*, 9, pp. 602–19.

Maxwell, N. [1988]: 'Quantum Propensiton Theory: A Testable Resolution of the Wave/Particle Dilemma', *British Journal for the Philosophy of Science*, 39, pp. 1–51.

Mellor, D. H. [1971]: *The Matter of Chance*, Cambridge: Cambridge University Press.

Milne, P. M. [1983]: 'A Note on Scale Invariance', *British Journal for the Philosophy of Science*, 34, pp. 49–55.

Milne, P. M. [1985]: 'Popper, Propensities and the Two-Slit Experiment', *British Journal for the Philosophy of Science*, 36, pp. 66–70.

Milne, P. M. [1993]: 'The Foundations of Probability and Quantum Mechanics', *Journal of Philosophical Logic*, 22, pp. 129–68.

Popper, K. R. [1959a]: *The Logic of Scientific Discovery*, London, Hutchinson.

Popper, K. R. [1959b]: 'The Propensity Interpretation of Probability', *British Journal for the Philosophy of Science*, 10, pp. 25–42.

Popper, K. R. [1967]: 'Quantum Mechanics without "The Observer"', in M. Bunge (*ed.*), *Quantum Theory and Reality*, New York, Springer.

Ramsey, F. P. [1926, 1928]: 'Truth and Probability' (1926), 'Chance' (1928), reprinted in D. H. Mellor (*ed.*), *Philosophical Papers*, Cambridge, Cambridge University Press [1990].

Redhead, M. L. G. [1974]: 'On Neyman's Paradox and the Theory of Statistical Tests', *British Journal for the Philosophy of Science*, 25, pp. 265–71.

Reichenbach, H. [1949]: *The Theory of Probability*, Berkeley, University of California Press.

Rosenkrantz, R. [1977]: *Inference, Method and Decision: Towards a Bayesian Philosophy of Science*, Dordrecht, Reidel.

Rosenkrantz, R. [1981]: *Foundations and Applications of Inductive Probability*, Atascadero, Ridgeview.

212

Salmon, W. C. [1971]: 'Statistical Explanation', *Statistical Explanation and Statistical Relevance*, Pittsburgh, University of Pittsburgh Press.

Savage, L. J. [1954]: *The Foundations of Statistics*, New York, John Wiley.

Savage, L. J. [1971]: 'Elicitation of Personal Probabilities and Expectations', *Journal of the American Statistical Association*, **66**, pp. 783–801.

Schick, F. [1986]: 'Dutch Bookies and Money Pumps', *Journal of Philosophy*, **83**, pp. 112–19.

Shafer, G. [1976]: *A Mathematical Theory of Evidence*, Princeton, Princeton University Press.

Shortliffe, E. H. and Buchanan, B. G. [1975]: 'A Model of Inexact Reasoning in Medicine', *Mathematical Biosciences*, **23**, pp. 351–79.

Skyrms, B. [1984]: *Pragmatics and Empiricism*, New Haven, Yale University Press.

Skyrms, B. [1987]: *Choice and Chance*, Belmont, Wadsworth.

Skyrms, B. [1991]: 'Stability and Chance', in W. Spohn *et al.* (*eds*), *Existence and Explanation*, pp. 149–63.

Smith, C. A. B. [1961]: 'Consistency in Statistical Inference and Decision', *Journal of the Royal Statistical Society*, B, **23**, pp. 1–25.

Spohn, W. [1990]: 'A General Non-Probabilistic Theory of Inductive Reasoning', in R. D. Schachter, T. S. Levitt, J. Lemmer, and N. Kanal (*eds*), *Uncertainty in Artificial Intelligence 4*, Amsterdam, Elsevier.

Teller, P. [1973]: 'Conditionalisation and Observation', *Synthese*, **26**, pp. 218–58.

Van Fraassen, B. C. [1980]: *The Scientific Image*, Oxford: The Clarendon Press.

Van Fraassen, B. C. [1984]: 'Belief and the Will', *Journal of Philosophy*, **81**, pp. 235–56.

Van Fraassen, B. C. [1989]: *Laws and Symmetry*, Oxford: The Clarendon Press.

Van Fraassen, B. C. [1991]: *Quantum Mechanics: An Empiricist View*, Oxford, The Clarendon Press.

Ville, J. [1939]: *Étude Critique de la Notion de Collectif*, Paris, Gauthier-Villars.

Von Mises, R. [1939]: *Probability, Statistics and Truth*, London, George Allen & Unwin.

Von Mises, R. [1964]: *Mathematical Theory of Probability and Statistics*, New York, Academic Press.

Von Plato, J. [1982]: 'The Significance of the Ergodic Decomposition of Stationary Measures for the Interpretation of Probability', *Synthese*, **53**, pp. 419–32.

Wald, A. [1938]: 'Die Widerspruchsfreiheit des Kollektivbegriffes', in *Wald: Selected Papers in Statistics and Probability*, New York, McGraw-Hill [1955].

Walley, P. [1991] *Statistical Reasoning with Imprecise Probabilities*, London, Chapman & Hall.

Williams, P. M. [1978]: 'On a New Theory of Epistemic Probability' (review of G. Shafer, *A Mathematical Theory of Evidence*), *British Journal of the Philosophy of Science*, **29**, pp. 375–87.

Williams, P. M. [1980]: 'Bayesian Conditionalisation and the Principle of Minimum Information', *British Journal for the Philosophy of Science*, **31**, pp. 131–44.

WESLEY SALMON

Vindication of induction [1]

1. The Problem of Induction In this paper I should like to discuss one of the many ways in which philosophers have attempted to deal with Hume's classic problem of the justification of induction—namely, the so-called *pragmatic justification* of induction.[2] I shall not argue against other approaches to the same problem; [3] rather, I shall try to show that this approach can be strengthened by supplementing the arguments which have been previously given by exponents of the pragmatic justification. This aim will be carried out by applying two criteria to three broad classes of inductive rules in order to show that in each of these three classes there is one rule which is superior to all other rules of that class with respect to the purpose the rules of that class are designed to serve. In my opinion, the argument, if correct, constitutes a pragmatic justification or *vindication* [4] of each of the three rules. However, even if the argument does not *justify* these inductive rules, it does, if correct, supply an important elucidation of the nature

[1] The author wishes to express his gratitude to several individuals who have made helpful suggestions and criticisms, in particular, Miss Mary L. Lind and Mr. Keith Lehrer of the Department of Philosophy, and Professor Frank Stewart of the Department of Mathematics, Brown University. The author is especially grateful to Professor Rudolf Carnap for a set of detailed comments which included pointing out a crucial error in a previous version of the paper.
[2] H. Reichenbach has been the chief exponent of this approach. See his *Experience and Prediction*. Chicago: Univ. of Chicago Press, 1938, Section 42. Also *Theory of Probability*. Berkeley: Univ. of California Press, 1949, Section 87.
[3] These reasons are given in W. Salmon, "Should We Attempt to Justify Induction," *Philosophical Studies*, April, 1957.
[4] The term "vindication" is used in the sense introduced by H. Feigl in "De Principiis non Disputandum . . . ?" in Max Black, ed., *Philosophical Analysis*. Ithaca: Cornell Univ. Press, 1950.

245

of the rules of these three broad classes and their relations to the purposes they are designed to serve. The three classes of rules to be considered are the following: [5]

1. *Rules of Predictive Inference:* Rules for inferring from one finite sample to another nonoverlapping finite sample of a population. (Sample-to-sample inference.)
2. *Rules of Inverse Inference:* Rules for inferring from a finite sample to the whole population. If the population is infinite, these are rules for inferring from an initial section of a sequence to the limit of the relative frequency. (Sample-to-population inference.)
3. *Rules of Direct Inference:* Rules for inferring from the whole population to a finite sample of that population. If the population is infinite, these are rules for inferring from the limit to the relative frequency in an initial section. (Population-to-sample inference.)

The two criteria to be applied will be:

1. The criterion of convergence.
2. The criterion of invariance with respect to purely linguistic transformations (hereinafter called "the criterion of linguistic invariance").

The *criterion of convergence* is essentially the criterion involved in Reichenbach's justification of induction. It may be stated roughly as follows:

No inductive rule is acceptable if its persistent use will necessarily lead to inductive inferences which become and remain incorrect even with respect to infinite populations in which relevant limits of relative frequencies exist.

As far as I know, the *criterion of linguistic invariance* has not previously been stated or discussed. It may be formulated roughly as follows:

No inductive rule is acceptable if the results it yields are functions of the arbitrary features of the choice of language.

I hope to illustrate the application of this criterion later in the paper and I shall attempt a more precise formulation of the criterion at the conclusion.

2. Confirmation Functions and the Predictive Inference It is a well-known fact that unrestricted use of the *principle of indifference*

[5] Names of the types of inference are taken from R. Carnap, *Logical Foundations of Probability.* Chicago: Univ. of Chicago Press, 1950, pp. 207–8.

in probability theory leads to paradoxes. The famous Bertrand paradox [6] is illustrated by the following example. Suppose it is known that an automobile travels 1 mile in a time between 1 minute and 2 minutes, and that nothing further is known about the time taken. The principle of indifference would allow us to conclude that there is a probability of 0.5 that the time was between 1 and $1\frac{1}{2}$ minutes and a probability of 0.5 that the time was between $1\frac{1}{2}$ and 2 minutes. The same situation may be approached in a different way. If the time of the trip was between 1 and 2 minutes, it follows that the average speed was between 60 and 30 miles per hour. Another use of the principle of indifference would allow us to infer that there is a probability of 0.5 that the average speed was between 60 and 45 miles per hour and a probability of 0.5 that the average speed was between 45 and 30 miles per hour. This result, however, conflicts with the previous one. Translating times into average speeds, the previous use of the principle of indifference yielded the result that there is a probability of 0.5 that the average speed was between 60 and 40 miles per hour and a probability of 0.5 that the average speed was between 40 and 30 miles per hour. The principle of indifference may be applied to the same problem in two different ways to yield results which contradict each other; there is no rule which tells which application, if any, is correct.

Carnap's theory of probability is designed to avoid this difficulty without entirely relinquishing the principle of indifference. His confirmation function c^* utilizes the principle of indifference, first with respect to structure descriptions, and second with respect to the state descriptions within a given structure description.[7] It seems to me that Carnap's theory gives rise to a difficulty analogous to the Bertrand paradox.

First, let us consider the language RWB which contains the three predicates "red," "white," and "blue," and two individual names "*a*" and "*b*." This language does *not* meet Carnap's requirement that the three predicates be logically independent. In fact, we shall stipulate that they are mutually exclusive and exhaustive within the universe we wish to describe. There are nine possible "states" of this universe:

1.	*Ra.Rb*	6.	*Ra.Bb*
2.	*Wa.Wb*	7.	*Ba.Rb*
3.	*Ba.Bb*	8.	*Wa.Bb*
4.	*Ra.Wb*	9.	*Ba.Wb*
5.	*Wa.Rb*		

---

[6] This paradox is frequently mentioned in contemporary literature, for example, R. von Mises, *Probability, Statistics, and Truth*, 2d English ed. New York: The Macmillan Co., 1958, pp. 77–79.

[7] c^* is defined and discussed in R. Carnap, *op. cit.*, Appendix.

Second, let us introduce three languages which apply to this same universe and which satisfy Carnap's independence requirement. Each of these languages contains the same two individual names "a" and "b." L_1 contains only "R" as a predicate, L_2 contains only "B" as a predicate, and L_3 contains only "C" as a predicate (where "C" means "colored" and is equivalent to "$R \vee B$"). In each of these languages we have four state descriptions and three structure descriptions; each of the state descriptions in each of the languages L_1, L_2, and L_3 corresponds to one or more of the state descriptions of RWB, as follows:

		State Description	*Weight*	*RWB State Descriptions*
L_1:	1.	$Ra.Rb$	$\frac{1}{3}$	1
	2.	$Ra.\sim Rb$	$\frac{1}{6}$	4, 6
	3.	$\sim Ra.Rb$	$\frac{1}{6}$	5, 7
	4.	$\sim Ra.\sim Rb$	$\frac{1}{3}$	2, 3, 8, 9
L_2:	1.	$Ba.Bb$	$\frac{1}{3}$	3
	2.	$Ba.\sim Bb$	$\frac{1}{6}$	7, 9
	3.	$\sim Ba.Bb$	$\frac{1}{6}$	6, 8
	4.	$\sim Ba.\sim Bb$	$\frac{1}{3}$	1, 2, 4, 5
L_3:	1.	$Ca.Cb$	$\frac{1}{3}$	1, 3, 6, 7
	2.	$Ca.\sim Cb$	$\frac{1}{6}$	4, 9
	3.	$\sim Ca.Cb$	$\frac{1}{6}$	5, 8
	4.	$\sim Ca.\sim Cb$	$\frac{1}{3}$	2

The question is: Is it possible to assign weights to the state descriptions in RWB which satisfy the requirements for weighting imposed by Carnap's function c^* as they are given above? The answer is negative. By considering L_1 we see that RWB state description 1 must be weighted $\frac{1}{3}$; looking at L_3 we see that the combined weights of RWB descriptions 1, 3, 6, 7 must be $\frac{1}{3}$. Since weights cannot be negative, the weights of 3, 6, 7 must each be zero. However, we see from L_2 that the weight of RWB description 3 must be $\frac{1}{3}$. We have found a contradiction. This contradiction has arisen by taking a language which is stronger than any of the languages L_1, L_2, L_3, but which is capable of consistently describing possible states of affairs. Any of the languages L_1, L_2, L_3, can be used to describe the same states of affairs, although none of these languages can describe these states of affairs as completely as RWB. Depending on which of the languages L_1, L_2, L_3 is chosen, we arrive at different weightings for the same state description in RWB. This means that the probability on tautological evidence of a given state of affairs changes from language to language. Similar difficulties arise when degree of confirmation on nontautological evidence is considered, but it is not necessary to illustrate this point here because degree of confirmation always depends upon the weighting of the state descriptions. We have seen that this cannot be done uniquely. Therefore the probabilities of the occurrences of objective facts change with the language in which we choose to talk about them. For this reason, Carnap's confirmation function $c*$ violates the *criterion of linguistic invariance*.

I should, perhaps, make it quite clear at this point that the foregoing considerations do not show any internal inconsistency in Carnap's theory. The difficulty has to do only with the adequacy of the theory for dealing with facts. The *RWB* language has been introduced in order to show that L_1, L_2, and L_3 yield serious discrepancies in dealing with facts describable in *RWB*. *RWB* can be regarded as part of a metalanguage containing the semantics of L_1, L_2, and L_3.

The diagnosis of the foregoing difficulties is relatively straightforward, thanks to the admirable clarity with which Carnap has presented his theory. Using $c*$ as the definition of degree of confirmation, the degree of confirmation (upon given evidence) of a hypothesis that an event of a certain type will occur depends upon two things, an empirical factor and a logical factor. The logical factor is determined by the choice of basic predicates of the language in question. The difficulties cited in this paper result from the fact that it is possible to choose different languages with different, but related, basic predicates. These different sets of predicates yield conflicting results with respect to the logical factor which in turn determines in part the degree of confirmation. This shows why the difficulties cited are analogous to the Bertrand paradox. The Bertrand paradox arises because the principle of indifference needs to be supplemented by rules telling which predicates it may be applied to. In Carnap's system, the principle of indifference is restricted to application to a basic set of predicates. But the difficulty reappears because the choice of a language and consequently the choice of basic predicates is largely arbitrary. The fundamental difficulty is this: In Carnap's system arbitrary features of the choice of language influence predictions concerning empirical fact.

So far, I have discussed only $c*$. However, it seems clear that the same fundamental difficulty would arise in connection with any confirmation function or inductive rule which admits the logical factor in the determination of degree of confirmation, for in any such case the arbitrary features of the choice of language will influence prediction of empirical fact. Carnap has shown that all of the inductive methods he discusses in *The Continuum of Inductive Methods*, except the *straight rule*, admit the logical factor.[8] I take it that the foregoing considerations constitute, therefore, an argument in favor of the *straight rule*.[9]

3. Inverse Inference and Reichenbach's Asymptotic Rules Reichenbach's famous pragmatic justification of induction justifies, according to him, an infinite class of "asymptotic rules" from which he selects his *rule of induction* on grounds of *descriptive simplicity*.[10] The consideration of descriptive simplicity is unconvincing because, although there is a sense in which all the asymptotic rules are equivalent in the long run, they diverge completely for any finite amount of evidence.[11] Hence, Reichenbach's justification of induction equally justifies an infinite set of conflicting rules.

The class of asymptotic rules may be characterized as follows. A rule is asymptotic if, and only if, given a sample of the class A consisting of n elements of which m are B, the rule yields $m/n + f$ as the inferred value of the limit of the relative frequency of Bs among As, where f is a function which converges to zero as n goes to infinity. (Reichenbach's *rule of induction* is the asymptotic rule in which f is always zero.)

Reichenbach's pragmatic justification of the asymptotic rules rests upon the following argument: It follows from the definitions of the terms "limit of a sequence" and "asymptotic rule" that all and only the asymptotic rules have the property that, if a sequence of relative frequencies has a limit, then the persistent use of an asymptotic rule will yield inferred values of the limit which become and remain accurate to any predetermined degree. If the sequence has no limit, however, no rule will enable us to determine what it is. Unfortunately, Reichenbach's justification leaves the choice of a rule from the class of asymptotic rules completely arbitrary.

By invoking the *criterion of linguistic invariance* we may show how to make the choice among the infinite class of asymptotic rules. This is done by examining the function f which appears in the definition of the asymptotic rules. To carry out this examination we consider the form the rule must assume if it is to be applied with respect to a set of mutually exclusive and exhaustive properties. This approach is help-

---

[8] See R. Carnap, *The Continuum of Inductive Methods*. Chicago: Univ. of Chicago Press, 1952.

[9] In *The Continuum of Inductive Methods,* Carnap has offered a number of arguments against the *straight rule;* unfortunately, space does not permit a detailed discussion of them in this paper.

[10] H. Reichenbach, *Experience and Prediction, loc. cit.,* Sections 39 and 42; *Theory of Probability, loc. cit.,* Section 87.

[11] This is shown in detail in W. Salmon, "The Predictive Inference," *Philosophy of Science,* April, 1957.

ful because it enables us to utilize certain *normalizing conditions* which we know must hold.

Let $B_1, \ldots, B_k$ be a set of properties which are mutually exclusive and exhaustive within A (the reference class under consideration). For any i ($i = 1, \ldots, k$) let m_i be the number of elements among the first n members of A which have the property B_i; that is, m_i/n is the relative frequency of B_i in a sample containing n members. Because of the mutually exclusive and exhaustive character of the properties B_i we know that every element of the sample must have just one of these properties and, therefore,

$$\sum_{i=1}^{k} \frac{m_i}{n} = 1$$

for each value of n. Furthermore, we know that the sum of the limits of the relative frequencies (if they exist) of the properties B_i and A must equal one; hence, we must impose the condition that

$$\sum_{i=1}^{k} \left(\frac{m_i}{n} + f \right) = 1$$

for each value of n. These are the *normalizing conditions*.

Now, it is easy to see that the function f cannot be a function of n only unless it identically vanishes for all n.[12] If f were a nonvanishing function of n for some n we would have

$$\sum_{i=1}^{k} \left(\frac{m_i}{n} + f \right) = 1 + kf$$

and this would violate the *normalizing conditions*. This shows that if, for some n, f is not identically zero, then, for that n, f must be nonconstant; that is, f must be a function of some number besides n. Let us then consider how, for some fixed n, f might vary.

First, f might be a function of i, the index of the set of predicates. If so, we would have a set of k functions f_i, some of which would differ from others. Our rule would then direct us to infer that the limit of the relative frequency of B_i in A is $m_i/n + f_i$, for each different i. If this were the case, however, our rule would immediately violate the *criterion of linguistic invariance*, for a mere reordering of the predicates

---

[12] For a discussion of the *normalizing conditions* and inductive rules which violate them see W. Salmon, "Regular Rules of Induction," *Philosophical Review*, July, 1956.

would lead to changes in the inferences of the limits of the relative frequencies of the various properties on the same data.

Second, f might be a function of k, the number of mutually exclusive and exhaustive properties in our set. This again, however, will lead to a violation of the *criterion of linguistic invariance*, because it is easy to make an arbitrary change in k. For example, we could define a new pair of mutually exclusive predicates whose disjunction is equivalent to one of the original predicates, thereby increasing k by one and introducing changes in our predictions of the limits of the relative frequencies of the properties with which we are dealing.

It is perhaps worth mentioning explicitly that the term "predicate" is being used as the name of a property. The observed frequency m_i/n of a property is not a function of the name of that property, nor is the limit of the relative frequency. The violations of the *criterion of linguistic invariance* referred to in the preceding two paragraphs consist in making the inference to the limit of the relative frequency depend upon the name of the property in question.

Third, the function f might be a function of the number m_i alone, or—since we are considering n fixed it comes to the same thing—m_i/n alone. It can be shown that if f is a function of m_i/n only, then f must be identically zero.[13] This means that if f is not identically zero, then any rule involving f will violate the *criterion of linguistic invariance*.

Reichenbach's *rule of induction* is the only asymptotic rule for which f is identically zero; hence, it is the only one which is not rejected by the foregoing considerations. The *criterion of convergence* selects the class of asymptotic rules from the class of all rules of inverse inference, and the *criterion of linguistic invariance* selects Reichenbach's *rule of induction* from the class of asymptotic rules.

4. Direct Inference and Short Run Rules Philosophers have long

[13] This is easily proved as follows: First, let $n = k$ and $m_i = 1$ for each i. Then

$$\sum_{i=1}^{k} f\left(\frac{m_i}{n}\right) = \sum_{i=1}^{k} f\left(\frac{1}{n}\right) = 0$$

Hence,

$$f\left(\frac{1}{n}\right) = 0$$

Next, let r be any integer such that $0 \leqq r \leqq n$. Let $k = n - r + 1$, let $m_1 = r$, and let $m_i = 1$ for each $i \neq 1$. Then

$$\sum_{i=1}^{k} f\left(\frac{m_i}{n}\right) = f\left(\frac{r}{n}\right) + \sum_{i=2}^{k} f\left(\frac{1}{n}\right) = f\left(\frac{r}{n}\right) = 0$$

Q.E.D.

realized the difficulties involved in attempting to infer the limit of the relative frequency from the relative frequency in an observed finite initial section of a sequence of empirically given events (the inverse inference). They have not as universally been aware of the difficulties involved in attempting to infer the relative frequency in a finite initial section from a knowledge of the limit of the relative frequency in the sequence (the direct inference). Yet, as C. S. Peirce pointed out, the latter problem is as difficult as, if not more difficult than, the former problem.[14] The latter problem is essentially the problem of the short run; it is the problem of determining how we may with justification apply knowledge of long-run probabilities in the short run.

In an earlier paper I attempted to give a pragmatic justification of a short-run rule analogous to Reichenbach's rule of induction by an argument analogous to his justification of induction.[15] The analogy was so strong that I succeeded, at best, in equally justifying an infinite class of asymptotic rules.[16] I do believe, however, that the argument in that paper was sufficient to eliminate all nonasymptotic rules of direct (short-run) inference.

The rule of inference I attempted to justify may be roughly stated as follows:

Infer that the relative frequency in the short run approximates the limit of the relative frequency as nearly as possible.

Let us call this rule the *short-run rule*. It may be more precisely formulated:

If p is the limit of the relative frequency of B in the sequence A, then infer that the relative frequency of B in the first n As in m/n, where m is an integer chosen so that the difference between m/n and p is as small as possible. (If two integers are available which make this difference equally small, the choice between them is arbitrary.)

The class of asymptotic rules may be characterized as follows:

A rule is asymptotic if, and only if, given that p is the limit of the relative frequency of Bs among As, the rule yields m/n as the value of the relative frequency of Bs among the first n As, where m is an integer chosen so that the difference between m/n and $p + f$ is as small as possible and where f is a function which converges to zero as n goes to infinity.

My *short-run rule* is the asymptotic rule for which the function f is identically zero.

[14] Charles Hartshorne and Paul Weiss, eds., *Collected Papers of Charles Sanders Peirce*. Cambridge: Harvard Univ. Press, 1931–35, 2.652.
[15] W. Salmon, "The Short Run," *Philosophy of Science*, July, 1955.
[16] W. Salmon, "The Predictive Inference," *loc. cit.*

All and only the asymptotic rules satisfy the *criterion of convergence*, for if we persistently use an asymptotic rule to infer relative frequencies in longer and longer initial sections, we shall eventually arrive at inferences which are accurate to any desired degree. Nonasymptotic rules do not have this property. The justification for preferring asymptotic to nonasymptotic rules lies in the fact that for any predetermined degree of accuracy, we do not know how long our short run must be in order that our asymptotic rule yield a value accurate to that degree. With respect to a given sequence of events, we do not know that any value yielded by the rule will be inaccurate—that is, a short run containing only one member may be sufficiently long. If we choose a nonasymptotic rule, however, we know that, for some degrees of accuracy and some lengths of initial section, it will yield inaccurate results. In other words if we use an asymptotic rule in connection with a given sequence, it may yield accurate results for all short runs in that sequence and it must yield accurate results for some; if we use a nonasymptotic rule, it may yield inaccurate results for all short runs but it must yield inaccurate results for some.[17]

The same kind of argument that was used in the previous section to select Reichenbach's *rule of induction* can be used to select my *short-run rule* from the class of asymptotic rules of direct inference. As before, in order to make use of *normalizing conditions*, we consider the application of the rules with reference to sets of mutually exclusive and exhaustive properties.

Let $B_1, \ldots, B_k$ be a set of properties which are mutually exclusive and exhaustive within A. For any i $(i = 1, \ldots, k)$ let p_i be the limit of the relative frequency of B_i in A. As *normalizing conditions* we have

$$\sum_{i=1}^{k} p_i = 1 \text{ and } \sum_{i=1}^{k} (p_i + f) = 1$$

As before, we point out that if f is not identically zero for all n, then f is not a function of n alone. If it were, we would have for some n,

$$\sum_{i=1}^{k} (p_i + f) = 1 + kf$$

which would violate the *normalizing conditions*.

Once more, if f is not identically zero, we choose a value of n for

---

[17] This argument is given more fully in W. Salmon, "The Short Run," *loc. cit.*

which it does not vanish and examine the manner in which it varies. As before, we see that if f is a function of i or k, the rule embodying f would violate the *criterion of linguistic invariance*. In addition, it can be shown that if, for any given n, f is a function of p_i only, then f must be identically zero.[18] It follows in this case as in the former that the only acceptable rule in the class of asymptotic rules is that in which the function f is identically zero. Therefore the *criterion of linguistic invariance* rejects all asymptotic rules of direct inference except the *short run rule*.

5. Vindication of Inductive Rules   I have attempted to show how the two criteria may be applied to three kinds of inductive rules. Carnap's confirmation functions serve as rules of predictive inference, among other things. Our considerations have shown that the *straight rule* is the preferred rule for this purpose. Reichenbach's *rule of induction* has been selected as the preferred rule of inverse inference. My *short-run rule* has been taken as the preferred rule of direct inference. There is a second manner of handling the predictive inference. A combination of the *rule of induction* and the *short-run rule* accomplishes a predictive inference. From a finite sample we may infer the limit of the relative frequency by the *rule of induction*, and from the limit of the relative frequency we may infer the relative frequency in another finite sample by the *short-run rule*. There is no cause for alarm here, for the combined use of the *rule of induction* and the *short-run rule* yields precisely the same result as does the use of the *straight rule*. In view of this fact, we may be content to discuss rules of inverse inference and rules of direct inference only.

Application of the two criteria give what I believe to be a vindication of induction. It will be recalled that a vindication, in the sense explained by Feigl, consists of an argument to show that a rule is better suited to fulfill the purposes it is designed to serve than is any other rule.[19] I think we may regard the ascertainment of limits of relative frequencies and the ascertainment of short-run frequencies as worthy enterprises. Since relative frequencies and their limits (if they exist)

[18] This may be proved for all rational values of p_i by the proof given in footnote 13, regarding p_i as a ratio m_i/n. I do not know whether the normalizing conditions are sufficient to prove the statement for all real values of p_i. However, if we add the condition (which seems legitimate) that $f(p_i)$ be continuous, the desired result follows immediately. Were we to deny continuity we would have the consequence that from arbitrarily close pairs of values of the limit of the relative frequency we would have to infer pairs of values of the short-run frequency which are not arbitrarily close. This would seem to be unreasonable.

[19] H. Feigl, *loc. cit.*

are objective facts and not linguistic conventions, it seems clear that any rule which makes inferences to relative frequencies and their limits functions of linguistic conventions is ill suited to the tasks at hand. In other words, if the task to be accomplished is the prediction of objective fact, then we do not want to adopt a rule which reflects the arbitrary features of the choice of language in its results. These considerations indicate briefly my reasons for regarding the *criterion of linguistic invariance* as an appropriate component of a vindication of inductive rules. The reasons for considering the *criterion of convergence* as another component have been sketched earlier in this paper and they have been given more fully elsewhere.[20]

It is not my purpose to argue that there are no other purposes of scientific inference besides those which can be reduced to inferring relative frequencies and their limits. Thus, I do not mean to say that no other rules of inductive (nondemonstrative) inference can be acceptable. I do not believe, however, that a successful vindication of any inductive (nondemonstrative) rule has previously been given, and I hope this paper provides a successful vindication of the three selected rules. If this hope is realized, I believe it constitutes some progress in dealing with the problem of induction Hume raised.

6. The Criterion of Linguistic Invariance   I now wish to give a more precise formulation of the *criterion of linguistic invariance*. It will be given in two parts, one for application to confirmation functions, the other for application to inductive rules.

1. Let $c(h/e)$ be a confirmation function defined for languages $L_1, \ldots, L_r$. Let h_i and e_i be two sentences of L_i ($1 \leq i \leq r$) such that $c(h_i/e_i) = c_1$ in L_i. To satisfy the criterion of linguistic invariance the confirmation function $c(h/e)$ must be such that it is impossible, using the values of $c(h/e)$ for any sentences h and e in any of the languages $L_1, \ldots, L_r$ and using the semantical and syntactical rules of $L_1, \ldots, L_r$, to prove that $c(h_i/e_i) = c_2$, where $c_1 \neq c_2$.

2. Let $e(A_i, B_i)$ be an inferred value according to a rule R of either a short-run frequency or a limit of a sequence of relative frequencies. Let A_j and B_j be terms of any language such that, according to the semantical and syntactical rules of the languages containing A_i, A_j, B_i, and B_j, $A_i = A_j$ and $B_i = B_j$. To satisfy the *criterion of linguistic invariance*, R must be such that $e(A_i, B_i) = e(A_j, B_j)$.

[20] H. Reichenbach, *Experience and Prediction* and *Theory of Probability;* W. Salmon, "Should We Attempt to Justify Induction?" and "The Short Run," *loc. cit.*

II. THE 'JUSTIFICATION' OF INDUCTION

7. We have seen something, then, of the nature of inductive reasoning; of how one statement or set of statements may support another statement, S, which they do not entail, with varying degrees of strength, ranging from being conclusive evidence for S to being only slender evidence for it; from making S as certain as the supporting statements, to giving it some slight probability. We have seen, too, how the question of degree of support is complicated by consideration of relative frequencies and numerical chances.

There is, however, a residual philosophical question which enters so largely into discussion of the subject that it must be discussed. It can be raised, roughly, in the following forms. What reason have we to place reliance on inductive procedures? Why should we suppose that the accumulation of instances of *A*s which are *B*s, however various the conditions in which they are observed, gives any good reason for expecting the next *A* we encounter to be a *B*? It is our habit to form expectations in this way; but can the habit be rationally justified? When this doubt has entered our minds it may be difficult to free ourselves from it. For the doubt has its source in a confusion; and some attempts to resolve the doubt preserve the confusion; and other attempts to show that the doubt is senseless seem altogether too facile. The root-confusion is easily described; but simply to describe it seems an inadequate remedy against it. So the doubt must be examined again and again, in the light of different attempts to remove it.

If someone asked what grounds there were for supposing that deductive reasoning was valid, we might answer that there were in fact no grounds for supposing that deductive reasoning was always valid; sometimes people made valid inferences, and sometimes they were guilty of logical fallacies. If he said that we had misunderstood his question, and that what he wanted to know was what grounds there were for regarding deduction *in general* as a valid method of argument, we should have to answer that his question was without sense, for to say that an argument, or a form or method of argument, was valid or invalid would *imply* that it was deductive; the concepts of validity and invalidity had application only to individual deductive arguments or forms of deductive argument. Similarly, if a man asked what grounds there were for thinking it reasonable to hold beliefs arrived at inductively, one might at first answer that there were good and bad inductive arguments, that sometimes it was reasonable to hold a belief arrived at inductively and sometimes it was not. If he, too, said that his question had been misunderstood, that he wanted to know whether induction in general was a reasonable method of inference, then we might well think his question senseless in the same way as the question whether deduction is in general valid; for to call a particular belief reasonable or unreasonable is to apply inductive standards, just as to call a particular argument valid or invalid is to apply deductive standards. The parallel is not wholly convincing;

R 2

for words like ' reasonable ' and ' rational ' have not so precise and technical a sense as the word ' valid '. Yet it is sufficiently powerful to make us wonder how the second question could be raised at all, to wonder why, in contrast with the corresponding question about deduction, it should have seemed to constitute a genuine problem.

Suppose that a man is brought up to regard formal logic as the study of the science and art of reasoning. He observes that all inductive processes are, by deductive standards, invalid; the premises never entail the conclusions. Now inductive processes are notoriously important in the formation of beliefs and expectations about everything which lies beyond the observation of available witnesses. But an *invalid* argument is an *unsound* argument; an *unsound* argument is one in which *no good reason* is produced for accepting the conclusion. So if inductive processes are invalid, if all the arguments we should produce, if challenged, in support of our beliefs about what lies beyond the observation of available witnesses are unsound, then we have no good reason for any of these beliefs. This conclusion is repugnant. So there arises the demand for a justification, not of this or that particular belief which goes beyond what is entailed by our evidence, but a justification of induction in general. And when the demand arises in this way it is, in effect, the demand that induction shall be shown to be really a kind of deduction; for nothing less will satisfy the doubter when this is the route to his doubts.

Tracing this, the most common route to the general doubt about the reasonableness of induction, shows how the doubt seems to escape the absurdity of a demand that induction in general shall be justified by inductive standards. The demand is that induction should be shown to be a rational process; and this turns out to be the demand that one kind of reasoning should be shown to be another and different kind. Put thus crudely, the demand seems to escape one absurdity only to fall into another. Of course, inductive arguments are not deductively valid; if they were, they would be deductive arguments. Inductive reasoning must be assessed, for soundness, by inductive standards. Nevertheless, fantastic as the wish for induction to be deduction may seem, it is only in terms of it that we can understand some of the attempts that have been made to justify induction.

8. The first kind of attempt I shall consider might be called the search for the supreme premise of inductions. In its primitive form it is quite a crude attempt; and I shall make it cruder by caricature. We have already seen that for a particular inductive step, such as ' The kettle has been on the fire for ten minutes, so it will be boiling by now ', we can substitute a deductive argument by introducing a generalization (e.g., ' A kettle always boils within ten minutes of being put on the fire ') as an additional premise. This manœuvre shifted the emphasis of the problem of inductive support on to the question of how we established such generalizations as these, which rested on grounds by which they were not entailed. But suppose the manœuvre could be repeated. Suppose we could find one supremely general proposition, which taken in conjunction with the evidence for any accepted generalization of science or daily life (or at least of science) would entail that generalization. Then, so long as the status of the supreme generalization could be satisfactorily explained, we could regard all sound inductions to unqualified general conclusions as, at bottom, valid deductions. The justification would be found, for at least these cases. The most obvious difficulty in this suggestion is that of formulating the supreme general proposition in such a way that it shall be precise enough to yield the desired entailments, and yet not obviously false or arbitrary. Consider, for example, the formula : ' For all f, g, wherever n cases of $f.g$, and no cases of $f. \sim g$, are observed, then all cases of f are cases of g.' To turn it into a sentence, we have only to replace ' n ' by some number. But what number? If we take the value of ' n ' to be 1 or 20 or 500, the resulting statement is obviously false. Moreover, the choice of any number would seem quite arbitrary ; there is no privileged number of favourable instances which we take as decisive in establishing a generalization. If, on the other hand, we phrase the proposition vaguely enough to escape these objections—if, for example, we phrase it as ' Nature is uniform '—then it becomes too vague to provide the desired entailments. It should be noticed that the impossibility of framing a general proposition of the kind required is really a special case of the impossibility of framing precise rules for the assessment of evidence. If we could frame a rule which would tell us precisely when we had *conclusive* evidence for a generaliza-

tion, then it would yield just the proposition required as the supreme premise.

Even if these difficulties could be met, the question of the status of the supreme premise would remain. How, if a non-necessary proposition, could it be established? The appeal to experience, to inductive support, is clearly barred on pain of circularity. If, on the other hand, it were a necessary truth and possessed, in conjunction with the evidence for a generaliza-tion, the required logical power to entail the generalization (e.g., if the latter were the conclusion of a hypothetical syllogism, of which the hypothetical premise was the necessary truth in ques-tion), then the evidence would entail the generalization indepen-dently, and the problem would not arise : a conclusion unbear-ably paradoxical. In practice, the extreme vagueness with which candidates for the role of supreme premise are expressed prevents their acquiring such logical power, and at the same time renders it very difficult to classify them as analytic or synthetic : under pressure they may tend to tautology; and, when the pressure is removed, assume an expansively synthetic air.

In theories of the kind which I have here caricatured the ideal of deduction is not usually so blatantly manifest as I have made it. One finds the ' Law of the Uniformity of Nature ' presented less as the suppressed premise of crypto-deductive inferences than as, say, the ' presupposition of the validity of inductive reasoning '. I shall have more to say about this in my last section.

9. I shall next consider a more sophisticated kind of attempt to justify induction : more sophisticated both in its interpreta-tion of this aim and in the method adopted to achieve it. The aim envisaged is that of proving that the probability of a gener-alization, whether universal or proportional, increases with the number of instances for which it is found to hold. This clearly is a realistic aim : for the proposition to be proved does state, as we have already seen, a fundamental feature of our criteria for assessing the strength of evidence. The method of proof proposed is mathematical. Use is to be made of the arithmetical calculation of chances. This, however, seems less realistic : for we have already seen that the prospect of analysing the notion of support in these terms seems poor.

I state the argument as simply as possible; but, even so, it will be necessary to introduce and explain some new terms. Suppose we had a collection of objects of different kinds, some with some characteristics and some with others. Suppose, for example, we had a bag containing 100 balls, of which 70 were white and 30 black. Let us call such a collection of objects a *population*; and let us call the way it is made up (e.g., in the case imagined, of 70 white and 30 black balls) the *constitution* of the population. From such a population it would be possible to take *samples* of various sizes. For example, we might take from our bag a sample of 30 balls. Suppose each ball in the bag had an individual number. Then the collection of balls numbered 10 to 39 inclusive would be one sample of the given size; the collection of balls numbered 11 to 40 inclusive would be another and different sample of the same size; the collection of balls numbered 2, 4, 6, 8 . . . 58, 60 would be another such sample; and so on. Each possible collection of 30 balls is a different sample of the same size. Some different samples of the same size will have the same constitutions as one another; others will have different constitutions. Thus there will be only one sample made up of 30 black balls. There will be many different samples which share the constitution : 20 white and 10 black. It would be a simple matter of mathematics to work out the number of possible samples of the given size which had any one possible constitution. Let us say that a sample *matches* the population if, allowing for the difference between them in size, the constitution of the sample corresponds, within certain limits, to that of the population. For example, we might say that any possible sample consisting of, say, 21 white and 9 black balls matched the constitution (70 white and 30 black) of the population, whereas a sample consisting of 20 white and 10 black balls did not. Now it is a proposition of pure mathematics that, given any population, the proportion of possible samples, all of the same size, which match the population, increases with the size of the sample.

We have seen that conclusions about the ratio of a subset of equally possible chances to the whole set of those chances may be expressed by the use of the word ' probability '. Thus of the 52 possible samples of one card from a population constituted like an orthodox pack, 16 are court-cards or aces. This fact we allow ourselves to express (under the conditions, inductively established, of equipossibility of draws) by saying that the probability of drawing a court-card or an ace was $\frac{4}{13}$. If we express the proposition referred to at the end of the last paragraph by means of this use of ' probability ' we shall obtain the result: The probability of a sample matching a given population increases with the size of the sample. It is tempting to try to derive from this result a general justification of the inductive procedure : which will not, indeed, show that any given inductive conclusion is entailed by the evidence for it, taken in conjunction with some universal premise, but will show that the multiplication of favourable instances of a generalization entails a proportionate increase in its probability. For, since *matching* is a symmetrical relation, it might seem a simple deductive step to move from

I. The probability of a sample matching a given population increases with the size of the sample

to

II. The probability of a population matching a given sample increases with the size of the sample.

II might seem to provide a guarantee that the greater the number of cases for which a generalization is observed to hold, the greater is its probability; since in increasing the number of cases we increase the size of the sample from whatever population forms the subject of our generalization. Thus pure mathematics might seem to provide the sought-for proof that the evidence for a generalization really does get stronger, the more favourable instances of it we find.

The argument is ingenious enough to be worthy of respect; but it fails of its purpose, and misrepresents the inductive situation. Our situation is not in the least like that of a man drawing a sample from a given, i.e., fixed and limited, population from which the drawing of any mathematically possible sample is equiprobable with that of any other. Our only datum is the sample. No limit is fixed beforehand to the diversity, and the possibilities of change, of the ' population ' from which it is drawn : or, better, to the multiplicity and variousness of different populations, each with different constitutions, any one of which might replace the present one before we make the next draw. Nor is there any *a priori* guarantee that different mathematically possible samples are equally likely to be drawn. If we have or can obtain any assurance on these points, then it is assurance derived inductively from our data, and cannot therefore be assumed at the outset of an argument designed to justify induction. So II, regarded as a justification of induction founded on purely mathematical considerations, is a fraud. The important shift of ' given ' from qualifying ' population ' in I to qualifying ' sample ' in II is illegitimate. Moreover, ' probability ', which means one thing in II (interpreted as giving the required guarantee) means something quite different in I (interpreted as a proposition of pure mathematics). In I probability is simply the measure of the ratio of one set of mathematically possible chances to another; in II it is the measure of the inductive acceptability of a generalization. As a mathematical proposition, I is certainly independent of the soundness of inductive procedures; and as a statement of one of the criteria we use in assessing the strength of evidence of a generalization, II is as certainly independent of mathematics.

It has not escaped the notice of those who have advocated a mathematical justification of induction, that certain assumptions are required to make the argument even seem to fulfil its purpose. Inductive reasoning would be of little use if it did not sometimes enable us to assign at least fairly high probabilities to certain conclusions. Now suppose, in conformity with the mathematical model, we represented the fact that the evidence for a proposition was conclusive by assigning to it the probability figure of 1; and the fact that the evidence for and against a proposition was evenly balanced by assigning to it the probability figure $\frac{1}{2}$; and so on. It is a familiar mathematical truth that, between any two fractions, say $\frac{1}{4}$ and $\frac{1}{6}$, there is an infinite number of intermediate quantities; that $\frac{1}{8}$ can be indefinitely increased without reaching equality to $\frac{1}{4}$. Even if we could regard II as mathematically established, therefore, it fails to give us what we require; for it fails to provide a guarantee that the probability of an inductive conclusion ever attains a degree at which it begins to be of use. It was accordingly necessary to buttress the purely mathematical argument by large, vague assumptions, comparable with the principles designed for the role of supreme premise in the first type of attempt. These

assumptions, like those principles, could never actually be used
to give a deductive turn to inductive arguments; for they could
not be formulated with precision. They were the shadows of
precise unknown truths, which, if one did know them, would
suffice, along with the data for our accepted generalizations, to
enable the probability of the latter to be assigned, after calculation, a precise numerical fraction of a tolerable size. So this
theory represents our inductions as the vague sublunary
shadows of deductive calculations which we cannot make.

10. Let us turn from attempts to justify induction to attempts
to show that the demand for a justification is mistaken. We
have seen already that what lies behind such a demand is often
the absurd wish that induction should be shown to be some kind
of deduction—and this wish is clearly traceable in the two
attempts at justification which we have examined. What
other sense could we give to the demand? Sometimes it is
expressed in the form of a request for proof that induction is a
reasonable or *rational* procedure, that we have *good grounds* for
placing reliance upon it. Consider the uses of the phrases
‘ good grounds ’, ‘ justification ’, ‘ reasonable ’, &c. Often we
say such things as ‘ He has *every justification* for believing that
p ’; ‘ I have *very good reasons* for believing it ’; ‘ There are
good grounds for the view that *q* ’; ‘ There is *good evidence* that
r ’. We often talk, in such ways as these, of justification, good
grounds or reasons or evidence for certain beliefs. Suppose
such a belief were one expressible in the form ‘ Every case of *f*
is a case of *g* ’. And suppose someone were asked what he
meant by saying that he had good grounds or reasons for holding
it. I think it would be felt to be a satisfactory answer if he
replied : ‘ Well, in all my wide and varied experience I’ve come
across innumerable cases of *f* and never a case of *f* which wasn’t
a case of *g*.’ In saying this, he is clearly claiming to have
inductive support, *inductive* evidence, of a certain kind, for his
belief; and he is also giving a perfectly proper answer to the
question, what he meant by saying that he had ample justification, good grounds, good reasons for his belief. It is an analytic
proposition that it is reasonable to have a degree of belief in a
statement which is proportional to the strength of the evidence
in its favour; and it is an analytic proposition, though not a

proposition of mathematics, that, other things being equal,[1] the evidence for a generalization is strong in proportion as the number of favourable instances, and the variety of circumstances in which they have been found, is great. So to ask whether it is reasonable to place reliance on inductive procedures is like asking whether it is reasonable to proportion the degree of one's convictions to the strength of the evidence. Doing this is what ' being reasonable ' *means* in such a context.

As for the other form in which the doubt may be expressed, viz., ' Is induction a justified, or justifiable, procedure? ', it emerges in a still less favourable light. No sense has been given to it, though it is easy to see why it seems to have a sense. For it is generally proper to inquire *of a particular belief*, whether its adoption is justified; and, in asking this, we are asking whether there is good, bad, or any, evidence for it. In applying or withholding the epithets ' justified ', ' well founded ', &c., in the case of specific beliefs, we are appealing to, and applying, inductive standards. But to what standards are we appealing when we ask whether the application of inductive standards is justified or well grounded? If we cannot answer, then no sense has been given to the question. Compare it with the question : Is the law legal? It makes perfectly good sense to inquire of a particular action, of an administrative regulation, or even, in the case of some states, of a particular enactment of the legislature, whether or not it is legal. The question is answered by an appeal to a legal system, by the application of a set of legal (or constitutional) rules or standards. But it makes no sense to inquire in general whether the law of the land, the legal system as a whole, is or is not legal. For to what legal standards are we appealing?

The only way in which a sense might be given to the question, whether induction is in general a justified or justifiable procedure, is a trival one which we have already noticed. We might interpret it to mean ' Are all conclusions, arrived at inductively, justified? ', i.e., ' Do people always have adequate evidence for the conclusions they draw? ' The answer to this question is easy, but uninteresting : it is that sometimes people have adequate evidence, and sometimes they do not.

[1] This phrase embodies the large abstractions referred to in Sections 5 and 6.

11. It seems, however, that this way of showing the request for a general justification of induction to be absurd is sometimes insufficient to allay the worry that produces it. And to point out that ' forming rational opinions about the unobserved on the evidence available ' and ' assessing the evidence by inductive standards ' are phrases which describe the same thing, is more apt to produce irritation than relief. The point is felt to be ' merely a verbal ' one; and though the point of this protest is itself hard to see, it is clear that something more is required. So the question must be pursued further. First, I want to point out that there is something a little odd about talking of ' the inductive method ', or even ' the inductive policy ', as if it were just one possible method among others of arguing from the observed to the unobserved, from the available evidence to the facts in question. If one asked a meteorologist what method or methods he used to forecast the weather, one would be surprised if he answered : ' Oh, just the inductive method.' If one asked a doctor by what means he diagnosed a certain disease, the answer ' By induction ' would be felt as an impatient evasion, a joke, or a rebuke. The answer one hopes for is an account of the tests made, the signs taken account of, the rules and recipes and general laws applied. When such a specific method of prediction or diagnosis is in question, one can ask whether the method is justified in practice; and here again one is asking whether its employment is inductively justified, whether it commonly gives correct results. This question would normally seem an admissible one. One might be tempted to conclude that, while there are many different specific methods of prediction, diagnosis, &c., appropriate to different subjects of inquiry, all such methods could properly be called ' inductive ' in the sense that their employment rested on inductive support; and that, hence, the phrase ' non-inductive method of finding out about what lies deductively beyond the evidence ' was a description without meaning, a phrase to which no sense had been given; so that there could be no question of justifying our selection of one method, called ' the inductive ', of doing this.

However, someone might object : ' Surely it is possible, though it might be foolish, to use methods utterly different from accredited scientific ones. Suppose a man, whenever he wanted to form an opinion about what lay beyond his observation or the observation of available witnesses, simply shut his eyes, asked himself the appropriate question, and accepted the first answer that came into his head. Wouldn't this be a non-inductive method?' Well, let us suppose this. The man is asked : 'Do you usually get the right answer by your method?' He might answer : 'You've mentioned one of its drawbacks ; I never do get the right answer; but it's an extremely easy method.' One might then be inclined to think that it was not a method of finding things out at all. But suppose he answered : Yes, it's usually (always) the right answer. Then we might be willing to call it a method of finding out, though a strange one. But, then, by the very fact of its success, it would be an inductively supported method. For each application of the method would be an application of the general rule, 'The first answer that comes into my head is generally (always) the right one'; and for the truth of this generalization there would be the inductive evidence of a long run of favourable instances with no unfavourable ones (if it were ' always '), or of a sustained high proportion of successes to trials (if it were ' generally ').

So every successful method or recipe for finding out about the unobserved must be one which has inductive support; for to say that a recipe is successful is to say that it has been repeatedly applied with success; and repeated successful application of a recipe constitutes just what we mean by inductive evidence in its favour. Pointing out this fact must not be confused with saying that ' the inductive method ' is justified by its success, justified because it works. This is a mistake, and an important one. I am not seeking to ' justify the inductive method ', for no meaning has been given to this phrase. *A fortiori*, I am not saying that induction is justified by its success in finding out about the unobserved. I am saying, rather, that any successsful method of finding out about the unobserved is necessarily justified by induction. This is an analytic proposition. The phrase ' successful method of finding things out which has no inductive support ' is self-contradictory. Having, or acquiring, inductive support is a necessary condition of the success of a method.

Why point this out at all? First, it may have a certain therapeutic force, a power to reassure. Second, it may counteract the tendency to think of ' the inductive method ' as some-

thing on a par with specific methods of diagnosis or prediction
and therefore, like them, standing in need of (inductive) justification.

12. There is one further confusion, perhaps the most powerful
of all in producing the doubts, questions, and spurious solutions
discussed in this Part. We may approach it by considering the
claim that induction is justified by its success in practice. The
phrase ' success of induction ' is by no means clear and perhaps
embodies the confusion of induction with some specific method
of prediction, &c., appropriate to some particular line of inquiry.
But, whatever the phrase may mean, the claim has an obviously
circular look. Presumably the suggestion is that we should
argue from the past ' successes of induction ' to the continuance
of those successes in the future; from the fact that it has
worked hitherto to the conclusion that it will continue to work.
Since an argument of this kind is plainly inductive, it will not
serve as a justification of induction. One cannot establish a
principle of argument by an argument which uses that principle.
But let us go a little deeper. The argument rests the justification of induction on a matter of fact (its ' past successes ').
This is characteristic of nearly all attempts to find a justification.
The desired premise of Section 8 was to be some fact about the
constitution of the universe which, even if it could not be used
as a suppressed premise to give inductive arguments a deductive
turn, was at any rate a ' presupposition of the validity of induction '. Even the mathematical argument of Section 9
required buttressing with some large assumption about the make-
up of the world. I think the source of this general desire to find
out some fact about the constitution of the universe which will
' justify induction ' or ' show it to be a rational policy ' is the
confusion, the running together, of two fundamentally different
questions : to one of which the answer is a matter of non-
linguistic fact, while to the other it is a matter of meanings.
     There is nothing self-contradictory in supposing that all the
uniformities in the course of things that we have hitherto observed and come to count on should cease to operate to-morrow;
that all our familiar recipes should let us down, and that we
should be unable to frame new ones because such regularities
as there were were too complex for us to make out. (We may

assume that even the expectation that all of us, in such circumstances, would perish, were falsified by someone surviving to observe the new chaos in which, roughly speaking, nothing foreseeable happens.) Of course, we do not believe that this will happen. We believe, on the contrary, that our inductively supported expectation-rules, though some of them will have, no doubt, to be dropped or modified, will continue, on the whole, to serve us fairly well; and that we shall generally be able to replace the rules we abandon with others similarly arrived at. We might give a sense to the phrase ' success of induction ' by calling this vague belief the belief that induction will continue to be successful. It is certainly a factual belief, not a necessary truth; a belief, one may say, about the constitution of the universe. We might express it as follows, choosing a phraseology which will serve the better to expose the confusion I wish to expose :

 I. (The universe is such that) induction will continue to be successful.

I is very vague : it amounts to saying that there are, and will continue to be, natural uniformities and regularities which exhibit a humanly manageable degree of simplicity. But, though it is vague, certain definite things can be said about it. (1) It is not a necessary, but a contingent, statement; for chaos is not a self-contradictory concept. (2) We have good inductive reasons for believing it, good inductive evidence for it. We believe that some of our recipes will continue to hold good because they have held good for so long. We believe that we shall be able to frame new and useful ones, because we have been able to do so repeatedly in the past. Of course, it would be absurd to try to use I to ' justify induction ', to show that it is a reasonable policy; because I is a conclusion inductively supported.

 Consider now the fundamentally different statement :

 II. Induction is rational (reasonable).

We have already seen that the rationality of induction, unlike its ' successfulness ', is not a fact about the constitution of the world. It is a matter of what we mean by the word ' rational ' in its application to any procedure for forming opinions about what lies outside our observations or that of available witnesses. For to have good reasons for any such opinion is to have good inductive support for it. The chaotic universe just envisaged, therefore, is not one in which induction would cease to be rational; it is simply one in which it would be impossible to form rational expectations to the effect that specific things would happen. It might be said that in such a universe it would at least be rational to refrain from forming specific expectations, to expect nothing but irregularities. Just so. But this is itself a higher-order induction : where irregularity is the rule, expect further irregularities. Learning not to count on things is as much learning an inductive lesson as learning what things to count on.

So it is a contingent, factual matter that it is sometimes possible to form rational opinions concerning what specifically happened or will happen in given circumstances (I); it is a non-contingent, *a priori* matter that the only ways of doing this must be inductive ways (II). What people have done is to run together, to conflate, the question to which I is answer and the quite different question to which II is an answer; producing the muddled and senseless questions : ' Is the universe such that inductive procedures are rational ? ' or ' What must the universe be like in order for inductive procedures to be rational ? ' It is the attempt to answer these confused questions which leads to statements like ' The uniformity of nature is a pre-supposition of the validity of induction '. The statement that nature is uniform might be taken to be a vague way of expressing what we expressed by I; and certainly this fact is a condition of, for it is identical with, the likewise contingent fact that we are, and shall continue to be, able to form rational opinions, of the kind we are most anxious to form, about the unobserved. But neither this fact about the world, nor any other, is a condition of the necessary truth that, if it is possible to form rational opinions of this kind, these will be inductively supported opinions. The discordance of the conflated questions manifests itself in an uncertainty about the status to be accorded to the alleged presupposition of the ' validity ' of induction. For it was dimly, and correctly, felt that the reasonableness of inductive procedures was not merely a contingent, but a necessary, matter; so any necessary condition of their reasonableness had likewise to be a necessary matter.   On the other hand, it was uncomfortably clear that chaos is not a self-contradictory concept; that the fact that some phenomena do exhibit a tolerable degree of simplicity and repetitiveness is not guaranteed by logic, but is a contingent affair.   So the presupposition of induction had to be both contingent and necessary : which is absurd.   And the absurdity is only lightly veiled by the use of the phrase 'synthetic *a priori*' instead of 'contingent necessary'.

VOL. LIV.   No. 213.]   [January, 1945.

MIND

A QUARTERLY REVIEW

OF

PSYCHOLOGY AND PHILOSOPHY

———

I.—STUDIES IN THE LOGIC OF CONFIRMATION (I.).

To the memory of my wife, Eva Ahrends Hempel.

BY CARL G. HEMPEL.

1. *Objective of the Study.*[1]—The defining characteristic of an empirical statement is its capability of being tested by a confrontation with experimental findings, *i.e.* with the results of suitable experiments or " focussed " observations. This feature distinguishes statements which have empirical content both from the statements of the formal sciences, logic and mathematics, which require no experiential test for their validation, and from

---

[1] The present analysis of confirmation was to a large extent suggested and stimulated by a co-operative study of certain more general problems which were raised by Dr. Paul Oppenheim, and which I have been investigating with him for several years. These problems concern the form and the function of scientific laws and the comparative methodology of the different branches of empirical science. The discussion with Mr. Oppenheim of these issues suggested to me the central problem of the present essay. The more comprehensive problems just referred to will be dealt with by Mr. Oppenheim in a publication which he is now preparing.

In my occupation with the logical aspects of confirmation, I have benefited greatly by discussions with several students of logic, including Professor R. Carnap, Professor A. Tarski, and particularly Dr. Nelson Goodman, to whom I am indebted for several valuable suggestions which will be indicated subsequently.

A detailed exposition of the more technical aspects of the analysis of confirmation presented in this article is included in my article " A Purely Syntactical Definition of Confirmation ", *The Journal of Symbolic Logic*, vol. 8 (1943).

1

the formulations of transempirical metaphysics, which do not admit of any

The testability here referred to has to be understood in the comprehensive sense of " testability in principle " ; there are many empirical statements which, for practical reasons, cannot be actually tested at present. To call a statement of this kind testable in principle means that it is possible to state just what experiential findings, if they were actually obtained, would constitute favourable evidence for it, and what findings or " data ", as we shall say for brevity, would constitute unfavourable evidence ; in other words, a statement is called testable in principle, if it is possible to describe the kind of data which would confirm or disconfirm it.

The concepts of confirmation and of disconfirmation as here understood are clearly more comprehensive than those of conclusive verification and falsification. Thus, e.g. no finite amount of experiential evidence can conclusively verify a hypothesis expressing a general law such as the law of gravitation, which covers an infinity of potential instances, many of which belong either to the as yet inaccessible future, or to the irretrievable past ; but a finite set of relevant data may well be " in accord with " the hypothesis and thus constitute confirming evidence for it. Similarly, an existential hypothesis, asserting, say, the existence of an as yet unknown chemical element with certain specified characteristics, cannot be conclusively proved false by a finite amount of evidence which fails to " bear out " the hypothesis ; but such unfavourable data may, under certain conditions, be considered as weakening the hypothesis in question, or as constituting disconfirming evidence for it.[1]

While, in the practice of scientific research, judgments as to the confirming or disconfirming character of experiential data obtained in the test of a hypothesis are often made without hesitation and with a wide consensus of opinion, it can hardly be said that these judgments are based on an explicit theory providing general criteria of confirmation and of disconfirmation. In this respect, the situation is comparable to the manner in which deductive inferences are carried out in the practice of scientific research : This, too, is often done without reference to an explicitly stated system of rules of logical inference. But while criteria of valid deduction can be and have been supplied by

---

[1] This point as well as the possibility of conclusive verification and conclusive falsification will be discussed in some detail in section 10 of the present paper.

formal logic, no satisfactory theory providing general criteria of confirmation and disconfirmation appears to be available so far.

In the present essay, an attempt will be made to provide the elements of a theory of this kind. After a brief survey of the significance and the present status of the problem, I propose to present a detailed critical analysis of some common conceptions of confirmation and disconfirmation and then to construct explicit definitions for these concepts and to formulate some basic principles of what might be called the logic of confirmation.

2. *Significance and Present Status of the Problem.*—The establishment of a general theory of confirmation may well be regarded as one of the most urgent desiderata of the present methodology of empirical science.[1] Indeed, it seems that a precise analysis of the concept of confirmation is a necessary condition for an adequate solution of various fundamental problems concerning the logical structure of scientific procedure. Let us briefly survey the most outstanding of these problems.

(*a*) In the discussion of scientific method, the concept of relevant evidence plays an important part. And while certain " inductivist " accounts of scientific procedure seem to assume that relevant evidence, or relevant data, can be collected in the context of an inquiry prior to the formulation of any hypothesis, it should be clear upon brief reflection that relevance is a relative concept ; experiential data can be said to be relevant or irrelevant only with respect to a given hypothesis ; and it is the hypothesis which determines what kind of data or evidence are relevant for it. Indeed, an empirical finding is relevant for a hypothesis if and only if it constitutes either favourable or unfavourable evidence for it ; in other words, if it either confirms or disconfirms the hypothesis. Thus, a precise definition of relevance presupposes an analysis of confirmation and disconfirmation.

(*b*) A closely related concept is that of instance of a hypothesis. The so-called method of inductive inference is usually presented as proceeding from specific cases to a general hypothesis of which each of the special cases is an " instance " in the sense that it " conforms to " the general hypothesis in question, and thus constitutes confirming evidence for it.

Thus, any discussion of induction which refers to the establishment of general hypotheses on the strength of particular instances is fraught with all those logical difficulties—soon to be expounded

---

[1] Or of the " logic of science ", as understood by R. Carnap ; *cf. The Logical Syntax of Language* (New York and London, 1937), sect. 72, and the supplementary remarks in *Introduction to Semantics* (Cambridge, Mass., 1942), p. 250.

4   CARL G. HEMPEL :

—which beset the concept of confirmation. A precise analysis of this concept is, therefore, a necessary condition for a clear statement of the issues involved in the problem complex of induction and of the ideas suggested for their solution—no matter what their theoretical merits or demerits may be.

(c) Another issue customarily connected with the study of scientific method is the quest for " rules of induction ". Generally speaking, such rules would enable us to " infer ", from a given set of data, that hypothesis or generalization which accounts best for all the particular data in the given set. Recent logical analyses have made it increasingly clear that this way of conceiving the problem involves a misconception : While the process of invention by which scientific discoveries are made is as a rule *psychologically guided and stimulated* by antecedent knowledge of specific facts, its results are *not logically determined* by them ; the way in which scientific hypotheses or theories are discovered cannot be mirrored in a set of general rules of inductive inference.[1] One of the crucial considerations which lead to this conclusion is the following : Take a scientific theory such as the atomic theory of matter. The evidence on which it rests may be described in terms referring to directly observable phenomena, namely to certain " macroscopic " aspects of the various experimental and observational data which are relevant to the theory. On the other hand, the theory itself contains a large number of highly abstract, non-observational terms such as " atom ", " electron ", " nucleus ", " dissociation ", " valence " and others, none of which figures in the description of the observational data. An adequate rule of induction would therefore have to provide, for this and for every conceivable other case, mechanically applicable criteria determining unambiguously, and without any reliance on the inventiveness or additional scientific knowledge of its user, all those new abstract concepts which need to be created for the formulation of the theory that will account for the given evidence. Clearly, this requirement cannot be satisfied by any set of rules, however ingeniously devised ; there can be no general rules of induction in the above sense ; the demand for them rests on a confusion of logical and psychological issues. What determines the soundness of a hypothesis is not the way it

[1] See the lucid presentation of this point in Karl Popper's *Logik der Forschung* (Wien, 1935), esp. sect. 1, 2, 3, and 25, 26, 27 ; *cf.* also Albert Einstein's remarks in his lecture *On the Method of Theoretical Physics* (Oxford, 1933,) pp. 11 and 12. Also of interest in this context is the critical discussion of induction by H. Feigl in " The Logical Character of the Principle of Induction, " *Philosophy of Science,* vol. 1 (1934).

is arrived at (it may even have been suggested by a dream or a hallucination), but the way it stands up when tested, *i.e.* when confronted with relevant observational data. Accordingly, the quest for rules of induction in the original sense of canons of scientific discovery has to be replaced, in the logic of science, by the quest for general objective criteria determining (A) whether, and—if possible—even (B) to what degree, a hypothesis H may be said to be corroborated by a given body of evidence E. This approach differs essentially from the inductivist conception of the problem in that it presupposes not only E, but also H as given and then seeks to determine a certain logical relationship between them. The two parts of this latter problem can be restated in somewhat more precise terms as follows :

(A) To give precise definitions of the two non-quantitative relational concepts of confirmation and of disconfirmation ; *i.e.* to define the meaning of the phrases " E confirms H " and " E disconfirms H ". (When E neither confirms nor disconfirms H, we shall say that E is neutral, or irrelevant, with respect to H.)

(B) (1) To lay down criteria defining a metrical concept " degree of confirmation of H with respect to E ", whose values are real numbers ; or, failing this,

(2) To lay down criteria defining two relational concepts, "more highly confirmed than " and " equally well confirmed with ", which make possible a non-metrical comparison of hypotheses (each with a body of evidence assigned to it) with respect to the extent of their confirmation.

Interestingly, problem B has received much more attention in methodological research than problem A ; in particular, the various theories of the " probability of hypotheses " may be regarded as concerning this problem complex ; we have here adopted [1] the more neutral term " degree of confirmation " instead of " probability " because the latter is used in science in a definite technical sense involving reference to the relative frequency of the occurrence of a given event in a sequence, and it is at least an open question whether the degree of confirmation of a hypothesis can generally be defined as a probability in this statistical sense.

The theories dealing with the probability of hypotheses fall into two main groups : the " logical " theories construe probability as a logical relation between sentences (or propositions ;

[1] Following R. Carnap's usage in Testability and Meaning, *Philosophy of Science*, vols. 3 (1936) and 4 (1937) ; esp. sect. 3 (in vol. 3).

6 CARL G. HEMPEL :

it is not always clear which is meant)[1] ; the " statistical " theories
interpret the probability of a hypothesis in substance as the
limit of the relative frequency of its confirming instances among
all relevant cases.[2] Now it is a remarkable fact that none of
the theories of the first type which have been developed so far
provides an explicit general definition of the probability (or
degree of confirmation) of a hypothesis H with respect to a body
of evidence E ; they all limit themselves essentially to the construction of an uninterpreted postulational system of logical
probability. For this reason, these theories fail to provide a
complete solution of problem B. The statistical approach, on
the other hand, would, if successful, provide an explicit numerical
definition of the degree of confirmation of a hypothesis ; this
definition would be formulated in terms of the numbers of confirming and disconfirming instances for H which constitute the
body of evidence E. Thus, a necessary condition for an adequate
interpretation of degrees of confirmation as statistical probabilities
is the establishment of precise criteria of confirmation and disconfirmation, in other words, the solution of problem A.

However, despite their great ingenuity and suggestiveness, the
attempts which have been made so far to formulate a precise
statistical definition of the degree of confirmation of a hypothesis
seem open to certain objections,[3] and several authors [4] have
expressed doubts as to the possibility of defining the degree of
confirmation of a hypothesis as a metrical magnitude, though

[1] This group includes the work of such writers as Janina Hosiasson-Lindenbaum (cf. for instance, her article "Induction et analogie: Comparàison
de leur fondement ", MIND, vol. L (1941) ; also see p. 21, n. 2), H.
Jeffreys, J. M. Keynes, B. O. Koopman, J. Nicod (sèe p. 9, n. 2),
St. Mazurkiewicz, F. Waismann. For a brief discussion of this conception
of probability, see Ernest Nagel, *Principles of the Theory of Probability*
(Internat. Encyclopedia of Unified Science, vol. i, no. 6, Chicago, 1939),
esp. sects. 6 and 8.
[2] The chief proponent of this view is Hans Reichenbach ; cf. especially
Ueber Induktion und Wahrscheinlichkeit, *Erkenntnis*, vol. v (1935), and
Experience and Prediction (Chicago, 1938), Ch. V.
[3] Cf. Karl Popper, *Logik der Forschung* (Wien, 1935), sect. 80 ; Ernest
Nagel, *l.c.*, sect. 8, and " Probability and the Theory of Knowledge ",
Philosophy of Science, vol. 6 (1939); C. G. Hempel, " Le problème de la
vérité ", *Theoria* (Göteborg), vol. 3 (1937), sect. 5, and " On the Logical
Form of Probability Statements ", *Erkenntnis*, vol. 7 (1937-38), esp. sect.
5. Cf. also Morton White, " Probability and Confirmation ", *The Journal
of Philosophy*, vol. 36 (1939).
[4] See, for example, J. M. Keynes, *A Treatise on Probability*, London,
1929, esp. Ch. III ; Ernest Nagel, *Principles of the Theory of Probability*
(cf. n. 1 above), esp. p. 70 ; compare also the somewhat less definitely
sceptical statement by Carnap, *l.c.* (see p. 5, n. 1), sect. 3, p. 427.

some of them consider it as possible, under certain conditions, to solve at least the less exacting problem B (2), *i.e.* to establish standards of non-metrical comparison between hypotheses with respect to the extent of their confirmation. An adequate comparison of this kind might have to take into account a variety of different factors[1]; but again the numbers of the confirming and of the disconfirming instances which the given evidence includes will be among the most important of those factors.

Thus, of the two problems, A and B, the former appears to be the more basic one, first, because it does not presuppose the possibility of defining numerical degrees of confirmation or of comparing different hypotheses as to the extent of their confirmation; and second because our considerations indicate that any attempt to solve problem B—unless it is to remain in the stage of an axiomatized system without interpretation—is likely to require a precise definition of the concepts of confirming and disconfirming instance of a hypothesis before it can proceed to define numerical degrees of confirmation, or to lay down non-metrical standards of comparison.

(*d*) It is now clear that an analysis of confirmation is of fundamental importance also for the study of the central problem of what is customarily called epistemology; this problem may be characterized as the elaboration of " standards of rational belief " or of criteria of warranted assertibility. In the methodology of empirical science this problem is usually phrased as concerning the rules governing the test and the subsequent acceptance or rejection of empirical hypotheses on the basis of experimental or observational findings, while in its " epistemological " version the issue is often formulated as concerning the validation of beliefs by reference to perceptions, sense data, or the like. But no matter how the final empirical evidence is construed and in what terms it is accordingly expressed, the theoretical problem remains the same : to characterize, in precise and general terms, the conditions under which a body of evidence can be said to confirm, or to disconfirm, a hypothesis of empirical character ; and that is again our problem A.

(*e*) The same problem arises when one attempts to give a precise statement of the empiricist and operationalist criteria for the empirical meaningfulness of a sentence ; these criteria, as is well known, are formulated by reference to the theoretical

---

[1] See especially the survey of such factors given by Ernest Nagel in *Principles of the Theory of Probability (cf.* p. 6, n. 1), pp. 66-73.

8 CARL G. HEMPEL:

testability of the sentence by means of experiential evidence [1] :
and the concept of theoretical testability, as was pointed out
earlier, is closely related to the concepts of confirmation and disconfirmation.[2]

Considering the great importance of the concept of confirmation,
it is surprising that no systematic theory of the non-quantitative
relation of confirmation seems to have been developed so far.
Perhaps this fact reflects the tacit assumption that the concepts
of confirmation and of disconfirmation have a sufficiently clear
meaning to make explicit definitions unnecessary or at least
comparatively trivial. And indeed, as will be shown below, there
are certain features which are rather generally associated with
the intuitive notion of confirming evidence, and which, at first,
seem well suited to serve as defining characteristics of confirmation. Closer examination will reveal the definitions thus obtainable to be seriously deficient and will make it clear that an
adequate definition of confirmation involves considerable difficulties.

Now the very existence of such difficulties suggests the question
whether the problem we are considering does not rest on a false
assumption : Perhaps there are no objective criteria of confirmation ; perhaps the decision as to whether a given hypothesis is
acceptable in the light of a given body of evidence is no more
subject to rational, objective rules than is the process of inventing
a scientific hypothesis or theory ; perhaps, in the last analysis,
it is a " sense of evidence ", or a feeling of plausibility in view of
the relevant data, which ultimately decides whether a hypothesis
is scientifically acceptable.[3] This view is comparable to the
opinion that the validity of a mathematical proof or of a logical
argument has to be judged ultimately by reference to a feeling
of soundness or convincingness ; and both theses have to be
rejected on analogous grounds : They involve a confusion of
logical and psychological considerations. Clearly, the occurrence

[1] Cf., for example, A. J. Ayer, *Language, Truth and Logic*, London and
New York, 1936, Ch. I ; R. Carnap, " Testability and Meaning " (cf.
p. 5, n. 1) sects. 1, 2, 3 ; H. Feigl, *Logical Empiricism* (in *Twentieth
Century Philosophy*, ed. by Dagobert D. Runes, New York, 1943) ;
P. W. Bridgman, *The Logic of Modern Physics*, New York, 1928.
[2] It should be noted, however, that in his essay "Testability and Meaning "
(cf. p. 5, n. 1) R. Carnap has constructed definitions of testability and
confirmability which avoid reference to the concept of confirming and of
disconfirming evidence ; in fact, no proposal for the definition of these
latter concepts is made in that study.
[3] A view of this kind has been expressed, for example, by M. Mandelbaum
in " Causal Analyses in History ", *Journal of the History of Ideas*, vol. 3
(1942) ; cf. esp. pp. 46-47.

or non-occurrence of a feeling of conviction upon the presentation of grounds for an assertion is a subjective matter which varies from person to person, and with the same person in the course of time ; it is often deceptive, and can certainly serve neither as a necessary nor as a sufficient condition for the soundness of the given assertion.[1] A rational reconstruction of the standards of scientific validation cannot, therefore, involve reference to a sense of evidence ; it has to be based on objective criteria. In fact, it seems reasonable to require that the criteria of empirical confirmation, besides being objective in character, should contain no reference to the specific subject-matter of the hypothesis or of the evidence in question ; it ought to be possible, one feels, to set up purely formal criteria of confirmation in a manner similar to that in which deductive logic provides purely formal criteria for the validity of deductive inferences.

With this goal in mind, we now turn to a study of the non-quantitative concept of confirmation. We shall begin by examining some current conceptions of confirmation and exhibiting their logical and methodological inadequacies ; in the course of this analysis, we shall develop a set of conditions for the adequacy of any proposed definition of confirmation ; and finally, we shall construct a definition of confirmation which satisfies those general standards of adequacy.

3. *Nicod's Criterion of Confirmation and its Shortcomings.*—We consider first a conception of confirmation which underlies many recent studies of induction and of scientific method. A very explicit statement of this conception has been given by Jean Nicod in the following passage : " Consider the formula or the law : *A entails B*. How can a particular proposition, or more briefly, a fact, affect its probability ? If this fact consists of the presence of B in a case of A, it is favourable to the law ' *A entails B* ' ; on the contrary, if it consists of the absence of B in a case of A, it is unfavourable to this law. It is conceivable that we have here the only two direct modes in which a fact can influence the probability of a law. . . . Thus, the entire influence of particular truths or facts on the probability of universal propositions or laws would operate by means of these two elementary relations which we shall call *confirmation* and *invalidation*." [2] Note that the applicability of this criterion is restricted to hypotheses of

[1] See Karl Popper's pertinent statement, *l.c.*, sect. 8.
[2] Jean Nicod, *Foundations of Geometry and Induction* (transl. by P. P. Wiener), London, 1930 ; p. 219 ; *cf.* also R. M. Eaton's discussion of " Confirmation and Infirmation ", which is based on Nicod's views ; it is included in Ch. III of his *General Logic*, New York, 1931.

the form " *A entails B* ". Any hypothesis *H* of this kind may be expressed in the notation of symbolic logic [1] by means of a universal conditional sentence, such as, in the simplest case,

$$(x)(P(x) \supset Q(x)),$$

i.e. " For any object *x* : if *x* is a *P*, then *x* is a *Q*," or also " Occurrence of the quality *P* entails occurrence of the quality *Q*." According to the above criterion this hypothesis is confirmed by an object *a*, if *a* is *P* and *Q* ; and the hypothesis is disconfirmed by *a* if *a* is *P*, but not *Q*. In other words, an object confirms a universal conditional hypothesis if and only if it satisfies both the antecedent (here : ' *P(x)* ') and the consequent (here : ' *Q(x)* ') of the conditional ; it disconfirms the hypothesis if and only if it satisfies the antecedent, but not the consequent of the conditional ; and (we add this to Nicod's statement) it is neutral, or irrelevant, with respect to the hypothesis if it does not satisfy the antecedent.

This criterion can readily be extended so as to be applicable also to universal conditionals containing more than one quantifier, such as " Twins always resemble each other ", or, in symbolic notation, ' $(x)(y)(\text{Twins}(x, y) \supset \text{Rsbl}(x, y))$ '. In these cases, a confirming instance consists of an ordered couple, or triple, etc., of objects satisfying the antecedent and the consequent of the conditional. (In the case of the last illustration, any two persons who are twins and resemble each other would confirm the hypothesis ; twins who do not resemble each other would disconfirm it ; and any two persons not twins—no matter whether they resemble each other or not—would constitute irrelevant evidence.)

We shall refer to this criterion as Nicod's criterion.[2] It states explicitly what is perhaps the most common tacit interpretation of the concept of confirmation. While seemingly quite adequate, it suffers from serious shortcomings, as will now be shown.

(*a*) First, the applicability of this criterion is restricted to hypotheses of universal conditional form; it provides no standards of confirmation for existential hypotheses (such as " There exists organic life on other stars ", or " Poliomyelitis is caused by some virus ") or for hypotheses whose explicit formulation calls for the use of both universal and existential quantifiers (such as

---

[1] In this paper, only the most elementary devices of this notation are used ; the symbolism is essentially that of *Principia Mathematica*, except that parentheses are used instead of dots, and that existential quantification is symbolized by ' (E) ' instead of by the inverted ' E '.

[2] This term is chosen for convenience, and in view of the above explicit formulation given by Nicod ; it is not, of course, intended to imply that this conception of confirmation originated with Nicod.

" Every human being dies some finite number of years after his birth ", or the psychological hypothesis, " You can fool all of the people some of the time and some of the people all of the time, but you cannot fool all of the people all of the time ", which may be symbolized by ' $(x)(Et)$Fl(x, t) . $(Ex)(t)$Fl(x, t) . $\sim (x)(t)$Fl(x, t) ', (where ' Fl(x, t) ' stands for " You can fool (person) x at time t "). We note, therefore, the desideratum of establishing a criterion of confirmation which is applicable to hypotheses of any form.[1]

(b) We now turn to a second shortcoming of Nicod's criterion. Consider the two sentences

$$S_1 : \text{ ' } (x)(\text{Raven}(x) \supset \text{Black}(x)) \text{ ' ;}$$
$$S_2 : \text{ ' } (x)(\sim \text{Black}(x) \supset \sim \text{Raven}(x)) \text{ '}$$

(i.e. " All ravens are black " and " Whatever is not black is not a raven "), and let a, b, c, d be four objects such that a is a raven and black, b a raven but not black, c not a raven but black, and d neither a raven nor black. Then, according to Nicod's criterion, a would confirm S_1, but be neutral with respect to S_2 ; b would disconfirm both S_1 and S_2 ; c would be neutral with respect to both S_1 and S_2, and d would confirm S_2, but be neutral with respect to S_1.

But S_1 and S_2 are logically equivalent ; they have the same content, they are different formulations of the same hypothesis. And yet, by Nicod's criterion, either of the objects a and d would be confirming for one of the two sentences, but neutral with respect to the other. This means that Nicod's criterion makes confirmation depend not only on the content of the hypothesis, but also on its formulation.[2]

One remarkable consequence of this situation is that every hypothesis to which the criterion is applicable—i.e. every universal conditional—can be stated in a form for which there cannot possibly exist any confirming instances. Thus, e.g. the sentence

$$(x)[(\text{Raven}(x) . \sim \text{Black}(x)) \supset (\text{Raven}(x) . \sim \text{Raven}(x)]$$

is readily recognized as equivalent to both S_1 and S_2 above ; yet no object whatever can confirm this sentence, i.e. satisfy both

---

[1] For a rigorous formulation of the problem, it is necessary first to lay down assumptions as to the means of expression and the logical structure of the language in which the hypotheses are supposed to be formulated ; the desideratum then calls for a definition of confirmation applicable to any hypothesis which can be expressed in the given language. Generally speaking, the problem becomes increasingly difficult with increasing richness and complexity of the assumed " language of science ".

[2] This difficulty was pointed out, in substance, in my article " Le problème de la vérité ", *Theoria* (Göteborg), vol. 3 (1937), esp. p. 222.

its antecedent and its consequent ; for the consequent is contradictory. An analogous transformation is, of course, applicable to any other sentence of universal conditional form.

4. *The Equivalence Condition.*—The results just obtained call attention to a condition which an adequately defined concept of confirmation should satisfy, and in the light of which Nicod's criterion has to be rejected as inadequate : *Equivalence condition :* Whatever confirms (disconfirms) one of two equivalent sentences, also confirms (disconfirms) the other.

Fulfilment of this condition makes the confirmation of a hypothesis independent of the way in which it is formulated ; and no doubt it will be conceded that this is a necessary condition for the adequacy of any proposed criterion of confirmation. Otherwise, the question as to whether certain data confirm a given hypothesis would have to be answered by saying : " That depends on which of the different equivalent formulations of the hypothesis is considered "—which appears absurd. Furthermore—and this is a more important point than an appeal to a feeling of absurdity —an adequate definition of confirmation will have to do justice to the way in which empirical hypotheses function in theoretical scientific contexts such as explanations and predictions ; but when hypotheses are used for purposes of explanation or prediction,[1] they serve as premises in a deductive argument whose conclusion is a description of the event to be explained or predicted. The deduction is governed by the principles of formal logic, and according to the latter, a deduction which is valid will remain so if some or all of the premises are replaced by different, but equivalent statements ; and indeed, a scientist will feel free, in any theoretical reasoning involving certain hypotheses, to use the latter in whichever of their equivalent formulations is most convenient for the development of his conclusions. But if we adopted a concept of confirmation which did not satisfy the equivalence condition, then it would be possible, and indeed necessary, to argue in certain cases that it was sound scientific procedure to base a prediction on a given hypothesis if formulated in a sentence S_1, because a good deal of confirming evidence had

[1] For a more detailed account of the logical structure of scientific explanation and prediction, *cf.* C. G. Hempel, " The Function of General Laws in History ", *The Journal of Philosophy*, vol. 39 (1942), esp. sects. 2, 3, 4. The characterization, given in that paper as well as in the above text, of explanations and predictions as arguments of a deductive logical structure, embodies an over-simplification : as will be shown in sect. 7 of the present essay, explanations and predictions often involve " quasi-inductive " steps besides deductive ones. This point, however, does not affect the validity of the above argument.

STUDIES IN THE LOGIC OF CONFIRMATION. 13

been found for S_1; but that it was altogether inadmissible to base the prediction (say, for convenience of deduction) on an equivalent formulation S_2, because no confirming evidence for S_2 was available. Thus, the equivalence condition has to be regarded as a necessary condition for the adequacy of any definition of confirmation.

5. *The " Paradoxes " of Confirmation.*—Perhaps we seem to have been labouring the obvious in stressing the necessity of satisfying the equivalence condition. This impression is likely to vanish upon consideration of certain consequences which derive from a combination of the equivalence condition with a most natural and plausible assumption concerning a sufficient condition of confirmation.

The essence of the criticism we have levelled so far against Nicod's criterion is that it certainly cannot serve as a necessary condition of confirmation ; thus, in the illustration given in the beginning of section 3, the object a confirms S_1 and should therefore also be considered as confirming S_2, while according to Nicod's criterion it is not. Satisfaction of the latter is therefore not a necessary condition for confirming evidence.

On the other hand, Nicod's criterion might still be considered as stating a particularly obvious and important sufficient condition of confirmation. And indeed, if we restrict ourselves to universal conditional hypotheses in one variable [1]—such as S_1

---

[1] This restriction is essential : In its general form, which applies to universal conditionals in any number of variables, Nicod's criterion cannot even be construed as expressing a sufficient condition of confirmation. This is shown by the following rather surprising example : Consider the hypothesis S_1: $(x)(y)[\sim (R(x, y)) \supset (R(x, y) . \sim R(y, x))]$.
Let a, b be two objects such that $R(a, b)$ and $\sim R(b, a)$. Then clearly, the couple (a, b) satisfies both the antecedent and the consequent of the universal conditional S_1 ; hence, if Nicod's criterion in its general form is accepted as stating a sufficient condition of confirmation, (a, b) constitutes confirming evidence for S_1. However, S_1 can be shown to be equivalent to

$$S_2 : (x)(y)R(x, y)$$

Now, by hypothesis, we have $\sim R(b, a)$; and this flatly contradicts S_2 and thus S_1. Thus, the couple (a, b), although satisfying both the antecedent and the consequent of the universal conditional S_1 actually constitutes disconfirming evidence of the strongest kind (conclusively disconfirming evidence, as we shall say later) for that sentence. This illustration reveals a striking and—as far as I am aware—hitherto unnoticed weakness of that conception of confirmation which underlies Nicod's criterion. In order to realize the bearing of our illustration upon Nicod's original formulation, let A and B be $\sim (R(x, y) . R(y, x))$ and $R(x, y) . \sim R(y, x)$ respectively. Then S_1 asserts that A entails B, and the couple (a, b) is a case of the presence of B in the presence of A ; this should, according to Nicod, be favourable to S_1.

and S_2 in the above illustration—then it seems perfectly reasonable to qualify an object as confirming such a hypothesis if it satisfies both its antecedent and its consequent. The plausibility of this view will be further corroborated in the course of our subsequent analyses.

Thus, we shall agree that if a is both a raven and black, then a certainly confirms S_1 : ' (x) (Raven(x) $\supset$ Black(x)) ', and if d is neither black nor a raven, d certainly confirms S_2 :

$$' (x) (\sim Black(x) \supset \sim Raven(x)).'$$

Let us now combine this simple stipulation with the equivalence condition : Since S_1 and S_2 are equivalent, d is confirming also for S_1 ; and thus, we have to recognize as confirming for S_1 any object which is neither black nor a raven. Consequently, any red pencil, any green leaf, and yellow cow, etc., becomes confirming evidence for the hypothesis that all ravens are black. This surprising consequence of two very adequate assumptions (the equivalence condition and the above sufficient condition of confirmation) can be further expanded : The following sentence can readily be shown to be equivalent to S_1 : S_3 : ' (x) [(Raven(x) v $\sim$ Raven(x)) $\supset$ ($\sim$ Raven(x) v Black(x))] ', $i.e.$ " Anything which is or is not a raven is either no raven or black ". According to the above sufficient condition, S_3 is certainly confirmed by any object, say e, such that (1) e is or is not a raven and, in addition, (2) e is not a raven or also black. Since (1) is analytic. these conditions reduce to (2). By virtue of the equivalence condition, we have therefore to consider as confirming for S_1 any object which is either no raven or also black (in other words: any object which is no raven at all, or a black raven).

Of the four objects characterized in section 3, a, c and d would therefore constitute confirming evidence for S_1, while b would be disconfirming for S_1. This implies that any non-raven represents confirming evidence for the hypothesis that all ravens are black.

We shall refer to these implications of the equivalence criterion and of the above sufficient condition of confirmation as the *paradoxes of confirmation.*

How are these paradoxes to be dealt with ? Renouncing the equivalence condition would not represent an acceptable solution, as is shown by the considerations presented in section 4. Nor does it seem possible to dispense with the stipulation that an object satisfying two conditions, C_1 and C_2, should be considered as confirming a general hypothesis to the effect that any object which satisfies C_1, also satisfies C_2.

But the deduction of the above paradoxical results rests on

one other assumption which is usually taken for granted, namely, that the meaning of general empirical hypotheses, such as that all ravens are black, or that all sodium salts burn yellow, can be adequately expressed by means of sentences of universal conditional form, such as ' (x) (Raven(x) ⊃ Black(x))' and ' (x) (Sod. Salt(x) ⊃ Burn Yellow (x))', etc. Perhaps this customary mode of presentation has to be modified ; and perhaps such a modification would automatically remove the paradoxes of confirmation ? If this is not so, there seems to be only one alternative left, namely to show that the impression of the paradoxical character of those consequences is due to misunderstanding and can be dispelled, so that no theoretical difficulty remains. We shall now consider these two possibilities in turn : The sub-sections 5.11 and 5.12 are devoted to a discussion of two different proposals for a modified representation of general hypotheses ; in subsection 5.2, we shall discuss the second alternative, *i.e.* the possibility of tracing the impression of paradoxicality back to a misunderstanding.

5.11. It has often been pointed out that while Aristotelian logic, in agreement with prevalent every day usage, confers " existential import " upon sentences of the form " All P's are Q's ", a universal conditional sentence, in the sense of modern logic, has no existential import ; thus, the sentence

$$' (x) \text{ (Mermaid}(x) \supset \text{Green}(x))'$$

does not imply the existence of mermaids ; it merely asserts that any object either is not a mermaid at all, or a green mermaid ; and it is true simply because of the fact that there are no mermaids. General laws and hypotheses in science, however—so it might be argued—are meant to have existential import ; and one might attempt to express the latter by supplementing the customary universal conditional by an existential clause. Thus, the hypothesis that all ravens are black would be expressed by means of the sentence S_1 : ' (x) (Raven(x) ⊃ Black(x)) . (Ex)Raven(x) ; and the hypothesis that no non-black things are ravens by S_2 : ' (x)($\sim$ Black(x) ⊃ $\sim$ Raven(x))) . (Ex) $\sim$ Black(x). Clearly, these sentences are not equivalent, and of the four objects a, b, c, d characterized in section 3, part (b), only a might reasonably be said to confirm S_1, and only d to confirm S_2. Yet this method of avoiding the paradoxes of confirmation is open to serious objections :

(a) First of all, the representation of every general hypothesis by a conjunction of a universal conditional and an existential sentence would invalidate many logical inferences which are generally accepted as permissible in a theoretical argument. Thus, for example, the assertions that all sodium salts burn yellow, and that whatever does not burn yellow is no sodium salt are logically equivalent according to customary understanding and usage ; and their representation by universal conditionals preserves this equivalence ; but if existential clauses are added, the two assertions are no longer equivalent, as is illustrated above by the analogous case of S_1 and S_2.

(b) Second, the customary formulation of general hypotheses in empirical science clearly does not contain an existential clause, nor does it, as a rule, even indirectly determine such a clause unambiguously. Thus, consider the hypothesis that if a person after receiving an injection of a certain test substance has a positive skin reaction, he has diphtheria. Should we construe the existential clause here as referring to persons, to persons receiving the injection, or to persons who, upon receiving the injection, show a positive skin reaction ? A more or less arbitrary decision has to be made ; each of the possible decisions gives a different interpretation to the hypothesis, and none of them seems to be really implied by the latter.

(c) Finally, many universal hypotheses cannot be said to imply an existential clause at all. Thus, it may happen that from a certain astrophysical theory a universal hypothesis is deduced concerning the character of the phenomena which would take place under certain specified extreme conditions. A hypothesis of this kind need not (and, as a rule, does not) imply that such extreme conditions ever were or will be realized ; it has no existential import. Or consider a biological hypothesis to the effect that whenever man and ape are crossed, the offspring will have such and such characteristics. This is a general hypothesis ; it might be contemplated as a mere conjecture, or as a consequence of a broader genetic theory, other implications of which may already have been tested with positive results ; but unquestionably the hypothesis does not imply an existential clause asserting that the contemplated kind of cross-breeding referred to will, at some time, actually take place.

While, therefore, the adjunction of an existential clause to the customary symbolization of a general hypothesis cannot be considered ·as an adequate *general* method of coping with the paradoxes of confirmation, there is a purpose which the use of an existential clause may serve very well, as was pointed out to me by Dr. Paul Oppenheim [1] : if somebody feels that objects of the

---

[1] This observation is related to Mr. Oppenheim's methodological studies referred to in p. 1, n. 1.

types c and d mentioned above are irrelevant rather than confirming for the hypothesis in question, and that qualifying them as confirming evidence does violence to the meaning of the hypothesis, then this may indicate that he is consciously or unconsciously construing the latter as having existential import ; and this kind of understanding of general hypotheses is in fact very common.   In this case, the " paradox " may be removed by pointing out that an adequate symbolization of the intended meaning requires the adjunction of an existential clause.   The formulation thus obtained is more restrictive than the universal conditional alone ; and while we have as yet set up no criteria of confirmation applicable to hypotheses of this more complex form, it is clear that according to every acceptable definition of confirmation objects of the types c and d will fail to qualify as confirming cases.   In this manner, the use of an existential clause may prove helpful in distinguishing and rendering explicit different possible interpretations of a given general hypothesis which is stated in non-symbolic terms.

5.12. Perhaps the impression of the paradoxical character of the cases discussed in the beginning of section 5 may be said to grow out of the feeling that the hypothesis that all ravens are black is about ravens, and not about non-black things, nor about all things.   The use of an existential clause was one attempt at expressing this presumed peculiarity of the hypothesis.   The attempt has failed, and if we wish to reflect the point in question, we shall have to look for a stronger device.   The idea suggests itself of representing a general hypothesis by the customary universal conditional, supplemented by the indication of the specific " field of application " of the hypothesis ; thus, we might represent the hypothesis that all ravens are black by the sentence ' (x) (Raven(x) ⊃ Black(x)) ' (or any one of its equivalents), plus the indication " Class of ravens " characterizing the field of application ; and we might then require that every confirming instance should belong to the field of application.   This procedure would exclude the objects c and d from those constituting confirming evidence and would thus avoid those undesirable consequences of the existential-clause device which were pointed out in 5.11 (c).   But apart from this advantage, the second method is open to objections similar to those which apply to the first : (a) The way in which general hypotheses are used in science never involves the statement of a field of application ; and the choice of the latter in a symbolic formulation of a given hypothesis thus introduces again a considerable measure of arbitrariness.   In particular, for a scientific hypothesis to the effect that

2

all P's are Q's, the field of application cannot simply be said to be the class of all P's ; for a hypothesis such as that all sodium salts burn yellow finds important applications in tests with negative results ; *i.e.* it may be applied to a substance of which it is not known whether it contains sodium salts, nor whether it burns yellow ; and if the flame does not turn yellow, the hypothesis serves to establish the absence of sodium salts. The same is true of all other hypotheses used for tests of this type. (*b*) Again, the consistent use of a domain of application in the formulation of general hypotheses would involve considerable logical complications, and yet would have no counterpart in the theoretical procedure of science, where hypotheses are subjected to various kinds of logical transformation and inference without any consideration that might be regarded as referring to changes in the fields of application. This method of meeting the paradoxes would therefore amount to dodging the problem by means of an *ad hoc* device which cannot be justified by reference to actual scientific procedure.

5.2. We have examined two alternatives to the customary method of representing general hypotheses by means of universal conditionals ; neither of them proved an adequate means of precluding the paradoxes of confirmation. We shall now try to show that what is wrong does not lie in the customary way of construing and representing general hypotheses, but rather in our reliance on a misleading intuition in the matter : The impression of a paradoxical situation is not objectively founded ; it is a psychological illusion.

(*a*) One source of misunderstanding is the view, referred to before, that a hypothesis of the simple form " Every P is a Q " such as " All sodium salts burn yellow ", asserts something about a certain limited class of objects only, namely, the class of all P's. This idea involves a confusion of logical and practical considerations : Our interest in the hypothesis may be focussed upon its applicability to that particular class of objects, but the hypothesis nevertheless asserts something about, and indeed imposes restrictions upon, *all* objects (within the logical type of the variable occurring in the hypothesis, which in the case of our last illustration might be the class of all physical objects). Indeed, a hypothesis of the form " Every P is a Q " forbids the occurrence of any objects having the property P but lacking the property Q ; *i.e.* it restricts all objects whatsoever to the class of those which either lack the property P or also have the property Q. Now, every object either belongs to this class or falls outside it, and thus, every object—and not only the P's—either conforms to the hypothesis or violates it ; there is no object which is not implicitly " referred to " by a hypothesis of this type. In particular, every object which either is no sodium salt or burns yellow conforms to, and thus " bears out " the hypothesis that all sodium salts burn yellow ; every other object violates that hypothesis.

The weakness of the idea under consideration is evidenced also by the observation that the class of objects about which a hypothesis is supposed to assert something is in no way clearly determined, and that it changes with the context, as was shown in 5.12 (a).

(b) A second important source of the appearance of paradoxicality in certain cases of confirmation is exhibited by the following consideration.

Suppose that in support of the assertion " All sodium salts burn yellow " somebody were to adduce an experiment in which a piece of pure ice was held into a colourless flame and did not turn the flame yellow. This result would confirm the assertion, " Whatever does not burn yellow is no sodium salt ", and consequently, by virtue of the equivalence condition, it would confirm the original formulation. Why does this impress us as paradoxical ? The reason becomes clear when we compare the previous situation with the case of an experiment where an object whose chemical constitution is as yet unknown to us is held into a flame and fails to turn it yellow, and where subsequent analysis reveals it to contain no sodium salt. This outcome, we should no doubt agree, is what was to be expected on the basis of the hypothesis that all sodium salts burn yellow—no matter in which of its various equivalent formulations it may be expressed ; thus, the data here obtained constitute confirming evidence for the hypothesis. Now the only difference between the two situations here considered is that in the first case we are told beforehand the test substance is ice, and we happen to " know anyhow " that ice contains no sodium salt ; this has the consequence that the outcome of the flame-colour test becomes entirely irrelevant for the confirmation of the hypothesis and thus can yield no new evidence for us. Indeed, if the flame should not turn yellow, the hypothesis requires that the substance contain no sodium salt— and we know beforehand that ice does not—and if the flame should turn yellow, the hypothesis would impose no further restrictions on the substance ; hence, either of the possible outcomes of the experiment would be in accord with the hypothesis.

The analysis of this example illustrates a general point : In

the seemingly paradoxical cases of confirmation, we are often
not actually judging the relation of the given evidence, E alone
to the hypothesis H (we fail to observe the "methodological
fiction", characteristic of every case of confirmation, that we
have no relevant evidence for H other than that included in E);
instead, we tacitly introduce a comparison of H with a body of
evidence which consists of E in conjunction with an additional
amount of information which we happen to have at our disposal;
in our illustration, this information includes the knowledge (1) that
the substance used in the experiment is ice, and (2) that ice contains no sodium salt. If we assume this additional information
as given, then, of course, the outcome of the experiment can add
no strength to the hypothesis under consideration. But if we
are careful to avoid this tacit reference to additional knowledge
(which entirely changes the character of the problem), and if we
formulate the question as to the confirming character of the
evidence in a manner adequate to the concept of confirmation
as used in this paper, we have to ask: Given some object a
(it happens to be a piece of ice, but this fact is not included
in the evidence), and given the fact that a does not turn
the flame yellow and is no sodium salt—does a then constitute
confirming evidence for the hypothesis? And now—no matter
whether a is ice or some other substance—it is clear that the
answer has to be in the affirmative; and the paradoxes
vanish.

So far, in section (b), we have considered mainly that type of
paradoxical case which is illustrated by the assertion that any
non-black non-raven constitutes confirming evidence for the hypothesis, "All ravens are black." However, the general idea just
outlined applies as well to the even more extreme cases exemplified
by the assertion that any non-raven as well as any black object
confirms the hypothesis in question. Let us illustrate this by
reference to the latter case. If the given evidence E—i.e. in the
sense of the required methodological fiction, all our data relevant
for the hypothesis—consists only of one object which, in addition,
is black, then E may reasonably be said to support even the
hypothesis that all objects are black, and a fortiori E supports the
weaker assertion that all ravens are black. In this case, again,
our factual knowledge that not all objects are black tends to
create an impression of paradoxicality which is not justified on
logical grounds. Other "paradoxical" cases of confirmation
may be dealt with analogously, and it thus turns out that the
"paradoxes of confirmation", as formulated above, are due to
a misguided intuition in the matter rather than to a logical flaw

in the two stipulations from which the "paradoxes" were derived.[1], [2]

[1] The basic idea of sect. (b) in the above analysis of the "paradoxes of confirmation" is due to Dr. Nelson Goodman, to whom I wish to reiterate my thanks for the help he rendered me, through many discussions, in clarifying my ideas on this point.

[2] The considerations presented in section (b) above are also influenced by, though not identical in content with, the very illuminating discussion of the "paradoxes" by the Polish methodologist and logician Janina Hosiasson-Lindenbaum ; cf. her article " On Confirmation ", *The Journal of Symbolic Logic*, vol. 5 (1940), especially sect. 4.   Dr. Hosiasson's attention had been called to the paradoxes by the article referred to in p. 11, n. 2, and by discussions with the author.   To my knowledge, hers has so far been the only publication which presents an explicit attempt to solve the problem.   Her solution is based on a theory of degrees of confirmation, which is developed in the form of an uninterpreted axiomatic system (cf. also p. 6, n. 1, and part (b) in sect. 1 of the present article), and most of her arguments presuppose that theoretical framework.   I have profited, however, by some of Miss Hosiasson's more general observations which proved relevant for the analysis of the paradoxes of the non-gradated relation of confirmation which forms the object of the present study.

One point in those of Miss Hosiasson's comments which rest on her theory of degrees of confirmation is of particular interest, and I should like to discuss it briefly.   Stated in reference to the raven-hypothesis, it consists in the suggestion that the finding of one non-black object which is no raven, while constituting confirming evidence for the hypothesis, would increase the degree of confirmation of the hypothesis by a smaller amount than the finding of one raven which is black.   This is said to be so because the class of all ravens is much less numerous than that of all non-black objects, so that—to put the idea in suggestive though somewhat misleading terms—the finding of one black raven confirms a larger portion of the total content of the hypothesis than the finding of one non-black non-raven. In fact, from the basic assumptions of her theory, Miss Hosiasson is able to derive a theorem according to which the above statement about the relative increase in degree of confirmation will hold provided that actually the number of all ravens is small compared with the number of all non-black objects.   But is this last numerical assumption actually warranted in the present case and analogously in all other "paradoxical" cases ? The answer depends in part upon the logical structure of the language of science.   If a "co-ordinate language" is used, in which, say, finite space-time regions figure as individuals, then the raven-hypothesis assumes some such form as " Every space-time region which contains a raven, contains something black " ; and even if the total number of ravens ever to exist is finite, the class of space-time regions containing a raven has the power of the continuum, and so does the class of space-time regions containing something non-black ; thus, for a co-ordinate language of the type under consideration, the above numerical assumption is not warranted.   Now the use of a co-ordinate language may appear quite artificial in this particular illustration ; but it will seem very appropriate in many other contexts, such as, e.g., that of physical field theories.   On the other hand, Miss Hosiasson's numerical assumption may well be justified on the basis of a " thing language ", in which physical objects of finite size function

6. *Confirmation Construed as a Relation between Sentences.*—Our
analysis of Nicod's criterion has so far led to two main results :
The rejection of that criterion in view of several deficiencies, and
the emergence of the equivalence condition as a necessary condition of adequacy for any proposed definition of confirmation.
Another aspect of Nicod's criterion requires consideration now.
In our formulation of the criterion, confirmation was construed
as a dyadic relation between an object or an ordered set of objects,
representing the evidence, and a sentence, representing the
hypothesis. This means that confirmation was conceived of as
a semantical relation [1] obtaining between certain extra-linguistic
objects [2] on one hand and certain sentences on the other. It is
possible, however, to construe confirmation in an alternative
fashion as a relation between two sentences, one describing the
given evidence, the other expressing the hypothesis. Thus, *e.g.*
instead of saying that an object *a* which is both a raven and
black (or the " fact " of *a* being both a raven and black) confirms
the hypothesis, " All ravens are black ", we may say that the
evidence sentence, " *a* is a raven, and *a* is black ", confirms the
hypothesis-sentence (briefly, the hypothesis), " All ravens are
black ". We shall adopt this conception of confirmation as a
relation between sentences here for the following reasons : First,
the evidence adduced in support or criticism of a scientific hypothesis is always expressed in sentences, which frequently have
the character of observation reports ; and second, it will prove
very fruitful to pursue the parallel, alluded to in section 2 above,
between the concepts of confirmation and of logical consequence.
And just as in the theory of the consequence relation, *i.e.* in
deductive logic, the premisses of which a given conclusion is a
consequence are construed as sentences rather than as " facts ",
so we propose to construe the data which confirm a given hypothesis as given in the form of sentences.

The preceding reference to observation reports suggests a
certain restriction which might be imposed on evidence sentences.
Indeed, the evidence adduced in support of a scientific hypothesis

as individuals. Of course, even on this basis, it remains an empirical
question, for every hypothesis of the form " All *P*'s are *Q*'s ", whether
actually the class of non-*Q*'s is much more numerous than the class of
P's ; and in many cases this question will be very difficult to decide.

[1] For a detailed account of this concept, see C. W. Morris, *Foundations
of the Theory of Signs* (Internat. Encyclopedia of Unified Science, vol. i,
no. 2, Chicago, 1938), and R. Carnap, *Introduction to Semantics* (Cambridge,
Mass., 1942), esp. sects. 4 and 37.

[2] Instead of making the first term of the relation an object or a sequence
of objects, we might construe it as a " state of affairs " (or perhaps as a
" fact ", or a " proposition ", as Nicod puts it), such as that state of affairs
which consists in *a* being a black raven, etc.

or theory consists, in the last analysis, in data accessible to what is loosely called " direct observation ", and such data are expressible in the form of " observation reports ".   In view of this considera-tion, we shall restrict the evidence sentences which form the domain of the relation of confirmation, to sentences of the charac-ter of observation reports.   In order to give a precise meaning to the concept of observation report, we shall assume that we are given a well-determined " language of science ", in terms of which all sentences under consideration, hypotheses as well as evidence sentences, are formulated.   We shall further assume that this language contains, among other terms, a clearly de-limited " observational vocabulary " which consists of terms designating more or less directly observable attributes of things or events, such as, say, " black ", " taller than ", " burning with a yellow light ", etc., but no theoretical constructs such as " aliphatic compound ", " circularly polarized light ", " heavy hydrogen ", etc.

We shall now understand by a hypothesis any sentence which can be expressed in the assumed language of science, no matter whether it is a generalized sentence, containing quantifiers, or a particular sentence referring only to a finite number of particular objects.   An observation report will be construed as a finite class (or a conjunction of a finite number) of observation sen-tences ; and an observation sentence as a sentence which either asserts or denies that a given object has a certain observable property (such as " a is a raven ", " d is not black "), or that a given sequence of objects stand in a certain observable relation (such as " a is between b and c ").

Now the concept of observability itself obviously is relative to the techniques of observation used.   What is unobservable to the unaided senses may well be observable by means of suitable devices such as telescopes, microscopes, polariscopes, lie-detectors, Gallup-polls, etc.   If by direct observation we mean such ob-servational procedures as do not make use of auxiliary devices, then such property terms as " black ", " hard ", " liquid ", " cool ", and such relation terms as " above ", " between ", " spatially coincident ", etc., might be said to refer to directly observable attributes ; if observability is construed in a broader sense, so as to allow for the use of certain specified instruments or other devices, the concept of observable attribute becomes more comprehensive.   If, in our study of confirmation, we wanted to analyze the manner in which the hypotheses and theories of empirical science are ultimately supported by " evidence of the senses ", then we should have to require that observation reports refer exclusively to directly observable attributes.   This view

was taken, for simplicity and concreteness, in the preceding parts
of this section. Actually, however, the general logical characteristics of that relation which obtains between a hypothesis and a
group of empirical statements which " support " it, can be studied
in isolation from this restriction to dirℓct observability. All we
will assume here is that in the context of the scientific test of
a given hypothesis or theory, certain specified techniques of
observation have been agreed upon ; these determine an observational vocabulary, namely a set of terms designating properties and relations observable by means of the accepted techniques.
For our purposes it is entirely sufficient that these terms, constituting the " observational vocabulary ", be given. An observation sentence is then defined simply as a sentence affirming
or denying that a given object, or sequence of objects, possesses
one of those observable attributes.[1]

Let it be noted that we do not require an observation sentence
to be true, nor to be accepted on the basis of actual observations ;
rather, an observation sentence expresses something that is de-
cidable by means of the accepted techniques of observation ;
in other words : An observation sentence describes a possible outcome of the accepted observational techniques ; it asserts something that might conceivably be established by means of those

---

[1] The concept of observation sentence has, in the context of our study,
a status and a logical function closely akin to that of the concepts of protocol statement or basis sentence, etc., as used in many recent studies of
empiricism. However, the conception of observation sentence which is
being proposed in the present study is more liberal in that it renders the
discussion of the logical problems of testing and confirmation independent
of various highly controversial epistemological issues ; thus, e.g. we do
not stipulate that observation reports must be about psychic acts, or about
sense perceptions (i.e. that they have to be expressed in terms of a vocabulary of phenomenology, or of introspective psychology). According to the
conception of observation sentence adopted in the present study, the
" objects " referred to in an observation sentence may be construed in any
one of the senses just referred to, or in various other ways ; for example,
they might be space-time regions, or again physical objects such as stones,
trees, etc. (most of the illustrations given throughout this article represent
observation sentences belonging to this kind of " thing-language ") ; all
that we require is that the few very general conditions stated above be
satisfied.

These conditions impose on observation sentences and on observation
reports certain restrictions with respect to their form ; in particular, neither
kind of sentence may contain any quantifiers. This stipulation recommends
itself for the purposes of the logical analysis here to be undertaken ; but
we do not wish to claim that this formal restriction is indispensable. On
the contrary, it is quite possible and perhaps desirable also to allow for
observation sentences containing quantifiers : our simplifying assumption
is introduced mainly in order to avoid considerable logical complications
in the definition of confirmation.

techniques. Possibly, the term "observation-type sentence" would be more suggestive ; but for convenience we give preference to the shorter term. An analogous comment applies, of course, to our definition of an observation report as a class or a conjunction of observation sentences. The need for this broad conception of observation sentences and observation reports is readily recognized : Confirmation as here conceived is a logical relationship between sentences, just as logical consequence is. Now whether a sentence S_2 is a consequence of a sentence S_1 does not depend on whether S_1 is true (or known to be true), or not ; and analogously, the criteria of whether a given statement, expressed in terms of the observational vocabulary, confirms a certain hypothesis cannot depend on whether the statements in the report are true, or based on actual experience, or the like. Our definition of confirmation must enable us to indicate what kind of evidence *would* confirm a given hypothesis *if* it were available ; and clearly the sentence characterizing such evidence can be required only to express something that might be observed, but not necessarily something that has actually been established by obsrvation.

It may be helpful to carry the analogy between confirmation and consequence one step further. The truth or falsity of S_1 is irrelevant for the question of whether S_2 is a consequence of S_1 (whether S_2 can be validly inferred from S_1) ; but in a logical inference which justifies a sentence S_2 by showing that it is a logical consequence of a conjunction of premises, S_1, we can be certain of the truth of S_2 only if we know S_1 to be true. Analogously, the question of whether an observation report stands in the relation of confirmation to a given hypothesis does not depend on whether the report states actual or fictitious observational findings ; but for a decision as to the soundness or acceptability of a hypothesis which is confirmed by a certain report, it is of course necessary to know whether the report is based on actual experience or not. Just as a conclusion of a logical inference, in order to be reliably true must be (a1) validly inferred from (a2) a set of true premises, so a hypothesis, in order to be scientifically acceptable, must be (b1) formally confirmed by (b2) reliable reports on observational findings.

The central problem of this essay is to establish general criteria for the formal relation of confirmation as referred to in (b1) ; the analysis of the concept of a reliable observation report, which belongs largely to the field of pragmatics,[1] falls outside the scope of the present study. One point, however, deserves mention here : A statement of the form of an observation report (for

[1] An account of the concept of pragmatics may be found in the publications listed in p. 22, n. 1.

example, about the position of the pointer of a certain thermograph at 3 a.m.) may be accepted or rejected in science either on the basis of direct observation, or because it is indirectly confirmed or disconfirmed by other accepted observation sentences (in the example, these might be sentences describing the curve traced by the pointer during the night), and because of this possibility of indirect confirmation, our study has a bearing also on the question of the acceptance of hypotheses which have themselves the form of observation reports.

The conception of confirmation as a relation between sentences analogous to that of logical consequence suggests yet another specification for the attempted definition of confirmation: While logical consequence has to be conceived of as a basically semantical relation between sentences, it has been possible, for certain languages, to establish criteria of logical consequence in purely syntactical terms.[1] Analogously, confirmation may be conceived of as a semantical relation between an observation report and a hypothesis ; but the parallel with the consequence relation suggests that it should be possible, for certain languages, to establish purely syntactical criteria of confirmation. The subsequent considerations will indeed eventuate in a definition of confirmation based on the concept of logical consequence and other purely syntactical concepts.

The interpretation of confirmation as a logical relation between sentences involves no essential change in the central problem of the present study. In particular, all the points made in the preceding sections can readily be rephrased in accordance with this interpretation. Thus, for example, the assertion that an object a which is a swan and white confirms the hypothesis '(x) (Swan(x) ⊃ White(x))' can be expressed by saying that the observation report 'Swan(a) . White(a)' confirms that hypothesis. Similarly, the equivalence condition can be reformulated as follows : If an observation report confirms a certain sentence, then it also confirms every sentence which is logically equivalent with the latter. Nicod's criterion as well as our grounds for rejecting it can be re-formulated along the same lines. We presented Nicod's concept of confirmation as referring to a relation between non-linguistic objects on one hand and sentences on the other because this approach seemed to approximate most closely Nicod's own formulations, and because it enabled us to avoid certain technicalities which are actually unnecessary in that context.

(To be concluded)

---

[1] *Cf.* especially the two publications by R. Carnap listed in p. 3, n. 1.

VOL. LIV. No. 214.]   [April, 1945.

MIND

A QUARTERLY REVIEW

OF

PSYCHOLOGY AND PHILOSOPHY

I.—STUDIES IN THE LOGIC OF CON-FIRMATION (II.).

BY CARL G. HEMPEL.

7. *The Prediction-criterion of Confirmation and its Short-comings.*—We are now in a position to analyze a second conception of confirmation which is reflected in many methodological discussions and which can claim a great deal of plausibility. Its basic idea is very simple : General hypotheses in science as well as in everyday usage are intended to enable us to anticipate future events ; hence, it seems reasonable to count any prediction which is borne out by subsequent observation as confirming evidence for the hypothesis on which it is based, and any prediction that fails as disconfirming evidence. To illustrate : Let H_1 be the hypothesis that all metals, when heated, expand ; symbolically : '(x) ((Metal (x) . Heated (x)) $\supset$ Exp(x)) '. If we are given an observation report to the effect that a certain object a is a metal and is heated, then by means of H_1 we can derive the prediction that a expands. Suppose that this is borne out by observation and described in an additional observation statement. We should then have the total observation report. {Metal(a), Heated(a), Exp.(a)}.[1] This report would be qualified as confirming evidence for H_1 because its last sentence bears out what could be predicted, or derived, from the first two by means of

---

[1] An (observation) report, it will be recalled, may be represented by a conjunction or by a class of observation sentences ; in the latter case, we characterize it by writing the sentences between braces ; the quotation marks which normally would be used are, for convenience, assumed to be absorbed by the braces.

7

98   CARL G. HEMPEL:

H_1; more explicitly : because the last sentence can be derived from the first two in conjunction with H_1.—Now let H_2 be the hypothesis that all swans are white ; symbolically : ' (x) (Swan $(x) \supset$ White(x)) ' ; and consider the observation report {Swan(a), ~ White(a)}. This report would constitute disconfirming evidence for H_2 because the second of its sentences contradicts (and thus fails to bear out) the prediction ' White(a) ' which can be deduced from the first sentence in conjunction with H_2 ; or, symmetrically, because the first sentence contradicts the consequence ' ~ Swan(a) ' which can be derived from the second in conjunction with H_2. Obviously, either of these formulations implies that H_2 is incompatible with the given observation report. These illustrations suggest the following general definition of confirmation as successful prediction :

Prediction-criterion of Confirmation : Let H be a hypothesis, B an observation report, *i.e.* a class of observation sentences. Then

(a) B is said to confirm H if B can be divided into two mutually exclusive subclasses B_1 and B_2 such that B_2 is not empty, and every sentence of B_2 can be logically deduced from B_1 in conjunction with H, but not from B_1 alone.

(b) B is said to disconfirm H if H logically contradicts B.[1]

(c) B is said to be neutral with respect to H if it neither confirms nor disconfirms H.[2]

But while this criterion is quite sound as a statement of sufficient conditions of confirmation for hypotheses of the type illustrated above, it is considerably too narrow to serve as a general definition of confirmation. Generally speaking, this criterion would serve its purpose if all scientific hypotheses could be construed as asserting regular connections of observable features in the subject-matter under investigation ; *i.e.* if they all were of

[1] It might seem more natural to stipulate that B disconfirms H if it can be divided into two mutually exclusive classes B_1 and B_2 such that the denial of at least one sentence in B_2 can be deduced from B_1 in conjunction with H ; but this condition can be shown to be equivalent to (b) above.

[2] The following quotations from A. J. Ayer's book *Language, Truth and Logic* (London, 1936) formulate in a particularly clear fashion the conception of confirmation as successful prediction (although the two are not explicitly identified by definition) : ". . . the function of an empirical hypothesis is to enable us to anticipate experience. Accordingly, if an observation to which a given proposition is relevant conforms to our expectations, . . . that proposition is confirmed " (*loc. cit.* pp. 142-143). ". . . it is the mark of a genuine factual proposition . . . that some experiential propositions can be deduced from it in conjunction with certain premises without being deducible from those other premises alone ". (*loc. cit.* p. 26).

the form " Whenever the observable characteristic P is present in an object or a situation, then the observable characteristic Q will also be present." But actually, most scientific hypotheses and laws are not of this simple type ; as a rule, they express regular connections of characteristics which are not observable in the sense of direct observability, nor even in a much more liberal sense. Consider, for example, the following hypothesis : " Whenever plane-polarized light of wave length λ traverses a layer of quartz of thickness d, then its plane of polarization is rotated through an angle α which is proportional to $\frac{d}{\lambda}$."—Let us assume that the observational vocabulary, by means of which our observation reports have to be formulated, contains exclusively terms referring to directly observable attributes. Then, since the question of whether a given ray of light is plane-polarized and has the wave length λ cannot be decided by means of direct observation, no observation report of the kind here admitted could include information of this type. This in itself would not be crucial if at least we could assume that the fact that a given ray of light is plane-polarized, etc., could be logically inferred from some possible observation report ; for then, from a suitable report of this kind, in conjunction with the given hypothesis, one would be able to predict a rotation of the plane of polarization ; and from this prediction, which itself is not yet expressed in exclusively observational terms, one might expect to derive further predictions in the form of genuine observation sentences. But actually, a hypothesis to the effect that a given ray of light is plane-polarized has to be considered as a general hypothesis which entails an unlimited number of observation sentences ; thus it cannot be logically inferred from, but at best be confirmed by, a suitable set of observational findings. The logically essential point can best be exhibited by reference to a very simple abstract case : Let us assume that R_1 and R_2 are two relations of a kind accessible to direct observation, and that the field of scientific investigation contains infinitely many objects. Consider now the hypothesis

$$(H) \qquad (x)((y)R_1(x, y) \supset (\mathsf{E}z)R_2(x, z)),$$

i.e. : Whenever an object x stands in R_1 to every object y, then it stands in R_2 to at least one object z.—This simple hypothesis has the following property : However many observation sentences may be given, H does not enable us to derive any new observation sentences from them. Indeed—to state the reason in suggestive though not formally rigorous terms—in order to make a prediction concerning some specific object a, we should first have to know that a stands in R_1 to every object ; and this necessary information clearly cannot be contained in any finite number, however large, of observation sentences, because a finite set of observation sentences can tell us at best for a finite number of objects that a stands in R_1 to them. Thus an observation report, which always involves only a finite number of observation sentences, can never provide a sufficiently broad basis for a prediction by means of H.[1]—Besides, even if we did know that a stood in R_1 to every object, the prediction derivable by means of H would not be an observation sentence ; it would assert that a stands in R_2 to some object, without specifying which, and where to find it. Thus, H would be an empirical hypothesis, containing, besides purely logical terms, only expressions belonging to the observational vocabulary, and yet the predictions which it renders possible neither start from nor lead to observation reports.

It is, therefore, a considerable over-simplification to say that scientific hypotheses and theories enable us to derive predictions of future experiences from descriptions of past ones. Unquestionably, scientific hypotheses do have a predictive function ; but the way in which they perform this function, the manner in which they establish logical connections between observation reports, is logically more complex than a deductive inference. Thus, in the last illustration, the predictive use of H may assume the following form : On the basis of a number of individual tests, which show that a does stand in R_1 to three objects b, c, and d, we may accept the hypothesis that a stands in R_1 to all objects ; or, in terms of our formal mode of speech : In view of the observation report $\{R_1(a, b), R_1(a, c), R_1(a, d)\}$, the hypothesis that $(y)R_1(a, y)$ is accepted as confirmed by, though not logically inferable from, that report.[2] This process might be referred to as quasi-induction.[3] From the hypothesis thus established we

[1] To illustrate : a might be an iron object which possibly is a magnet ; R_1 might be the relation of attracting ; the objects under investigation might be iron objects. Then a finite number of observation reports to the effect that a did attract a particular piece of iron is insufficient to infer that a will attract every piece of iron.

[2] Thus, in the illustration given in the preceding footnote, the hypothesis that the object a will attract every piece of iron might be accepted as sufficiently well substantiated by, though by no means derivable from, an observation report to the effect that in tests a did attract the iron objects b, c, and d.

[3] The prefix " quasi " is to contradistinguish the procedure in question from so-called induction, which is usually supposed to be a method of discovering, or arriving at, general regularities on the basis of a finite

can then proceed to derive, by means of H, the prediction that a stands in R_2 to at least one object. This again, as was pointed out above, is not an observation sentence ; and indeed no observation sentence can be derived from it ; but it can, in turn, be confirmed by a suitable observation sentence, such as ' $R_2(a, b)$ '. —In other cases, the prediction of actual observation sentences may be possible ; thus if the given hypothesis asserts that $(x)((y)R_1(x, y) \supset (z)R_2(x, z))$, then after quasi-inductively accepting, as above, that $(y)R_1(a, y)$, we can derive, by means of the given hypothesis, the sentence that a stands in R_2 to every object, and thence, we can deduce special predictions such as ' $R_2(a, b)$ ', etc., which do have the form of observation sentences.

Thus, the chain of reasoning which leads from given observational findings to the " prediction " of new ones actually involves, besides deductive inferences, certain quasi-inductive steps each of which consists in the acceptance of an intermediate statement on the basis of confirming, but usually not logically conclusive, evidence. In most scientific predictions, this general pattern occurs in multiple re-iteration ; an analysis of the predictive use of the hypothesis mentioned above, concerning plane-polarized light, could serve as an illustration. In the present context, however, this general account of the structure of scientific prediction is sufficient : it shows that a general definition of confirmation by reference to successful prediction becomes circular ; indeed, in order to make the original formulation of the prediction-criterion of confirmation sufficiently comprehensive, we should have to replace the phrase " can be logically deduced " by " can be obtained by a series of steps of deduction and quasi-induction " ; and the definition of " quasi-induction " in the above sense presupposes the concept of confirmation.

Let us note, as a by-product of the preceding consideration, the fact that an adequate analysis of scientific prediction (and analogously, of scientific explanation, and of the testing of empirical hypotheses) requires an analysis of the concept of confirmation. The reason for this fact may be restated in general terms as follows : Scientific laws and theories, as a rule, connect terms which lie on the level of abstract theoretical constructs rather than on that of direct observation ; and from observation sentences, no merely deductive logical inference leads

number of instances. In quasi-induction, the hypothesis is not " discovered " but has to be *given* in addition to the observation report ; the process consists in the acceptance of the hypothesis if it is deemed sufficiently confirmed by the observation report. *Cf.* also the discussion in section 1c, above.

to statements about those theoretical constructs which are the starting point for scientific predictions ; statements about logical constructs, such as " This piece of iron is magnetic " or " Here, a plane-polarized ray of light traverses a quartz crystal " can be confirmed, but not entailed, by observation reports, and thus, even though based on general scientific laws, the " prediction " of new observational findings on the basis of given ones is a process involving confirmation in addition to logical deduction.[1]

8. *Conditions of Adequacy for any Definition of Confirmation.*— The two most customary conceptions of confirmation, which were rendered explicit in Nicod's criterion and in the prediction criterion, have thus been found unsuitable for a general definition of confirmation. Besides this negative result, the preceding analysis has also exhibited certain logical characteristics of scientific prediction, explanation, and testing, and it has led to the establishment of certain standards which an adequate definition of confirmation has to satisfy. These standards include the equivalence condition and the requirement that the definition of confirmation be applicable to hypotheses of any degree of logical complexity, rather than to the simplest type of universal conditional only. An adequate definition of confirmation, however, has to satisfy several further logical requirements, to which we now turn.

First of all, it will be agreed that any sentence which is entailed by—*i.e.* a logical consequence of—a given observation report has to be considered as confirmed by that report : Entailment is a special case of confirmation. Thus, *e.g.*, we want to say that the observation report " *a* is black " confirms the sentence (hypothesis) " *a* is black or grey " ; and—to refer to one of the illustrations given in the preceding section—the observation sentence ' $R_2(a, b)$ ' should certainly be confirming evidence for the sentence ' $(Ez)R_2(a, z)$ '. We are therefore led to the stipulation that any adequate definition of confirmation must insure the fulfilment of the

[1] In the above sketch of the structure of scientific prediction, we have disregarded the fact that in practically every case where a prediction is said to be obtained by means of a certain hypothesis or theory, a considerable mass of auxiliary theories is used in addition ; thus, *e.g.* the prediction of observable effects of the deflection of light in the gravitational field of the sun on the basis of the general theory of relativity, requires such auxiliary theories as mechanics and optics. But an explicit consideration of this fact would not affect our result that scientific predictions, even when based on hypotheses or theories of universal form, still are not purely deductive in character, but involve quasi-inductive steps as well.

(8.1) *Entailment condition :* Any sentence which is entailed by an observation report is confirmed by it.[1]

This condition is suggested by the preceding consideration, but of course not proved by it. To make it a standard of adequacy for the definition of confirmation means to lay down the stipulation that a proposed definition of confirmation will be rejected as logically inadequate if it is not constructed in such a way that (8.1) is unconditionally satisfied. An analogous remark applies to the subsequently proposed further standards of adequacy.—

Second, an observation report which confirms certain hypotheses would invariably be qualified as confirming any consequence of those hypotheses. Indeed : any such consequence is but an assertion of all or part of the combined content of the original hypotheses and has therefore to be regarded as confirmed by any evidence which confirms the original hypotheses. This suggests the following condition of adequacy :

(8.2) *Consequence Condition :* If an observation report confirms every one of a class K of sentences, then it also confirms any sentence which is a logical consequence of K.

If (8.2) is satisfied, then the same is true of the following two more special conditions :

(8.21) *Special Consequence Condition :* If an observation report confirms a hypothesis H, then it also confirms every consequence of H.

(8.22) *Equivalence Condition :* If an observation report confirms a hypothesis H, then it also confirms every hypothesis which is logically equivalent with H.

(This follows from (8.21) in view of the fact that equivalent hypotheses are mutual consequences of each other.) Thus, the satisfaction of the consequence condition entails that of our earlier equivalence condition, and the latter loses its status of an independent requirement.

In view of the apparent obviousness of these conditions, it is interesting to note that the definition of confirmation in terms of successful prediction, while satisfying the equivalence condition, would violate the consequence condition. Consider, for example, the formulation of the prediction-criterion given in the earlier

---

[1] As a consequence of this stipulation, a contradictory observation report, such as {Black(a), $\sim$ Black(a)} confirms every sentence, because it has every sentence as a consequence. Of course, it is possible to exclude the possibility of contradictory observation reports altogether by a slight restriction of the definition of " observation report ". There is, however, no important reason to do so.

part of the preceding section. Clearly, if the observational findings B_2 can be predicted on the basis of the findings B_1 by means of the hypothesis H, the same prediction is obtainable by means of any equivalent hypothesis, but not generally by means of a weaker one.

On the other hand, any prediction obtainable by means of H can obviously also be established by means of any hypothesis which is stronger than H, i.e. which logically entails H. Thus, while the consequence condition stipulates in effect that whatever confirms a given hypothesis also confirms any weaker hypothesis, the relation of confirmation defined in terms of successful prediction would satisfy the condition that whatever confirms a given hypothesis, also confirms every stronger one.

But is this " converse consequence condition ", as it might be called, not reasonable enough, and should it not even be included among our standards of adequacy for the definition of confirmation ? The second of these two suggestions can be readily disposed of : The adoption of the new condition, in addition to (8.1) and (8.2), would have the consequence that any observation report B would confirm any hypothesis H whatsoever. Thus, e.g., if B is the report " a is a raven " and H is Hooke's law, then, according to (8.1), B confirms the sentence " a is a raven ", hence B would, according to the converse consequence condition, confirm the stronger sentence " a is a raven, and Hooke's law holds " ; and finally, by virtue of (8.2), B would confirm H, which is a consequence of the last sentence. Obviously, the same type of argument can be applied in all other cases.

But is it not true, after all, that very often observational data which confirm a hypothesis H are considered also as confirming a stronger hypothesis ? Is it not true, for example, that those experimental findings which confirm Galileo's law, or Kepler's laws, are considered also as confirming Newton's law of gravitation ?[1] This is indeed the case, but this does not justify the acceptance of the converse entailment condition as a general rule of the logic of confirmation ; for in the cases just mentioned, the weaker hypothesis is connected with the stronger one by a logical bond of a particular kind : it is essentially a substitution instance of the stronger one ; thus, e.g., while the law of gravitation refers to the force obtaining between any two bodies, Galileo's law is a specialization referring to the case where one of

---

[1] Strictly speaking, Galileo's law and Kepler's laws can be deduced from the law of gravitation only if certain additional hypotheses—including the laws of motion—are presupposed ; but this does not affect the point under discussion.

the bodies is the earth, the other an object near its surface. In the preceding case, however, where Hooke's law was shown to be confirmed by the observation report that a is a raven, this situation does not prevail; and here, the rule that whatever confirms a given hypothesis also confirms any stronger one becomes an entirely absurd principle. Thus, the converse consequence condition does not provide a sound general condition of adequacy.[1]

A third condition remains to be stated : [2]

(8.3) *Consistency Condition :* Every logically consistent observation report is logically compatible with the class of all the hypotheses which it confirms.

The two most important implications of this requirement are the following :

(8.31) Unless an observation report is self-contradictory,[3] it does not confirm any hypothesis with which it is not logically compatible.

(8.32) Unless an observation report is self-contradictory, it does not confirm any hypotheses which contradict each other.

The first of these corollaries will readily be accepted ; the second, however,—and consequently (8.3) itself—will perhaps be

---

[1] William Barrett, in a paper entitled " Discussion on Dewey's Logic " (*The Philosophical Review*, vol. 50, 1941, pp. 305 ff., esp. p. 312) raises some questions closely related to what we have called above the consequence condition and the converse consequence condition. In fact, he invokes the latter (without stating it explicitly) in an argument which is designed to show that " not every observation which confirms a sentence need also confirm all its consequences ", in other words, that the special consequence condition (8.21) need not always be satisfied. He supports his point by reference to " the simplest case : the sentence ' C ' is an abbreviation of ' A.B ', and the observation O confirms ' A ', *and so* ' C ', but is irrelevant to ' B ', which is a consequence of ' C '." (Italics mine.)

For reasons contained in the above discussion of the consequence condition and the converse consequence condition, the application of the latter in the case under consideration seems to us unjustifiable, so that the illustration does not prove the author's point ; and indeed, there seems to be every reason to preserve the unrestricted validity of the consequence condition. As a matter of fact, Mr. Barrett himself argues that " the degree of confirmation for the consequence of a sentence cannot be less than that of the sentence itself " ; this is indeed quite sound ; but it is hard to see how the recognition of this principle can be reconciled with a renunciation of the special consequence condition, since the latter may be considered simply as the correlate, for the non-gradated relation of confirmation, of the former principle which is adapted to the concept of degree of confirmation.

[2] For a fourth condition, see n. 1, p. 110.

[3] A contradictory observation report confirms every hypothesis (*cf.* n. 1, p. 103) and is, of course, incompatible with every one of the hypotheses it confirms.

felt to embody a too severe restriction. It might be pointed out, for example, that a finite set of measurements concerning the variation of one physical magnitude, x, with another, y, may conform to, and thus be said to confirm, several different hypotheses as to the particular mathematical function in terms of which the relationship of x and y can be expressed ; but such hypotheses are incompatible because to at least one value of x, they will assign different values of y.

No doubt it is possible to liberalize the formal standards of adequacy in line with these considerations. This would amount to dropping (8.3) and (8.32) and retaining only (8.31). One of the effects of this measure would be that when a logically consistent observation report B confirms each of two hypotheses, it does not necessarily confirm their conjunction ; for the hypotheses might be mutually incompatible, hence their conjunction self-contradictory ; consequently, by (8.31), B could not confirm it.—This consequence is intuitively rather awkward, and one might therefore feel inclined to suggest that while (8.3) should be dropped and (8.31) retained, (8.32) should be replaced by the requirement (8.33): If an observation sentence confirms each of two hypotheses, then it also confirms their conjunction. But it can readily be shown that by virtue of (8.2) this set of conditions entails the fulfilment of (8.32).

If, therefore, the condition (8.3) appears to be too rigorous, the most obvious alternative would seem to lie in replacing (8.3) and its corollaries by the much weaker condition (8.31) alone ; and it is an important problem whether an intuitively adequate definition of confirmation can be constructed which satisfies (8.1), (8.2) and (8.31), but not (8.3).—One of the great advantages of a definition which satisfies (8.3) is that it sets a limit, so to speak, to the strength of the hypotheses which can be confirmed by given evidence.[1]

The remainder of the present study, therefore, will be concerned exclusively with the problem of establishing a definition of confirmation which satisfies the more severe formal conditions represented by (8.1), (8.2), and (8.3) together.

The fulfilment of these requirements, which may be regarded as general laws of the logic of confirmation, is of course only a necessary, not a sufficient, condition for the adequacy of any proposed definition of confirmation. Thus, e.g., if " B confirms

---

[1] This was pointed out to me by Dr. Nelson Goodman. The definition later to be outlined in this essay, which satisfies conditions (8.1), (8.2) and (8.3), lends itself, however, to certain generalizations which satisfy only the more liberal conditions of adequacy just considered.

H " were defined as meaning " B logically entails *H* ", then the above three conditions would clearly be satisfied ; but the definition would not be adequate because confirmation has to be a more comprehensive relation than entailment (the latter might be referred to as the special case of *conclusive* confirmation). Thus, a definition of confirmation, to be acceptable, also has to be materially adequate : it has to provide a reasonably close approximation to that conception of confirmation which is implicit in scientific procedure and methodological discussion. That conception is vague and to some extent quite unclear, as I have tried to show in earlier parts of this paper ; therefore, it would be too much to expect full agreement as to the material adequacy of a proposed definition of confirmation ; on the other hand, there will be rather general agreement on certain points ; thus, *e.g.*, the identification of confirmation with entailment, or the Nicod criterion of confirmation as analyzed above, or any definition of confirmation by reference to a " sense of evidence ", will probably now be admitted not to be adequate approximations to that concept of confirmation which is relevant for the logic of science.

On the other hand, the soundness of the logical analysis (which, in a clear sense, always involves a logical reconstruction) of a theoretical concept cannot be gauged simply by our feelings of satisfaction at a certain proposed analysis ; and if there are, say, two alternative proposals for defining a term on the basis of a logical analysis, and if both appear to come fairly close to the intended meaning, then the choice has to be made largely by reference to such features as the logical properties of the two reconstructions, and the comprehensiveness and simplicity of the theories to which they lead.

9. *The Satisfaction Criterion of Confirmation.*—As has been mentioned before, a precise definition of confirmation requires reference to some definite " language of science ", in which all observation reports and all hypotheses under consideration are assumed to be formulated, and whose logical structure is supposed to be precisely determined. The more complex this language, and the richer its logical means of expression, the more difficult it will be, as a rule, to establish an adequate definition of confirmation for it. However, the problem has been solved at least for certain cases : With respect to languages of a comparatively simple logical structure, it has been possible to construct an explicit definition of confirmation which satisfies all of the above logical requirements, and which appears to be intuitively rather adequate. An exposition of the technical details of this

definition has been published elsewhere ; [1] in the present study, which is concerned with the general logical and methodological aspects of the problem of confirmation rather than with technical details, it will be attempted to characterize the definition of confirmation thus obtained as clearly as possible with a minimum of technicalities.

Consider the simple case of the hypothesis H : ' (x)(Raven(x) ⊃ Black(x)) ', where ' Raven ' and ' Black ' are supposed to be terms of our observational vocabulary. Let B be an observation report to the effect that Raven(a) . Black(a) . $\sim$ Raven(c) . Black(c) . $\sim$ Raven(d) . $\sim$ Black(d). Then B may be said to confirm H in the following sense : There are three objects altogether mentioned in B, namely a, c, and d ; and as far as these are concerned, B informs us that all those which are ravens (*i.e.* just the object a) are also black.[2] In other words, from the information contained in B we can infer that the hypothesis H does hold true within the finite class of those objects which are mentioned in B.

Let us apply the same consideration to a hypothesis of a logically more complex structure. Let H be the hypothesis " Everybody likes somebody " ; in symbols : ' $(x)(Ey)$Likes(x, y)',

[1] In my article referred to in n. 1, p. 1. The logical structure of the languages to which the definition in question is applicable is that of the lower functional calculus with individual constants, and with predicate constants of any degree. All sentences of the language are assumed to be formed exclusively by means of predicate constants, individual constants, individual variables, universal and existential quantifiers for individual variables, and the connective symbols of denial, conjunction, alternation, and implication. The use of predicate variables or of the identity sign is not permitted.

As to the predicate constants, they are all assumed to belong to the observational vocabulary, *i.e.* to denote a property or a relation observable by means of the accepted techniques. (" Abstract " predicate terms are supposed to be defined in terms of those of the observational vocabulary and then actually to be replaced by their *definientia*, so that they never occur explicitly.)

As a consequence of these stipulations, an observation report can be characterized simply as a conjunction of sentences of the kind illustrated by ' $P(a)$ ', ' $\sim P(b)$ ', ' $R(c, d)$ ', ' $\sim R(e, f)$', etc., where ' P ', ' R ', etc., belong to the observational vocabulary, and ' a ', ' b ', ' c ', ' d ', ' e ', ' f ', etc., are individual names, denoting specific objects. It is also possible to define an observation report more liberally as any sentence containing no quantifiers, which means that besides conjunctions also alternations and implication sentences formed out of the above kind of components are included among the observation reports.

[2] I am indebted to Dr. Nelson Goodman for having suggested this idea ; it initiated all those considerations which finally led to the definition to be outlined below.

i.e. for every (person) x, there exists at least one (not necessarily different person) y such that x likes y. (Here again, ' Likes ' is supposed to be a relation-term which occurs in our observational vocabulary.) Suppose now that we are given an observation report B in which the names of two persons, say ' e ' and ' f ', occur. Under what conditions shall we say that B confirms H ? The previous illustration suggests the answer : If from B we can infer that H is satisfied within the finite class $\{e, f\}$; *i.e.* that within $\{e, f\}$ everybody likes somebody. This in turn means that e likes e or f, and f likes e or f. Thus, B would be said to confirm H if B entailed the statement " e likes e or f, and f likes e or f ". This latter statement will be called the development of H for the finite class $\{e, f\}$.—

The concept of *development of a hypothesis, H, for a finite class of individuals, C*, can be defined in a general fashion ; the development of H for C states what H would assert if there existed exclusively those objects which are elements of C.—Thus, *e.g.*, the development of the hypothesis $H_1 =$ ' $(x)(P(x) \lor Q(x))$ ' (*i.e.* " Every object has the property P or the property Q ") for the class $\{a, b\}$ is ' $(P(a) \lor Q(a)) \cdot (P(b) \lor Q(b))$ ' (*i.e.* " a has the property P or the property Q, and b has the property P or the property Q ") ; the development of the existential hypothesis H_2 that at least one object has the property P, *i.e.* ' $(Ex)P(x)$ ', for $\{a, b\}$ is ' $P(a) \lor P(b)$ ' ; the development of a hypothesis which contains no quantifiers, such as H_3 : ' $P(c) \lor Q(c)$ ' is defined as that hypothesis itself, no matter what the reference class of individuals is.

A more detailed formal analysis based on considerations of this type leads to the introduction of a general relation of confirmation in two steps ; the first consists in defining a special relation of direct confirmation along the lines just indicated ; the second step then defines the general relation of confirmation by reference to direct confirmation.

Omitting minor details, we may summarize the two definitions as follows :

(9.1 Df.) An observation report B directly confirms a hypothesis H if B entails the development of H for the class of those objects which are mentioned in B.

(9.2 Df.) An observation report B confirms a hypothesis H if H is entailed by a class of sentences each of which is directly confirmed by B.

The criterion expressed in these definitions might be called the satisfaction criterion of confirmation because its basic idea consists in construing a hypothesis as confirmed by a given observation report if the hypothesis is satisfied in the finite class of those individuals which are mentioned in the report.—Let us now apply the two definitions to our last examples : The observation report B_1 : ' $P(a) . Q(b)$ ' directly confirms (and therefore also confirms) the hypothesis H_1, because it entails the development of H_1 for the class $\{a, b\}$, which was given above.—The hypothesis H_3 is not directly confirmed by B, because its development—i.e. H_3 itself—obviously is not entailed by B_1. However, H_3 is entailed by H_1, which is directly confirmed by B_1 ; hence, by virtue of (9.2), B_1 confirms H_3.

Similarly, it can readily be seen that B_1 directly confirms H_2.

Finally, to refer to the first illustration given in this section : The observation report ' Raven(a) . Black(a) . $\sim$ Raven(c) . $\sim$ Black(c) . $\sim$ Raven(d) . $\sim$ Black(d) ' confirms (even directly) the hypothesis ' (x)(Raven(x) $\supset$ Black(x)) ', for it entails the development of the latter for the class $\{a, c, d\}$, which can be written as follows : ' (Raven(a) $\supset$ Black(a)) . (Raven(c) $\supset$ Black(c)) . (Raven (d) $\supset$ Black(d)) '.

It is now easy to define disconfirmation and neutrality :

(9.3 Df.) An observation report B disconfirms a hypothesis H if it confirms the denial of H.

(9.4 Df.) An observation report B is neutral with respect to a hypothesis H if B neither confirms nor disconfirms H.

By virtue of the criteria laid down in (9.2), (9.3), (9.4), every consistent observation report, B, divides all possible hypotheses into three mutually exclusive classes : those confirmed by B, those disconfirmed by B, and those with respect to which B is neutral.

The definition of confirmation here proposed can be shown to satisfy all the formal conditions of adequacy embodied in (8.1), (8.2), and (8.3) and their consequences ; for the condition (8.2) this is easy to see ; for the other conditions the proof is more complicated.[1]

[1] For these proofs, see the article referred to in n. 1, p. 1. I should like to take this opportunity to point out and to remedy a certain defect of the definition of confirmation which was developed in that article, and which has been outlined above : this defect was brought to my attention by a discussion with Dr. Olaf Helmer.

It will be agreed that an acceptable definition of confirmation should satisfy the following further condition which might well have been included among the logical standards of adequacy set up in section 8 above :

(8.4). If B_1 and B_2 are logically equivalent observation reports and B_1 confirms (disconfirms, is neutral with respect to) a hypothesis H, then B_2, too, confirms (disconfirms, is neutral with respect to) H. This condition is indeed satisfied if observation reports are construed, as they have been in this article, as classes or conjunctions of observation sentences. As was indicated at the end of n. 1, p. 108, however, this restriction of observation

Furthermore, the application of the above definition of confirmation is not restricted to hypotheses of universal conditional form (as Nicod's criterion is, for example), nor to universal hypotheses in general ; it applies, in fact, to any hypothesis which can be expressed by means of property and relation terms of the observational vocabulary of the given language, individual names, the customary connective symbols for ' not ', ' and ', ' or ', ' if-then ', and any number of universal and existential quantifiers. Finally, as is suggested by the preceding illustrations as well as by the general considerations which underlie the establishment of the above definition, it seems that we have obtained a definition

reports to a conjunctive form is not essential ; in fact, it has been adopted here only for greater convenience of exposition, and all the preceding results, including especially the definitions and theorems of the present section, remain applicable without change if observation reports are given the more liberal interpretation characterized at the end of n. 1, p. 108. (In this case, if ' P ' and ' Q ' belong to the observational vocabulary, such sentences as ' $P(a) \lor Q(a)$ ', ' $P(a) \lor \sim Q(b)$ ', etc., would qualify as observation reports.) This broader conception of observation reports was therefore adopted in the article referred to in n. 1, p. 1 ; but it has turned out that in this case, the definition of confirmation summarized above does not generally satisfy the requirement (8.4). Thus, e.g., the observation reports, $B_1 = $ ' $P(a)$ ' and $B_2 = $ ' $P(a) \cdot (Q(b) \lor \sim Q(b))$ ' are logically equivalent, but while B_1 confirms (and even directly confirms) the hypothesis $H_1 = $ ' $(x)P(x)$ ', the second report does not do so, essentially because it does not entail ' $P(a) \cdot P(b)$ ', which is the development of H_1 for the class of those objects mentioned in B_2. This deficiency can be remedied as follows : The fact that B_2 fails to confirm H_1 is obviously due to the circumstance that B_2 contains the individual constant ' b ', without asserting anything about b : The object b is mentioned only in an analytic component of B_2. The atomic constituent ' $Q(b)$ ' will therefore be said to occur (twice) inessentially in B_2. Generally, an atomic constituent A of a molecular sentence S will be said to occur inessentially in S if by virtue of the rules of the sentential calculus S is equivalent to a molecular sentence in which A does not occur at all. Now an object will be said to be mentioned inessentially in an observation report if it is mentioned only in such components of that report as occur inessentially in it. The sentential calculus clearly provides mechanical procedures for deciding whether a given observation report mentions any object inessentially, and for establishing equivalent formulations of the same report in which no object is mentioned inessentially. Finally, let us say that an object is mentioned essentially in an observation report if it is mentioned, but not only mentioned inessentially, in that report. Now we replace 9.1 by the following definition :

(9.1a) An observation report B directly confirms a hypothesis H if B entails the development of H for the class of those objects which are mentioned essentially in B.

The concept of confirmation as defined by (9.1a) and (9.2) now satisfies (8.4) in addition to (8.1), (8.2), (8.3) even if observation reports are construed in the broader fashion characterized earlier in this footnote.

112   CARL G. HEMPEL:

of confirmation which also is materially adequate in the sense of being a reasonable approximation to the intended meaning of confirmation.

A brief discussion of certain special cases of confirmation might serve to shed further light on this latter aspect of our analysis.

10. *The Relative and the Absolute Concepts of Verification and Falsification.*—If an observation report entails a hypothesis H, then, by virtue of (8.1), it confirms H. This is in good agreement with the customary conception of confirming evidence ; in fact, we have here an extreme case of confirmation, the case where B *conclusively confirms* H ; this case is realized if, and only if, B entails H. We shall then also say that B *verifies* H. Thus, verification is a special case of confirmation ; it is a logical relation between sentences ; more specifically, it is simply the relation of entailment with its domain restricted to observation sentences.

Analogously, we shall say that B *conclusively disconfirms* H, or B *falsifies* H, if and only if B is incompatible with H ; in this case, B entails the denial of H and therefore, by virtue of (8.1) and (9.3), confirms the denial of H and disconfirms H. Hence, falsification is a special case of disconfirmation ; it is the logical relation of incompatibility between sentences, with its domain restricted to observation sentences.

Clearly, the concepts of *verification* and *falsification* as here defined are *relative;* a hypothesis can be said to be verified or falsified only with respect to some observation report ; and a hypothesis may be verified by one observation report and may not be verified by another. There are, however, hypotheses which cannot be verified and others which cannot be falsified by any observation report. This will be shown presently. We shall say that a given *hypothesis is verifiable (falsifiable)* if it is possible to construct an observation report which verifies (falsifies) the hypothesis. Whether a hypothesis is verifiable, or falsifiable, in this sense depends exclusively on its logical form. Briefly, the following cases may be distinguished :

(*a*) If a hypothesis does not contain the quantifier terms " all " and " some " or their symbolic equivalents, then it is both verifiable and falsifiable. Thus, e.g., the hypothesis " Object *a* turns blue or green " is entailed and thus verified by the report " Object *a* turns blue " ; and the same hypothesis is incompatible with, and thus falsified by, the report " Object *a* turns neither blue nor green ".

(*b*) A purely existential hypothesis (*i.e.* one which can be symbolized by a formula consisting of one or more existential quantifiers followed by a sentential function containing no quantifiers) is verifiable, but not falsifiable, if—as is usually assumed—the universe of discourse contains an infinite number of objects.—Thus, e.g., the hypothesis " There are blue roses " is verified by the observation report " Object a is a blue rose ", but no finite observation report can ever contradict and thus falsify the hypothesis.

(c) Conversely, a purely universal hypothesis (symbolized by a formula consisting of one or more universal quantifiers followed by a sentential function containing no quantifiers) is falsifiable but not verifiable for an infinite universe of discourse. Thus, e.g., the hypothesis " $(x)(\text{Swan}(x) \supset \text{White}(x))$ " is completely falsified by the observation report $\{\text{Swan}(a), \sim \text{White}(a)\}$; but no finite observation report can entail and thus verify the hypothesis in question.

(d) Hypotheses which cannot be expressed by sentences of one of the thfee types mentioned so far, and which in this sense require both universal and existential quantifiers for their formulation, are as a rule neither verifiable nor falsifiable.[1]   Thus, e.g., the hypothesis " Every substance is soluble in some solvent "— symbolically ' $(x)(Ey)\text{Soluble}(x, y)$ '—is neither entailed by, nor incompatible with any observation report, no matter how many cases of solubility or non-solubility of particular substances in particular solvents the report may list.   An analogous remark applies to the hypothesis " You can fool some of the people all of the time ", whose symbolic formulation ' $(Ex)(t)\text{Fl}(x,t)$ ' contains one existential and one universal quantifier.   But of course, all of the hypotheses belonging to this fourth class are capable of being confirmed or disconfirmed by suitable observation reports ; this was illustrated early in section 9 by reference to the hypothesis ' $(x)(Ey)\text{Likes}(x, y)$ '.

This rather detailed account of verification and falsification has been presented not only in the hope of further elucidating the meaning of confirmation and disconfirmation as defined above, but also in order to provide a basis for a sharp differentiation of two meanings of verification (and similarly of falsification) which have not always been clearly separated in recent discussions of the character of empirical knowledge.   One of the two meanings of verification which we wish to distinguish here is the relative concept just explained ; for greater clarity we shall sometimes

---

[1] A more precise study of the conditions of non-verifiability and non-falsifiability would involve technicalities which are unnecessary for the purposes of the present study.   Not all hypotheses of the type described in (d) are neither verifiable nor falsifiable ; thus, e.g., the hypothesis ' $(x)(Ey)(P(x) \lor Q(y))$ ' is verified by the report ' $Q(a)$ ', and the hypothesis ' $(x)(Ey)(P(x) . Q(y))$ ' is falsified by ' $\sim P(a)$ '.

8

refer to it as *relative verification*. The other meaning is what may be called *absolute or definitive verification*. This latter concept of verification does not belong to formal logic, but rather to pragmatics [1] : it refers to the acceptance of hypotheses by " observers " or " scientists ", etc., on the basis of relevant evidence. Generally speaking, we may distinguish three phases in the scientific test of a given hypothesis (which do not necessarily occur in the order in which they are listed here). The first phase consists in the performance of suitable experiments or observations and the ensuing acceptance of observation sentences, or of observation reports, stating the results obtained ; the next phase consists in confronting the given hypothesis with the accepted observation reports, *i.e.* in ascertaining whether the latter constitute confirming, disconfirming or irrelevant evidence with respect to the hypothesis ; the final phase consists either in accepting or rejecting the hypothesis on the strength of the confirming or disconfirming evidence constituted by the accepted observation reports, or in suspending judgment, awaiting the establishment of further relevant evidence.

The present study has been concerned almost exclusively with the second phase ; as we have seen, this phase is of a purely logical character ; the standards of evaluation here invoked— namely the criteria of confirmation, disconfirmation and neutrality—can be completely formulated in terms of concepts belonging to the field of pure logic.

The first phase, on the other hand, is of a pragmatic character ; it involves no logical confrontation of sentences with other sentences. It consists in performing certain experiments or systematic observations and noting the results. The latter are expressed in sentences which have the form of observation reports, and their acceptance by the scientist is connected (by causal, not by logical relations) with experiences occurring in those tests. (Of course, a sentence which has the form of an observation report may in certain cases be accepted not on the basis of direct observation, but because it is confirmed by other observation reports which were previously established ; but this process is illustrative of the second phase, which was discussed before. Here we are considering the case where a sentence is accepted directly " on the basis of experiential findings " rather than because it is supported by previously established statements.)

The third phase, too, can be construed as pragmatic, namely as consisting in a decision on the part of the scientist or a group of

---

[1] In the sense in which the term is used by Carnap in the work referred to in n. 1, p. 22.

scientists to accept (or reject, òr leave in suspense, as the case may be) a given hypothesis after ascertaining what amount of confirming or of disconfirming evidence for the hypothesis is contained in the totality of the accepted observation sentences. However, it may well be attempted to give a reconstruction of this phase in purely logical terms. This would require the establishment of general "rules of acceptance"; roughly speaking, these rules would state how well a given hypothesis has to be confirmed by the accepted observation reports to be scientifically acceptable itself;[1] *i.e.* the rules would formulate criteria for the acceptance or rejection of a hypothesis by reference to the kind and amount of confirming or disconfirming evidence for it embodied in the totality of accepted observation reports; possibly, these criteria would also refer to such additional factors as the "simplicity" of the hypothesis in question, the manner in which it fits into the system of previously accepted theories, etc. It is at present an open question to what extent a satisfactory system of such rules can be formulated in purely logical terms.[2]

[1] A stimulating discussion of some aspects of what we have called rules of acceptance is contained in an article by Felix Kaufmann, 'The logical rules of scientific procedure', *Philosophy and Phenomenological Research*, June, 1942.

If an explicit definition of the degree of confirmation of a hypothesis were available, then it might be possible- to formulate criteria of acceptance in terms of the degree to which the accepted observation reports confirm the hypothesis in question.

[2] The preceding division of the test of an empirical hypothesis into three phases of different character may prove useful for the clarification of the question whether or to what extent an empiricist conception of confirmation implies a "coherence theory of truth". This issue has recently been raised by Bertrand Russell, who, in ch. x of his *Inquiry into Meaning and Truth*, has levelled a number of objections against the views of Otto Neurath on this subject (*cf.* the articles mentioned in the next footnote), and against statements made by myself in articles published in *Analysis* in 1935 and 1936. I should like to add here a few, necessarily brief, comments on this issue.

(1) While, in the articles in *Analysis*, I argued in effect that the only possible interpretation of the phrase "Sentence *S* is true" is "*S* is highly confirmed by accepted observation reports", I should now reject this view. As the work of A. Tarski, R. Carnap, and others has shown, it is possible to define a semantical concept of truth which is not synonymous with that of strong confirmation, and which corresponds much more closely to what has customarily been referred to as truth, especially in logic, but also in other contexts. Thus, *e.g.*, if *S* is any empirical sentence, then either *S* or its denial is true in the semantical sense, but clearly it is possible that neither *S* nor its denial is highly confirmed by available evidence. To assert that a hypothesis is true is equivalent to asserting the hypothesis

At any rate, the acceptance of a hypothesis on the basis of a sufficient body of confirming evidence will as a rule be tentative, and will hold only " until further notice ", *i.e.* with the proviso that if new and unfavourable evidence should turn up (in other words, if new observation reports should be accepted which disconfirm the hypothesis in question) the hypothesis will be abandoned again.

Are there any exceptions to this rule ? Are there any empirical hypotheses which are capable of being established definitively, hypotheses such that we can be sure that once accepted on the basis of experiential evidence, they will never have to be revoked ? Hypotheses of this kind will be called absolutely or definitively verifiable ; and the concept of absolute or definitive falsifiability will be construed analogously.

While the existence of hypotheses which are relatively verifiable or relatively falsifiable is a simple logical fact, which was illustrated in the beginning of this section, the question of the existence of absolutely verifiable, or absolutely falsifiable, hypotheses is a highly controversial issue which has received a great deal of attention in recent empiricist writings.[1] As the problem

itself ; therefore the truth of an empirical hypothesis can be ascertained only in the sense in which the hypothesis itself can be established : *i.e.* the hypothesis—and thereby *ipso facto* its truth—can be more or less well confirmed by empirical evidence ; there is no other access to the question of the truth of a hypothesis.

In the light of these considerations, it seems advisable to me to reserve the term ' truth ' for the semantical concept ; I should now phrase the statements in the *Analysis* articles as dealing with confirmation. (For a brief and very illuminating survey of the distinctive characteristics of truth and confirmation, see R. Carnap, " Wahrheit and Bewährung," *Actes Ier Congrès Internat. de Philosophie Scientifique 1935*, vol. 4 ; Paris, 1936.)

(2) It is now clear also in what sense the test of a hypothesis is a matter of confronting sentences with sentences rather than with " facts ", or a matter of the " coherence " of the hypothesis and the accepted basic sentences : All the logical aspects of scientific testing, *i.e.* all the criteria governing the second and third of the three phases distinguished above, are indeed concerned only with certain relationships between the hypotheses under test and certain other sentences (namely the accepted observation reports) ; no reference to extra-linguistic " facts " is needed. On the other hand, the first phase, the acceptance of certain basic sentences in connection with certain experiments or observations, involves, of course, extra-linguistic procedures ; but this had been explicitly stated by the author in the articles referred to before. The claim that the views concerning truth and confirmation which are held by contemporary logical empiricism involve a coherence theory of truth is therefore mistaken.

[1] *Cf.* especially A. Ayer, *The Foundations of Empirical Knowledge* (New York, 1940) ; see also the same author's article, " Verification and Experience ", *Proceedings of the Aristotelian Society* for 1937 ; R. Carnap,

is only loosely connected with the subject of this essay, we shall restrict ourselves here to a few general observations.

Let it be assumed that the language of science has the general structure characterized and presupposed in the previous discussions, especially in section 9. Then it is reasonable to expect that only such hypotheses can possibly be absolutely verifiable as are relatively verifiable by suitable observation reports ; hypotheses of universal form, for example, which are not even capable of relative verification, certainly cannot be expected to be absolutely verifiable : In however many instances such a hypothesis may have been borne out by experiential findings, it is always possible that new evidence will be obtained which disconfirms the hypothesis. Let us, therefore, restrict our search for absolutely verifiable hypotheses to the class of those hypotheses which are relatively verifiable.

Suppose now that *H* is a hypothesis of this latter type, and that it is relatively verified, *i.e.* logically entailed, by an observation report *B*, and that the latter is accepted in science as an account of the outcome of some experiment or observation. Can we then say that *H* is absolutely confirmed, that it will never be revoked ? Clearly, that depends on whether the report *B* has been accepted irrevocably, or whether it may conceivably suffer the fate of being disavowed later. Thus the question as to the existence of absolutely verifiable hypotheses leads back to the question of whether all, or at least some, observation reports become irrevocable parts of the system of science once they have been accepted in connection with certain observations or experiments. This question is not simply one of fact ; it cannot adequately be answered by a descriptive account of the research behaviour of scientists. Here, as in all other cases of logical analysis of science, the problem calls for a " rational reconstruction " of scientific procedure, *i.e.* for the construction of a consistent and comprehensive theoretical model of scientific inquiry, which is then to serve as a system of reference, or a standard, in the examination of any particular scientific research. The

" Ueber Protokollsätze ", *Erkenntnis*, vol. 3 (1932), and § 82 of the same author's *The Logical Syntax of Language* (see n. 1, p. 3). O. Neurath, " Protokollsätze ", *Erkenntnis*, vol. 3 (1932) ; " Radikaler Physikalismus und ' wirkliche Welt ' ", *Erkenntnis*, vol. 4 (1934) ; " Pseudo-rationalismus der Falsifikation ", *Erkenntnis*, vol. 5 (1935). K. Popper, *Logik der Forschung* (see n. 1, p. 4). H. Reichenbach, *Experience and Prediction* (Chicago, 1938), ch. iii. Bertrand Russell, *An Inquiry into Meaning and Truth* (New York, 1940), especially chs. x and xi. M. Schlick, " Ueber das Fundament der Erkenntnis ", *Erkenntnis*, vol. 4 (1934).

construction of the theoretical model has, of course, to be oriented by the characteristics of actual scientific procedure, but it is not determined by the latter in the sense in which a descriptive account of some scientific study would be. Indeed, it is generally agreed that scientists sometimes infringe the standards of sound scientific procedure ; besides, for the sake of theoretical comprehensiveness and systematization, the abstract model will have to contain certain idealized elements which cannot possibly be determined in detail by a study of how scientists actually work. This is true especially of observation reports : A study of the way in which laboratory reports, or descriptions of other types of observational findings, are formulated in the practice of scientific research is of interest for the choice of assumptions concerning the form and the status of observation sentences in the model of a " language of science " ; but clearly, such a study cannot completely determine what form observation sentences are to have in the theoretical model, nor whether they are to be considered as irrevocable once they are accepted.

Perhaps an analogy may further elucidate this view concerning the character of logical analysis : Suppose that we observe two persons whose language we do not understand playing a game on some kind of chess board ; and suppose that we want to " reconstruct " the rules of the game. A mere descriptive account of the playing-behaviour of the individuals will not suffice to do this ; indeed, we should not even necessarily reject a theoretical reconstruction of the game which did not always characterize accurately the actual moves of the players : we should allow for the possibility of occasional violations of the rules. Our reconstruction would rather be guided by the objective of obtaining a consistent and comprehensive system of rules which are as simple as possible, and to which the observed playing behaviour conforms at least to a large extent. In terms of the standard thus obtained, we may then describe and critically analyze any concrete performance of the game.

The parallel is obvious ; and it appears to be clear, too, that in both cases the decision about various features of the theoretical model will have the character of a convention, which is influenced by considerations of simplicity, consistency, and comprehensiveness, and not only by a study of the actual procedure of scientists at work.[1]

[1] A clear account of the sense in which the results of logical analysis represent conventions can be found in §§ 9-11 and 25-30 of K. Popper's *Logik der Forschung.* An illustration of the considerations influencing the

This remark applies in particular to the specific question under consideration, namely whether " there are " in science any irrevocably accepted observation reports (all of whose consequences would then be absolutely verified empirical hypotheses). The situation becomes clearer when we put the question into this form : Shall we allow, in our rational reconstruction of science, for the possibility that certain observation reports may be accepted as irrevocable, or shall the acceptance of all observation reports be subject to the "until further notice " clause ? In comparing the merits of the alternative stipulations, we should have to investigate the extent to which each of them is capable of elucidating the structure of scientific inquiry in terms of a simple, consistent theory. We do not propose to enter into a discussion of this question here except for mentioning that various considerations militate in favour of the convention that no observation report is to be accepted definitively and irrevocably.[1] If this alternative is chosen, then not even those hypotheses which are entailed by accepted observation reports are absolutely verified, nor are those hypotheses which are found incompatible with accepted observation reports thereby absolutely falsified : in fact, in this case, no hypothesis whatsoever would be absolutely verifiable or absolutely falsifiable. If, on the other hand, some—or even all—observation sentences are declared irrevocable once they have been accepted, then those hypotheses entailed by or incompatible with irrevocable observation sentences will be absolutely verified, or absolutely falsified, respectively.

It should now be clear that the concepts of absolute and of relative verifiability (and falsifiability) are of an entirely different character. Failure to distinguish them has caused considerable misunderstanding in recent discussions on the nature of scientific knowledge. Thus, *e.g.*, K. Popper's proposal to admit as scientific hypotheses exclusively sentences which are (relatively) falsifiable by suitable observation reports has been criticized by means of arguments which, in effect, support the claim that scientific hypotheses should not be construed as being absolutely falsifiable—a point that Popper had not denied.—As can be seen from our earlier discussion of relative falsifiability, however, Popper's proposal to limit scientific hypotheses to the form of (relatively) falsifiable sentences involves a very severe restriction

determination of various features of the theoretical model is provided by the discussion in n. 1, p. 24.

[1] *Cf.* especially the publications by Carnap, Neurath, and Popper mentioned in n. 1, p. 116 ; also Reichenbach, *loc. cit.*, ch. ii, § 9.

of the possible forms of scientific hypotheses [1] ; in particular, it rules out all purely existential hypotheses as well as most hypotheses whose formulation requires both universal and existential quantification ; and it may be criticized on this account ; for in terms of this theoretical reconstruction of science it seems difficult or altogether impossible to give an adequate account of the status and function of the more complex scientific hypotheses and theories.—

With these remarks let us conclude our study of the logic of confirmation. What has been said above about the nature of the logical analysis of science in general, applies to the present analysis of confirmation in particular : It is a specific proposal for a systematic and comprehensive logical reconstruction of a concept which is basic for the methodology of empirical science as well as for the problem area customarily called " epistemology ". The need for a theoretical clarification of that concept was evidenced by the fact that no general theoretical account of confirmation has been available so far, and that certain widely accepted conceptions of confirmation involve difficulties so serious that it might be doubted whether a satisfactory theory of the concept is at all attainable.

It was found, however, that the problem can be solved : A general definition of confirmation, couched in purely logical terms, was developed for scientific languages of a specified and relatively simple logical character. The logical model thus obtained appeared to be satisfactory in the sense of the formal and material standards of adequacy that had been set up previously.

I have tried to state the essential features of the proposed analysis and reconstruction of confirmation as explicitly as possible in the hope of stimulating a critical discussion and of facilitating further inquiries into the various issues pertinent to this problem area. Among the open questions which seem to deserve careful consideration, I should like to mention the exploration of concepts of confirmation which fail to satisfy the general consistency condition ; the extension of the definition of confirmation to the case where even observation sentences containing quantifiers are permitted ; and finally the development of

---

[1] This was pointed out by R. Carnap ; *cf.* his review of Popper's book in *Erkenntnis*, vol. 5 (1935), and " Testability and Meaning " (see n. 1, p. 5) §§ 25, 26. For the discussion of Popper's falsifiability criterion, see for example H. Reichenbach, " Ueber Induktion and Wahrscheinlichkeit ", *Erkenntnis*, vol. 5 (1935) ; O. Neurath, " Pseudorationalismus der Falsifikation ", *Erkenntnis*, vol. 5 (1935).

a definition of confirmation for languages of a more complex logical structure than that incorporated in our model.[1] Languages of this kind would provide a greater variety of means of expression and would thus come closer to the high logical complexity of the language of empirical science.

[1] The languages to which our definition is applicable have the structure of the lower functional calculus without identity sign (cf. n. 1, 1, p. 108) ; it would be highly desirable so to broaden the general theory of confirmation as to make it applicable to the lower functional calculus with identity sign, or even to the higher functional calculus ; for it seems hardly possible to give a precise formulation of more complex scientific theories without the logical means of expression provided by the higher functional calculus.

Confirmation and Relevance

Item: One of the earliest surprises to emerge from Carnap's precise and systematic study of confirmation was the untenability of the initially plausible Wittgenstein confirmation function c†. Carnap's objection rested on the fact that c† precludes "learning from experience" because it fails to incorporate suitable relevance relations. Carnap's alternative confirmation function c* was offered as a distinct improvement because it does sustain the desired relevance relations.[1]

Item: On somewhat similar grounds, it has been argued that the heuristic appeal of the concept of partial entailment, which is often used to explain the basic idea behind inductive logic, rests upon a confusion of relevance with nonrelevance relations. Once this confusion is cleared up, it seems, the apparent value of the analogy between full and partial entailment vanishes.[2]

Item: In a careful discussion, based upon his detailed analysis of relevance, Carnap showed convincingly that Hempel's classic conditions of adequacy for any explication of the concept of confirmation are vitiated by another confusion of relevance with nonrelevance relations.[3]

Item: A famous controversy, in which Popper charges that Carnap's theory of confirmation contains a logical inconsistency, revolves around the same issue. As a result of this controversy, Carnap acknowledged in the preface to the second edition of Logical Foundations of Probability

AUTHOR'S NOTE: I wish to express gratitude to the National Science Foundation for support of research on inductive logic and probability. Some of the ideas in this paper were discussed nontechnically in my article "Confirmation," Scientific American, 228, 5 (May 1973), 75–83.

[1] Rudolf Carnap, Logical Foundations of Probability (Chicago: University of Chicago Press, 1950), sec. 110A.
[2] Wesley C. Salmon, "Partial Entailment as a Basis for Inductive Logic," in Nicholas Rescher, ed., Essays in Honor of Carl G. Hempel (Dordrecht: Reidel, 1969), and "Carnap's Inductive Logic," Journal of Philosophy, 64 (1967), 725–39.
[3] Carnap, Logical Foundations, secs. 86–88.

Wesley C. Salmon

that the first edition had been unclear with regard to this very distinction between relevance and nonrelevance concepts.[4]

Item: A new account of statistical explanation, based upon relations of relevance, has recently been proposed as an improvement over Hempel's well-known account, which is based upon relations of high degree of confirmation.[5]

Item: The problem of whether inductive logic can embody rules of acceptance — i.e., whether there are such things as inductive inferences in the usual sense — has been a source of deep concern to inductive logicians since the publication of Carnap's *Logical Foundations of Probability* (1950). Risto Hilpinen has proposed a rule of inductive inference which, he claims, avoids the "lottery paradox," thus overcoming the chief obstacle to the admission of rules of acceptance. Hilpinen's rule achieves this feat by incorporating a suitable combination of relevance and high confirmation requirements.[6]

The foregoing enumeration of issues is designed to show the crucial importance of the relations between confirmation and relevance. Yet, in spite of the fact that many important technical results have been available at least since the publication of Carnap's *Logical Foundations of Probability* in 1950, it seems to me that their consequences for the concept of confirmation have not been widely acknowledged or appreciated. In the first three sections of this paper, I shall summarize some of the most important facts, mainly to make them available in a single concise and relatively nontechnical presentation. All the material of these sections is taken from the published literature, much of it from the latter chapters of Carnap's book. In section 4, in response to questions raised by Adolf Grünbaum,[7] I shall apply some of these considerations to the Duhemian problems, where, to the best of my knowledge, they have not previously been brought to bear. In section 5, I shall attempt to pinpoint the source of some apparent difficulties with relevance concepts, and in the final section, I shall try to draw some morals from these results. All in all, I believe that many of the facts are shocking and counterintuitive

[4] Karl R. Popper, *The Logic of Scientific Discovery* (New York: Basic Books, 1959), appendix 9; Carnap, *Logical Foundations of Probability*, preface to the 2nd ed., 1962.

[5] Wesley C. Salmon, *Statistical Explanation and Statistical Relevance* (Pittsburgh: University of Pittsburgh Press, 1971).

[6] Risto Hilpinen, *Rules of Acceptance and Inductive Logic*, in Acta Philosophica Fennica, XXII (Amsterdam: North-Holland, 1968).

[7] In private conversation. At the same time he reported that much of his stimulus for raising these questions resulted from discussions with Professor Laurens Laudan.

and that they have considerable bearing upon current ideas about confirmation.

1. Carnap and Hempel

As Carnap pointed out in *Logical Foundations of Probability*, the concept of confirmation is radically ambiguous. If we say, for example, that the special theory of relativity has been confirmed by experimental evidence, we might have either of two quite distinct meanings in mind. On the one hand, we may intend to say that the special theory has become an accepted part of scientific knowledge and that it is very nearly certain in the light of its supporting evidence. If we admit that scientific hypotheses can have numerical degrees of confirmation, the sentence, on this construal, says that the degree of confirmation of the special theory on the available evidence is high. On the other hand, the same sentence might be used to make a very different statement. It may be taken to mean that some particular evidence — e.g., observations on the lifetimes of mesons — renders the special theory more acceptable or better founded than it was in the absence of this evidence. If numerical degrees of confirmation are again admitted, this latter construal of the sentence amounts to the claim that the special theory has a higher degree of confirmation on the basis of the new evidence than it had on the basis of the previous evidence alone.

The discrepancy between these two meanings is made obvious by the fact that a hypothesis h, which has a rather low degree of confirmation on prior evidence e, may have its degree of confirmation raised by an item of positive evidence i without attaining a high degree of confirmation on the augmented body of evidence $e.i$. In other words, a hypothesis may be confirmed (in the second sense) without being confirmed (in the first sense). Of course, we may believe that hypotheses can achieve high degrees of confirmation by the accumulation of many positive instances, but that is an entirely different matter. It is initially conceivable that a hypothesis with a low degree of confirmation might have its degree of confirmation increased repeatedly by positive instances, but in such a way that the confirmation approaches ¼ (say) rather than 1. Thus, it may be possible for hypotheses to be repeatedly confirmed (in the second sense) without ever getting confirmed (in the first sense). It can also work the other way. A hypothesis h that already has a high degree

299

of confirmation on evidence *e* may still have a high degree of confirmation on evidence *e.i.*, even though the addition of evidence *i* does not raise the degree of confirmation of *h*. In this case, *h* is confirmed (in the first sense) without being confirmed (in the second sense) on the basis of additional evidence *i*.

If we continue to speak in terms of numerical degrees of confirmation, as I shall do throughout this paper, we can formulate the distinction between these two senses of the term "confirm" clearly and concisely. For uniformity of formulation, let us assume some background evidence *e* (which may, upon occasion, be the tautological evidence *t*) as well as some additional evidence *i* on the basis of which degrees of confirmation are to be assessed. We can define "confirm" in the first (absolute; nonrelevance) sense as follows:

D1. Hypothesis *h* is confirmed (in the absolute sense) by evidence *i* in the presence of background evidence $e =_{df} c(h,e.i) > b$, where *b* is some chosen number, presumably close to 1.

This concept is absolute in that it makes no reference to the degree of confirmation of *h* on any other body of evidence.[8] The second (relevance) sense of "confirm" can be defined as follows:

D2. Hypothesis *h* is confirmed (in the relevance sense) by evidence *i* in the presence of background evidence $e =_{df} c(h,e.i) > c(h,e)$.

This is a relevance concept because it embodies a relation of change in degree of confirmation. Indeed, Carnap's main discussion of this distinction follows his technical treatment of relevance relations, and the second concept of confirmation is explicitly recognized as being identical with the concept of positive relevance.[9]

It is in this context that Carnap offers a detailed critical discussion of Hempel's criteria of adequacy for an explication of confirmation.[10] Car-

---

[8] The term "absolute probability" is sometimes used to refer to probabilities that are not relative or conditional. E.g., Carnap's null confirmation $c_0(h)$ is an absolute probability, as contrasted with $c(h,e)$ in which the degree of confirmation of *h* is relative to, or conditional upon, *e*. The distinction I am making between the concepts defined in D1 and D2 is quite different. It is a distinction between two different types of confirmation, where one is a conditional probability and the other is a relevance relation defined in terms of conditional probabilities. In this paper, I shall not use the concept of absolute probability at all; in place of null confirmation I shall always use the confirmation $c(h,t)$ on tautological evidence, which is equivalent to the null confirmation, but which is a conditional or relative probability.

[9] Carnap, *Logical Foundations*, sec. 86.

[10] These conditions of adequacy are presented in Carl G. Hempel, "Studies in the

nap shows conclusively, I believe, that Hempel has conflated the two concepts of confirmation, with the result that he has adopted an indefensible set of conditions of adequacy. As Carnap says, he is dealing with two explicanda, not with a single unambiguous one. The incipient confusion is signaled by his characterization of

the quest for general objective criteria determining (A) whether, and — if possible — even (B) to what degree, a hypothesis h may be said to be corroborated by a given body of evidence e. . . . The two parts of this . . . problem can be related in somewhat more precise terms as follows:
   (A) To give precise definitions of the two nonquantitative relational concepts of confirmation and disconfirmation; *i.e.* to define the meaning of the phrases 'e confirms h' and 'e disconfirms h'. (When e neither confirms nor disconfirms h, we shall say that e is neutral, or irrelevant, with respect to h.)
   (B) (1) To lay down criteria defining a metrical concept "degree of confirmation of h with respect to e," whose values are real numbers . . .[11]

The parenthetical remark under (A) makes it particularly clear that a relevance concept of confirmation is involved there, while a nonrelevance concept of confirmation is obviously involved in (B).

The difficulties start to show up when Hempel begins laying down conditions of adequacy for the concept of confirmation (A) (as opposed to degree of confirmation (B)). According to the very first condition "entailment is a special case of confirmation." This condition states:

   H-8.1 Entailment Condition. Any sentence which is entailed by an observation report is confirmed by it.[12]

If we are concerned with the absolute concept of confirmation, this condition is impeccable, for $c(h,e) = 1$ if e entails h. It is not acceptable, however, as a criterion of adequacy for a relevance concept of confirmation. For suppose our hypothesis h is "$(\exists x)Fx$" while evidence e is "Fa" and evidence i is "Fb." In this case, i entails h, but i does not confirm h in the relevance sense, for $c(h,e.i) = 1 = c(h,e)$.

Carnap offers further arguments to show that the following condition has a similar defect:

Logic of Confirmation," Mind, 54 (1945), 1–26, 97–121. Reprinted, with a 1964 postscript, in Carl G. Hempel, Aspects of Scientific Explanation (New York: Free Press, 1965). Page references in succeeding notes will be to the reprinted version.
   [11] Hempel, Aspects of Scientific Explanation, p. 6. Hempel's capital letters "H" and "E" have been changed to lowercase for uniformity with Carnap's notation.
   [12] Ibid., p. 31. Following Carnap, an "H" is attached to the numbers of Hempel's conditions.

Wesley C. Salmon

H-8.3 Consistency Condition. Every logically consistent observation report is logically compatible with the class of all the hypotheses which it confirms.[13]

This condition, like the entailment condition, is suitable for the absolute concept of confirmation, but not for the relevance concept. For, although no two incompatible hypotheses can have high degrees of confirmation on the same body of evidence, an observation report can be positively relevant to a number of different and incompatible hypotheses, provided that none of them has too high a prior degree of confirmation on the background evidence e. This happens typically when a given observation is compatible with a number of incompatible hypotheses — when, for example, a given bit of quantitative data fits several possible curves.

The remaining condition Hempel wished to adopt is as follows:

H-8.2 Consequence Condition. If an observation report confirms every one of a class k of sentences, then it also confirms any sentence which is a logical consequence of k.[14]

It will suffice to look at two conditions that follow from it:[15]

H-8.21 Special Consequence Condition. If an observation report confirms a hypothesis h, then it also confirms every consequence of h.

H-8.22 Equivalence Condition. If an observation report confirms a hypothesis h, then it also confirms every hypothesis that is logically equivalent with h.

The equivalence condition must hold for both concepts of confirmation. Within the formal calculus of probability (which Carnap's concept of degree of confirmation satisfies) we can show that, if h is equivalent to h', then $c(h,e) = c(h',e)$, for any evidence e whatever. Thus, if h has a high degree of confirmation on $e.i$, h' does also. Likewise, if i increases the degree of confirmation of h, it will also increase the degree of confirmation of h'.

The special consequence condition is easily shown to be valid for the nonrelevance concept of confirmation. If h entails k, then $c(k,e) \geqq c(h,e)$; hence, if $c(h,e.i) > b$, then $c(k,e.i) > b$. But here, I think, our intuitions mislead us most seductively. It turns out, as Carnap has shown with great clarity, that the special consequence condition fails for the

[13] *Ibid.*, p. 33.
[14] *Ibid.*, p. 31.
[15] *Ibid.*

302

relevance concept of confirmation. It is entirely possible for i to be positively relevant to h without being positively relevant to some logical consequence k. We shall return in section 3 to a more detailed discussion of this fact.

The net result of the confusion of the two different concepts is that obviously correct statements about confirmation relations of one type are laid down as conditions of adequacy for explications of concepts of the other type, where, upon examination, they turn out to be clearly unsatisfactory. Carnap showed how the entailment condition could be modified to make it acceptable as a condition of adequacy.[16] As long as we are dealing with the relevance concept of confirmation, it looks as if the consistency condition should simply be abandoned. The equivalence condition appears to be acceptable as it stands. The special consequence condition, surprisingly enough, cannot be retained.

Hempel tried to lay down conditions for a nonquantitative concept of confirmation, and we have seen some of the troubles he encountered. After careful investigation of this problem, Carnap came to the conclusion that it is best to establish a quantitative concept of degree of confirmation and then to make the definition of the two nonquantitative concepts dependent upon it, as we have done in D1 and D2 above.[17] Given a quantitative concept and the two definitions, there is no need for conditions of adequacy like those advanced by Hempel. The nonquantitative concepts of confirmation are fully determined by those definitions, but we may, if we wish, see what general conditions such as H-8.1, H-8.2, H-8.21, H-8.22, H-8.3 are satisfied. In a 1964 postcript to the earlier article, Hempel expresses general agreement with this approach of Carnap.[18] Yet, he does so with such equanimity that I wonder whether he, as well as many others, recognize the profound and far-reaching consequences of the fact that the relevance concept of confirmation fails to satisfy the special consequence condition and other closely related conditions (which will be discussed in section 3).

2. Carnap and Popper

Once the fundamental ambiguity of the term "confirm" has been pointed out, we might suppose that reasonably well-informed authors

---

[16] Carnap, *Logical Foundations*, p. 473.
[17] *Ibid.*, p. 467.
[18] Hempel, *Aspects of Scientific Explanation*, p. 50.

Wesley C. Salmon

could easily avoid confusing the two senses. Ironically, even Carnap himself did not remain entirely free from this fault. In the preface to the second edition of *Logical Foundations of Probability* (1962), he acknowledges that the first edition was not completely unambiguous. In the new preface, he attempts to straighten out the difficulty.

In the first edition, Carnap had distinguished a triplet of confirmation concepts:[19]

1. Classificatory — e confirms h.

2. Comparative — e confirms h more than e' confirms h'.

3. Quantitative — the degree of confirmation of h on e is u

In the second edition, he sees the need for two such triplets of concepts.[20] For this purpose, he begins by distinguishing what he calls "concepts of firmness" and "concepts of increase of firmness." The concept of confirmation we defined above in D1, which was called an "absolute" concept, falls under the heading "concepts of firmness." The concept of confirmation we defined in D2, and called a "relevance" concept, falls under the heading "concepts of increase of firmness." Under each of these headings, Carnap sets out a triplet of classificatory, comparative, and quantitative concepts:

I. Three Concepts of Firmness

I-1. h is firm on the basis of e. $c(h,e) > b$, where b is some fixed number.

I-2. h is firmer on e than is h' on e'. $c(h,e) > c(h',e')$.

I-3. The degree of firmness of h on e is u. $c(h,e) = u$.

To deal with the concepts of increase of firmness, Carnap introduces a simple relevance measure $D(h,i) =_{df} c(h,i) - c(h,t)$. This is a measure of what might be called "initial relevance," where the tautological evidence t serves as background evidence. The second triplet of concepts is given as follows:

II. Three Concepts of Increase of Firmness

II-1. h is made firmer by i. $D(h,i) > 0$.

II-2. The increase in firmness of h due to i is greater than the increase of firmness of h' due to i'. $D(h,i) > D(h',i')$.

[19] Carnap, *Logical Foundations*, sec. 8.
[20] Carnap, *Logical Foundations*, 2nd ed., pp. xv–xvi.

II-3. The amount of increase of firmness of h due to i is w.

$$D(h,i) = w.$$

Given the foregoing arrays of concepts, any temptation we might have had to identify the absolute (nonrelevance) concept of confirmation with the original classificatory concept, and to identify the relevance concept of confirmation with the original comparative concept, while distinguishing both from the original quantitative concept of degree of confirmation, can be seen to be quite mistaken. What we defined above in D1 as the absolute concept of confirmation is clearly seen to coincide with Carnap's new classificatory concept I-1, while our relevance concept of confirmation defined in D2 obviously coincides with Carnap's other new classificatory concept II-1. Carnap's familiar concept of degree of confirmation (probability$_1$) is obviously his quantitative concept of firmness I-3, while his new quantitative concept II-3 coincides with the concept of degree of relevance. Although we shall not have much occasion to deal with the comparative concepts, it is perhaps worth mentioning that the new comparative concept II-2 has an important use. When we compare the strengths of different tests of the same hypothesis we frequently have occasion to say that a test that yields evidence i is better than a test that yields evidence i'; sometimes, at least, this means that i is more relevant to h than is i' — i.e., the finding i would increase the degree of confirmation of h by a greater amount than would the finding i'.

It is useful to have a more general measure of relevance than the measure D of initial relevance. We therefore define the relevance of evidence i to hypothesis h on (in the presence of) background evidence e as follows:[21]

D3. $R(i,h,e) = c(h,e.i) - c(h,e)$.

Then we can say:

D4. i is positively relevant to h on $e =_{df} R(i,h,e) > 0$.

i is negatively relevant to h on $e =_{df} R(i,h,e) < 0$.

i is irrelevant to h on $e =_{df} R(i,h,e) = 0$.

[21] Carnap introduces the simple initial relevance measure D for temporary heuristic purposes in the preface to the second edition. In ch. 6 he discusses both our relevance measure R and Keynes's relevance quotient $c(h,e.i)/c(h,e)$, but for his own technical purposes he adopts a more complicated relevance measure $r(i,h,e)$. For purposes of this paper, I prefer the simpler and more intuitive measure R, which serves as well as Carnap's measure r in the present context. Since this measure differs from that used by Carnap in ch. 6 of *Logical Foundations*, I use a capital "R" to distinguish it from Carnap's lowercase symbol.

Wesley C. Salmon

Hence, the classificatory concept of confirmation in the relevance sense can be defined by the condition $R(i,h,e) > 0$. Using these relevance concepts, we can define a set of concepts of confirmation in the relevance sense as follows:

D5. i confirms h given $e =_{df} i$ is positively relevant to h on e.

i disconfirms h given $e =_{df} i$ is negatively relevant to h on e.

i neither confirms nor disconfirms h given $e =_{df} i$ is irrelevant to h on e.

Having delineated his two triplets of concepts, Carnap then acknowledges that his original triplet in the first edition was a mixture of concepts of the two types; in particular, he had adopted the classificatory concept of increase of firmness II-1, along with the comparative and quantitative concepts of firmness I-2 and I-3. This fact, plus some misleading informal remarks about these concepts, led to serious confusion.[22]

As Carnap further acknowledges, Popper called attention to these difficulties, but he fell victim to them as well.[23] By equivocating on the admittedly ambiguous concept of confirmation, he claimed to have derived a contradiction within Carnap's formal theory of probability₁. He offers the following example:

Consider the next throw with a homogeneous die. Let x be the statement 'six will turn up'; let y be its negation, that is to say, let $y = \bar{x}$; and let z be the information 'an even number will turn up'.
We have the following absolute probabilities:

$$p(x) = 1/6;\ p(y) = 5/6;\ p(z) = 1/2.$$

Moreover, we have the following relative probabilities:

$$p(x,z) = 1/3;\ p(y,z) = 2/3.$$

We see that x is supported by the information z, for z raises the probability of x from 1/6 to 2/6 = 1/3. We also see that y is undermined by z, for z lowers the probability of y by the same amount from 5/6 to 4/6 = 2/3. Nevertheless, we have $p(x,z) < p(y,z)$.[24]

From this example, Popper quite correctly draws the conclusion that there are statements x, y, and z such that z confirms x, z disconfirms y, and y has a higher degree of confirmation on z than x has. As Popper points out quite clearly, this result would be logically inconsistent if we were to take the term "confirm" in its nonrelevance sense. It would be self-contradictory to say,

[22] Carnap, Logical Foundations, 2nd ed., pp. xvii–xix.
[23] Ibid., p. xix, fn.
[24] Popper, The Logic of Scientific Discovery, p. 390.

The degree of confirmation of x on z is high, the degree of confirmation of y on z is not high, and the degree of confirmation of y on z is higher than the degree of confirmation of x on z.

The example, of course, justifies no such statement; there is no way to pick the number b employed by Carnap in his definition of confirmation in the (firmness) sense I-1 according to which we could say that the degree of confirmation of x on z is greater than b and the degree of confirmation of y on z is not greater than b. The proper formulation of the situation is:

The evidence z is positively relevant to x, the evidence z is negatively relevant to y, and the degree of confirmation of x on z is less than the degree of confirmation of y on z.

There is nothing even slightly paradoxical about this statement.

Popper's example shows clearly the danger of equivocating between the two concepts of confirmation, but it certainly does not show any inherent defect in Carnap's system of inductive logic, for this system contains both degree of confirmation $c(h,e)$ and degree of relevance $r(i,h,e)$. The latter is clearly and unambiguously defined in terms of the former, and there are no grounds for confusing them.[25] The example shows, however, the importance of exercising great care in our use of English language expressions in talking about these exact concepts.

3. The Vagaries of Relevance

It can be soundly urged, I believe, that the verb "to confirm" is used more frequently in its relevance sense than in the absolute sense. When we say that a test confirmed a hypothesis, we would normally be taken to mean that the result was positively relevant to the hypothesis. When we say that positive instances are confirming instances, it seems that we are characterizing confirming evidence as evidence that is positively relevant to the hypothesis in question. If we say that several investigators independently confirmed some hypothesis, it would seem sensible to understand that each of them had found positively relevant evidence. There is no need to belabor this point. Let us simply assert that the term "confirm" is often used in its relevance sense, and we wish to investigate some of the properties of this concept. In other words, let us agree for now to use the term "confirm" solely in its relevance sense

[25] Carnap, *Logical Foundations*, sec. 67.

Wesley C. Salmon

(unless some explicit qualification indicates otherwise), and see what we will be committed to.

It would be easy to suppose that, once we are clear on the two senses of the term *confirm* and once we resolve to use it in only one of these senses in a given context, it would be a simple matter to tidy up our usage enough to avoid such nasty equivocations as we have already discussed. This is not the case, I fear. For, as Carnap has shown by means of simple examples and elegant arguments, the relevance concept of confirmation has some highly counterintuitive properties.

Suppose, for instance, that two scientists are interested in the same hypothesis *h*, and they go off to their separate laboratories to perform tests of that hypothesis. The tests yield two positive results, *i* and *j*. Each of the evidence statements is positively relevant to *h* in the presence of common background information *e*. Each scientist happily reports his positive finding to the other. Can they now safely conclude that the net result of both tests is a confirmation of *h*? The answer, amazingly, is no! As Carnap has shown, two separate items of evidence can each be positively relevant to a hypothesis, while their conjunction is negative to that very same hypothesis. He offers the following example:[26]

Example 1. Let the prior evidence *e* contain the following information. Ten chess players participate in a chess tournament in New York City; some of them are local people, some from out of town; some are junior players, some are seniors; some are men (M), some are women (W). Their distribution is known to be as follows:

	Local players	Out-of-towners
Juniors	M, W, W	M, M
Seniors	M, M	W, W, W

Table 1

Furthermore, the evidence *e* is supposed to be such that on its basis each of the ten players has an equal chance of becoming the winner, hence 1/10 . . . It is assumed that in each case [of evidence that certain players have been eliminated] the remaining players have equal chances of winning.

Let *h* be the hypothesis that a man wins. Let *i* be the evidence that a local player wins; let *j* be the evidence that a junior wins. Using the equiprobability information embodied in the background evidence *e*, we can read the following values directly from table 1:

[26] *Ibid.*, pp. 382–83. Somewhat paraphrased for brevity.

308

$$c(h,e) = 1/2 \quad c(h,e.i) \quad = 3/5 \quad R(i,h,e) \quad = \quad 1/10$$
$$c(h,e.j) \quad = 3/5 \quad R(j,h,e) \quad = \quad 1/10$$
$$c(h,e.i.j) = 1/3 \quad R(i.j,h,e) = -1/6$$

Thus, i and j are each positively relevant to h, while the conjunction $i.j$ is negatively relevant to h. In other words, i confirms h and j confirms h but $i.j$ disconfirms h.

The setup of example 1 can be used to show that a given piece of evidence may confirm each of two hypotheses individually, while that same evidence disconfirms their conjunction.[27]

Example 2. Let the evidence e be the same as in example 1. Let h be the hypothesis that a local player wins; let k be the hypothesis that a junior wins. Let i be evidence stating that a man wins. The following values can be read directly from table 1:

$$c(h,e) \quad = 1/2 \quad c(h,e.i) \quad = 3/5 \quad R(i,h,e) \quad = \quad 1/10$$
$$c(k,e) \quad = 1/2 \quad c(k,e.i) \quad = 3/5 \quad R(i,k,e) \quad = \quad 1/10$$
$$c(h.k,e) = 3/10 \quad c(h.k,e.i) = 1/5 \quad R(i,h.k,e) = -1/10$$

Thus, i confirms h and i confirms k, but i disconfirms $h.k$.

In the light of this possibility it might transpire that a scientist has evidence that supports the hypothesis that there is gravitational attraction between any pair of bodies when at least one is of astronomical dimensions and the hypothesis of gravitational attraction between bodies when both are of terrestrial dimensions, but which disconfirms the law of universal gravitation! In the next section we shall see that this possibility has interesting philosophical consequences.

A further use of the same situation enables us to show that evidence can be positive to each of two hypotheses, and nevertheless negative to their disjunction.[28]

Example 3. Let the evidence e be the same as in example 1. Let h be the hypothesis that an out-of-towner wins; let k be the hypothesis that a senior wins. Let i be evidence stating that a woman wins. The following values can be read directly from table 1:

$$c(h,e) \quad = 1/2 \quad c(h,e.i) \quad = 3/5 \quad R(i,h,e) \quad = \quad 1/10$$
$$c(k,e) \quad = 1/2 \quad c(k,e.i) \quad = 3/5 \quad R(i,k,e) \quad = \quad 1/10$$
$$c(h \vee k,e) = 7/10 \quad c(h \vee k,e.i) = 3/5 \quad R(i,h \vee k,e) = -1/10$$

Thus, i confirms h and i confirms k, but i nevertheless disconfirms $h \vee k$.

[27] Ibid., pp. 394–95.
[28] Ibid., p. 384.

Wesley C. Salmon

Imagine the following situation:[29] a medical researcher finds evidence confirming the hypothesis that Jones is suffering from viral pneumonia and also confirming the hypothesis that Jones is suffering from bacterial pneumonia — yet this very same evidence disconfirms the hypothesis that Jones has pneumonia! It is difficult to entertain such a state of affairs, even as an abstract possibility.

These three examples illustrate a few members of a large series of severely counterintuitive situations that can be realized:

i. Each of two evidence statements may confirm a hypothesis, while their conjunction disconfirms it. (Example 1.)

ii. Each of two evidence statements may confirm a hypothesis, while their disjunction disconfirms it. (Example 2a, Carnap, *Logical Foundations*, p. 384.)

iii. A piece of evidence may confirm each of two hypotheses, while it disconfirms their conjunction. (Example 2.)

iv. A piece of evidence may confirm each of two hypotheses, while it disconfirms their disjunction. (Example 3.)

This list may be continued by systematically interchanging positive relevance (confirmation) and negative relevance (disconfirmation) throughout the preceding statements. Moreover, a large number of similar possibilities obtain if irrelevance is combined with positive or negative relevance. Carnap presents a systematic inventory of all of these possible relevance situations.[30]

In section 1, we mentioned that Hempel's special consequence condition does not hold for the relevance concept of confirmation. This fact immediately becomes apparent upon examination of statement iv above. Since h entails $h \vee k$, and since i may confirm h while disconfirming $h \vee k$, we have an instance in which evidence confirms a statement but fails to confirm one of its logical consequences. Statement ii, incidentally, shows that the converse consequence condition, which Hempel discusses but does not adopt,[31] also fails for the relevance concept of confirmation. Since $h.k$ entails h, and since i may confirm h without confirming $h.k$, we have an instance in which evidence confirms a hypothesis without confirming at least one statement from which that hypothesis follows. The failure of the special consequence condition and the converse consequence condition appears very mild when compared with the much

[29] This example is adapted from *ibid.*, p. 367.
[30] *Ibid.*, secs. 69, 71.

stronger results i–iv, and analogous ones. While one might, without feeling too queasy, give up the special consequence condition — the converse consequence condition being unsatisfactory on the basis of much more immediate and obvious considerations — it is much harder to swallow possibilities like i–iv without severe indigestion.

4. Duhem and Relevance

According to a crude account of scientific method, the testing of a hypothesis consists in deducing observational consequences and seeing whether or not the facts are as predicted. If the prediction is fulfilled we have a positive instance; if the prediction is false the result is a negative instance. There is a basic asymmetry between verification and falsification. If, from hypothesis h, an observational consequence o can be deduced, then the occurrence of a fact o' that is incompatible with o (o' entails $\sim o$) enables us to infer the falsity of h by good old modus tollens. If, however, we find the derived observational prediction fulfilled, we still cannot deduce the truth of h, for to do so would involve the fallacy of affirming the consequent.

There are many grounds for charging that the foregoing account is a gross oversimplification. One of the most familiar, which was emphasized by Duhem, points out that hypotheses are seldom, if ever, tested in isolation; instead, auxiliary hypotheses a are normally required as additional premises to make it possible to deduce observational consequences from the hypothesis h that is being tested. Hence, evidence o' (incompatible with o) does not entail the falsity of h, but only the falsity of the conjunction $h.a$. There are no deductive grounds on which we can say that h rather than a is the member of the conjunction that has been falsified by the negative outcome. To whatever extent auxiliary hypotheses are invoked in the deduction of the observational consequence, to that extent the alleged deductive asymmetry of verification and falsification is untenable.

At this point, a clear warning should be flashed, for we recall the strange things that happen when conjunctions of hypotheses are considered. Example 2 of the previous section showed that evidence that disconfirms a conjunction $h.a$ can nevertheless separately confirm each of the conjuncts. Is it possible that a *negative* result of a test of the hy-

[21] Hempel, *Aspects of Scientific Explanation*, pp. 32–33.

Wesley C. Salmon

pothesis h, in which auxiliary hypotheses a were also involved, could result in the confirmation of the hypothesis of interest h and in a confirmation of the auxiliary hypotheses a as well?

It might be objected, at this point, that in the Duhemian situation o' is not merely negatively relevant to $h.a$; rather,

(1) o' entails $\sim (h.a)$.

This objection, though not quite accurate, raises a crucial question. Its inaccuracy lies in the fact that h and a together do not normally entail o'; in the usual situation some initial conditions are required along with the main hypothesis and the auxiliary hypotheses. If this is the case, condition (1) does not obtain. We can deal with this trivial objection to (1), however, by saying that, since the initial conditions are established by observation, they are to be taken as part of the background evidence e which figures in all of our previous investigations of relevance relations. Thus, we can assert that, in the presence of background evidence e, o can be derived from $h.a$. This allows us to reinstate condition (1).

Unfortunately, condition (1) is of no help to us. Consider the following situation:

Example 4. The evidence e contains the same equiprobability assumptions as the evidence in example 1, except for the fact that the distribution of players is as indicated in the following table:

	Local players	Out-of-towners
Juniors	W	M, M
Seniors	M, M	M, W, W, W, W

Table 2

Let h be the hypothesis that a local player wins; let k be the hypothesis that a junior wins. Let i be evidence stating that a man wins. In this case, condition (1) is satisfied; the evidence i is logically incompatible with the conjunction $h.k$. The following values can be read directly from the table:

$c(h,e) = 0.3$ $c(h,e.i) = 0.4$ $R(i,h,e) = 0.1$
$c(k,e) = 0.3$ $c(k,e.i) = 0.4$ $R(i,k,e) = 0.1$
$c(h.k,e) = 0.1$ $c(h.k,e.i) = 0$ $R(i,h.k,e) = -0.1$

This example shows that evidence i, even though it conclusively refutes the conjunction $h.k$, nevertheless confirms both h and k taken individually.

312

Here is the situation. Scientist Smith comes home late at night after a hard day at the lab. "How did your work go today, dear?" asks his wife.

"Well, you know the Smith hypothesis, h_s, on which I have staked my entire scientific reputation? And you know the experiment I was running today to test my hypothesis? Well, the result was negative."

"Oh, dear, what a shame! Now you have to give up your hypothesis and watch your entire reputation go down the drain!"

"Not at all. In order to carry out the test, I had to make use of some auxiliary hypotheses."

"Oh, what a relief — saved by Duhem! Your hypothesis wasn't refuted after all," says the philosophical Mrs. Smith.

"Better than that," Smith continues, "I actually confirmed the Smith hypothesis."

"Why that's wonderful, dear," replies Mrs. Smith, "you found you could reject an auxiliary hypothesis, and that in so doing, you could establish the fact that the test actually confirmed your hypothesis? How ingenious!"

"No," Smith continues, "it's even better. I found I had confirmed the auxiliary hypotheses as well!"

This is the Duhemian thesis reinforced with a vengeance. Not only does a negative test result fail to refute the test hypothesis conclusively — it may end up confirming both the test hypothesis and the auxiliary hypotheses as well.

It is very tempting to suppose that much of the difficulty might be averted if only we could have sufficient confidence in our auxiliary hypotheses. If a medical researcher has a hypothesis about a disease which entails the presence of a certain microorganism in the blood of our favorite victim Jones, it would be outrageous for him to call into question the laws of optics as applied to microscopes as a way of explaining failure to find the bacterium. If the auxiliary hypotheses are well enough established beforehand, we seem to know where to lay the blame when our observational predictions go wrong. The question is how to establish the auxiliary hypotheses in the first place, for if the Duhemian is right, no hypotheses are ever tested in isolation. To test any hypothesis, according to this view, it is necessary to utilize auxiliary hypotheses; consequently, to establish our auxiliary hypotheses a for use in the tests of h, we would need some other auxiliary hypotheses a' to carry out the tests of a. A vicious regress threatens.

Wesley C. Salmon

A more contextual approach might be tried.[32] Suppose that a has been used repeatedly as an auxiliary hypothesis in the *successful* testing of other hypotheses j, k, l, etc. Suppose, that is, that the conjunctions j.a, k.a, l.a, etc., have been tested and repeatedly confirmed — i.e., all test results have been positively relevant instances. Can we say that a has been highly confirmed as a result of all of these successes? Initially, we might have been tempted to draw the affirmative conclusion, but by now we know better. Examples similar to those of the previous section can easily be found to show that evidence positively relevant to a conjunction of two hypotheses can nevertheless be negative to each conjunct.[33] It is therefore logically possible for each confirmation of a conjunction containing a to constitute a disconfirmation of a — and indeed a disconfirmation of the other conjunct as well in each such case.

There is one important restriction that applies to the case in which new observational evidence refutes a conjunction of two hypotheses, namely, hypotheses that are incompatible on evidence e.i can have, at most, probabilities that add to one. If e.i entails $\sim (h.k)$

$$c(h,e.i) + c(k,e.i) \leq 1.$$

Since we are interested in the case in which i is positively relevant to both h and k, these hypotheses must also satisfy the condition

$$c(h,e) + c(k,e) < 1.$$

We have here, incidentally, what remains of the asymmetry between confirmation and refutation. If evidence i refutes the conjunction h.k, that fact places an upper limit on the sum of the probabilities of h and k relative to e.i. If, however, evidence i confirms a conjunction h.k while disconfirming each of the conjuncts, there is no lower bound (other than zero) on the sum of their degrees of confirmation on i.

In this connection, let us recall our ingenious scientist Smith, who turned a refuting test result into a positive confirmation of both his pet hypothesis h_8 and his auxiliary hypotheses a. We see that he must have been working with a test hypothesis or auxiliaries (or both) which had rather low probabilities. We might well question the legitimacy of using hypotheses with degrees of confirmation appreciably less than one as

---

[32] This paper was motivated by Grünbaum's questions concerning this approach. See his "Falsifiability and Rationality" to be published in a volume edited by Joseph J. Kockelmans (proceedings of an international conference held at Pennsylvania State University).

[33] Carnap, *Logical Foundations*, pp. 394-95, 3b, is just such an example.

314

auxiliary hypotheses. If Smith's auxiliaries a had decent degrees of confirmation, his own hypothesis h_s must have been quite improbable. His clever wife might have made some choice remarks about his staking an entire reputation on so improbable a hypothesis. But I should not get carried away with dramatic license. If we eliminate all the unnecessary remarks about staking his reputation on h_s, and regard it rather as a hypothesis he finds interesting, then its initial improbability may be no ground for objection. Perhaps every interesting general scientific hypothesis starts its career with a very low prior probability. Knowing, as we do, that a positively relevant instance may disconfirm both our test hypothesis and our auxiliaries, while a negative instance may confirm them both, there remains a serious, and as yet unanswered, question how any hypothesis ever can become either reasonably well confirmed or reasonably conclusively disconfirmed (in the absolute sense). It obviously is still an open question how we could ever get any well-confirmed hypotheses to serve as auxiliaries for the purpose of testing other hypotheses.

Suppose, nevertheless, that we have a hypothesis h to test and some auxiliaries a that will be employed in conducting the test and that somehow we have ascertained that a has a higher prior confirmation than h on the initial evidence e:

$$c(a,e) > c(h,e).$$

Suppose, further, that as the result of the test we obtain negative evidence o' which refutes the conjunction $h.a$, but which confirms both h and a. Thus, o' entails $\sim(h.a)$ and

$$c(h,e.o') > c(h,e) \qquad c(a,e.o') > c(a,e).$$

We have already seen that this can happen (example 4). But now we ask the further question, is it possible that the posterior confirmation of h is greater than the posterior confirmation of a? In other words, can the negative evidence o' confirm both conjuncts and do so in a way that reverses the relation between h and a? A simple example will show that the answer is affirmative.

Example 5. The Department of History and Philosophy of Science at Polly Tech had two openings, one in history of science and the other in philosophy of science. Among the 1000 applicants for the position in history, 100 were women. Among the 2000 applicants for the position

Wesley C. Salmon

in philosophy, 100 were women. Let h be the hypothesis that the history job was filled by a woman; let k be the hypothesis that the philosophy job was filled by a woman. Since both selections were made by the use of a fair lottery device belonging to the inductive logician in the department,

$$c(h,e) = .1$$
$$c(k,e) = .05$$
$$c(h,e) > c(k,e).$$

Let i be the evidence that the two new appointees were discovered engaging in heterosexual intercourse with each other in the office of the historian. It follows at once that

$$c(h.k,e.i) = 0$$
$$c(h,e.i) + c(k,e.i) = 1$$

i.e., one appointee was a woman and the other a man, but we do not know which is which. Since it is considerably more probable, let us assume, that the office used was that of the male celebrant, we assign the values

$$c(h,e.i) = .2 \qquad c(k,e.i) = .8$$

with the result that

$$c(h,e.i) < c(k,e.i).$$

This illustrates the possibility of a reversal of the comparative relation between the test hypothesis and auxiliaries as a result of refuting evidence. It shows that a's initial superiority to h is no assurance that it will still be so subsequent to the refuting evidence. If, prior to the negative test result, we had to choose between h and a, we would have preferred a, but after the negative outcome, h is preferable to a.

There is one significant constraint that must be fulfilled if this reversal is to occur in the stated circumstances. If our auxiliary hypotheses a are initially better confirmed than our test hypothesis h, and if the conjunction $h.a$ is refuted by evidence o' that is positively relevant to both h and a, and if the posterior confirmation of h is greater than the posterior confirmation of a, then the prior confirmation of a must have been less than $1/2$. For,

$$c(h,e.o') + c(a,e.o') \leqq 1$$

and

$$c(h,e.o') > c(a,e.o').$$

316

Hence,

$$c(a,e.o') < 1/2.$$

Moreover,

$$c(a,e) < c(a,e.o').$$

Therefore,

$$c(a,e) < 1/2.$$

It follows that if a is initially more probable than h and also initially more probable than its own negation ~a, then it is impossible for a refuting instance o' which confirms both h and a to render h more probable than a. Perhaps that is some comfort. If our auxiliaries are more probable than not, and if they are better established before the test than our test hypothesis h, then a refuting test outcome which confirms both h and a cannot make h preferable to a.

But this is not really the tough case. The most serious problem is whether a refutation of the conjunction h.a can render h more probable than a by being positively relevant to h and negatively relevant to a, even when a is initially much more highly confirmed than h. You will readily conclude that this is possible; after all of the weird outcomes we have discussed, this situation seems quite prosaic. Consider the following example:

Example 6. Let

    e = Brown is an adult American male
    h = Brown is a Roman Catholic
    k = Brown is married

and suppose the following degrees of confirmation to obtain:

    $c(h,e) = .3$
    $c(k,e) = .8$
    $c(h.k,e) = .2.$

Let i be the information that Brown is a priest — that is, an ordained clergyman of the Roman Catholic, Episcopal, or Eastern Orthodox church. Clearly, i refutes the conjunction h.k, so

$$c(h.k,e.i) = 0.$$

Since the overwhelming majority of priests in America are Roman Catholic, let us assume that

$$c(h,e.i) = .9$$

317

and since some, but not all, non–Roman Catholic priests marry, let $c(k,e.i) = .05$.

We see that i is strongly relevant to both h and k; in particular, it is positively relevant to h and negatively relevant to k. Moreover, while k has a much higher degree of confirmation than h relative to the prior evidence e, h has a much higher degree of confirmation than k on the posterior evidence $e.i$. Thus, the refuting evidence serves to reverse the preferability relation between h and k.

It might be helpful to look at this situation diagrammatically and to think of it in terms of class ratios or frequencies. Since class ratios satisfy the mathematical calculus of probabilities, they provide a useful device for establishing the possibility of certain probability relations. With our background evidence e let us associate a reference class A, and with our hypotheses h and k let us associate two overlapping subclasses B and C respectively. With our additional evidence i let us associate a further subclass D of A. More precisely, let

$$e = x \in A, h = x \in B, k = x \in C, i = x \in D.$$

Since we are interested in the case in which the prior confirmation of k is high and the prior confirmation of h is somewhat lower, we want most of A to be included in C and somewhat less of A to be included in B. Moreover, since our hypotheses h and k should not be mutually exclusive on the prior evidence e alone, B and C must overlap. However, neither B nor C can be a subset of the other; they must be mutually exclusive within D, since h and k are mutually incompatible on additional evidence i. Moreover, because we are not considering cases in which $e.i$ entails either h or k alone, the intersections of D with B and C must both be nonnull. We incorporate all of these features in Figure 1. In order to achieve the desired result — that is, to have the posterior confirmation of h greater than the posterior confirmation of k — it is only necessary to draw D so that its intersection with B is larger than its intersection with C. This is obviously an unproblematic condition to fulfill. Indeed, there is no difficulty in arranging it so that the proportion of D occupied by its intersection with B is larger than the proportion of A occupied by its intersection with C. When this condition obtains, we can not only say that the evidence i has made the posterior confirmation of h greater than the posterior confirmation of k (thus

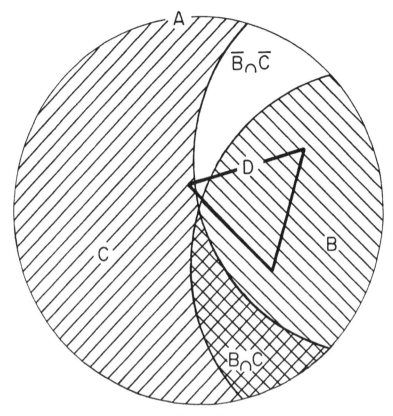

Figure 1

reversing their preferability standing), but also that the posterior confirmation of h is greater than the prior confirmation of k. Translated into the Duhemian situation, this means that not only can the refuting evidence o' disconfirm the initially highly probable auxiliary hypotheses a, but it can also confirm the test hypothesis h to the extent that its posterior confirmation makes it more certain than were the auxiliaries before the evidence o' turned up. This set of relationships is easily seen to be satisfied by example 6 if we let

$A =$ American men $B =$ Roman Catholics
$C =$ married men $D =$ priests.

It is evident from Figure 1 that C can be extremely probable relative to A without presenting any serious obstacle to the foregoing outcome, provided D is very much smaller than A. There seems little ground for

Wesley C. Salmon

assurance that a refuting experimental result will generally leave the auxiliaries intact, rather than drastically disconfirming them and radically elevating the degree of confirmation of the test hypothesis. This state of affairs can apparently arise with distressing ease, and there are no evident constraints that need to be fulfilled in order for it to happen. There seems to be no basis for confidence that it does not happen frequently in real scientific situations. If this is so, then whenever our auxiliaries have an initial degree of confirmation that falls ever so slightly short of certainty, we cannot legitimately infer with confidence that the test hypothesis h, rather than the auxiliaries, is disconfirmed by the refuting instance. Thus, no matter how high the initial probability of the auxiliaries a (provided it is somewhat less than certainty), it is still possible for a finding that entails the falsity of the conjunction $h.a$ to constitute a confirmation for either the one or the other. We certainly cannot say that the negative finding disconfirms h rather than a on the ground that a is more probable initially than h.

A parallel remark can be made about an instance that confirms the conjunction $h.a$. Such an instance might disconfirm either conjunct, and we have no way of saying which. In view of these dismal facts, we may well repeat, with even greater emphasis, the question posed earlier: how can any hypothesis (including those we need as auxiliary hypotheses for the testing of other hypotheses) ever be reasonably well confirmed or disconfirmed (in the absolute sense)?

The (to my mind) shocking possibilities that we have been surveying arise as consequences of the fact that we have been construing "confirm" in the relevance sense. What happens, I believe, is that we acknowledge with Carnap and Hempel that the classificatory concept of confirmation is best understood, in most contexts, as a concept of positive relevance defined on the basis of some quantitative degree of confirmation function. Then, in contexts such as the discussion of the Duhemian thesis, we proceed to talk casually about confirmation, forgetting that our intuitive notions are apt to be very seriously misleading. The old ambiguity of the absolute and the relevance sense of confirmation infects our intuitions, with the result that all kinds of unwarranted suppositions insinuate themselves. We find it extraordinarily difficult to keep a firm mental grasp upon such strange possibilities as we have seen in this section and the previous one.

320

5. Analysis of the Anomalies

There is, of course, a striking contrast between the "hypotheses" and "evidence" involved in our contrived examples, on the one hand, and the genuine hypotheses and evidence to be found in actual scientific practice, on the other. This observation might easily lead to the charge that the foregoing discussion is not pertinent to the logic of actual scientific confirmation, as opposed to the theory of confirmation constructed by Carnap on highly artificial and oversimplified languages. This irrelevance is demonstrated by the fact, so the objection might continue, that the kinds of problems and difficulties we have been discussing simply do not arise when real scientists test serious scientific hypotheses.

This objection, it seems to me, is wide of the mark. I am prepared to grant that such weird possibilities as we discussed in previous sections do not arise in scientific practice; at least, I have no concrete cases from the current or past history of science to offer as examples of them. This is, however, a statement of the problem rather than a solution. Carnap has provided a number of examples that, on the surface at least, seem to make a shambles of confirmation; why do they not also make a shambles of science itself? There can be no question that, for example, one statement can confirm each of two other statements separately while at the same time disconfirming their disjunction or conjunction. If that sort of phenomenon never occurs in actual scientific testing, it must be because we know something more about our evidence and hypotheses than merely that the evidence confirms the hypotheses. The problem is to determine the additional factors in the actual situation that block the strange results we can construct in the artificial case. In this section, I shall try to give some indications of what seems to be involved.

The crux of the situation seems to be the fact that we have said very little when we have stated merely that a hypothesis h has been confirmed by evidence i. This statement means, of course, that i raises the degree of confirmation of h, but that in itself provides very little information. It is by virtue of this paucity of content that we can go on and say that this same evidence i confirms hypothesis k as well, without being justified in saying anything about the effect of i upon the disjunction or the conjunction of h with k.

This state of affairs seems strange to intuitions that have been thoroughly conditioned on the extensional relations of truth-functional logic.

Wesley C. Salmon

Probabilities are not extensional in the same way. Given the truth-values of h and k we can immediately ascertain the truth-values of the disjunction and the conjunction. The degrees of confirmation ("probability values") of h and k do not, however, determine the degree of confirmation of either the disjunction or the conjunction. This fact follows immediately from the addition and multiplication rules of the probability calculus:

(2) $\quad c(h \vee k,e) = c(h,e) + c(k,e) - c(h.k,e)$

(3) $\quad c(h.k,e) = c(h,e) \times c(k,h.e) = c(k,e) \times c(h,k.e)$.

To determine the probability of the disjunction, we need, in addition to the values of the probabilities of the disjuncts, the probability of the conjunction. The disjunctive probability is the sum of the probabilities of the two disjuncts if they are mutually incompatible in the presence of evidence e, in which case $c(h.k,e) = 0$.[34] The probability of the conjunction, in turn, depends upon the probability of one of the conjuncts alone and the conditional probability of the other conjunct given the first.[35] If

(4) $\quad c(k,h.e) = c(k,e)$

the multiplication rule assumes the special form

(5) $\quad c(h.k,e) = c(h,e) \times c(k,e)$

in which case the probability of the conjunction is simply the product of the probabilities of the two conjuncts. When condition (4) is fulfilled, h and k are said to be independent of one another.[36] Independence, as thus defined, is obviously a relevance concept, for (4) is equivalent to the statement that h is irrelevant to k, i.e, $R(h,k,e) = 0$.

We can now see why strange things happen with regard to confirmation in the relevance sense. If the hypotheses h and k are mutually exclusive in the presence of e (and a fortiori in the presence of $e.i$), then

[34] The condition $c(h.k,e) = 0$ is obviously sufficient to make the probability of the disjunction equal to the sum of the probabilities of the disjuncts, and this is a weaker condition than e entails $\sim(h.k)$. Since the difference between these conditions has no particular import for the discussion of this paper, I shall, in effect, ignore it.

[35] Because of the commutativity of conjunction, it does not matter whether the probability of h conditional only on e or the probability of k conditional only on e is taken. This is shown by the double equality in formula (3).

[36] Independence is a symmetric relation; if h is independent of k then k will be independent of h.

(6) $c(h \lor k,e) = c(h,e) + c(k,e)$

(7) $c(h \lor k,e.i) = c(h,e.i) + c(k,e.i)$

so that if

(8) $c(h,e.i) > c(h,e)$ and $c(k,e.i) > c(k,e)$

it follows immediately that

(9) $c(h \lor k,e.i) > c(h \lor k,e)$.

Hence, in this special case, if i confirms h and i confirms k, then i must confirm their disjunction. This results from the fact that the relation between h and k is the same in the presence of $e.i$ as it is in the presence of e alone.[37]

If, however, h and k are not mutually exclusive on evidence e we must use the general formulas

(10) $c(h \lor k,e) = c(h,e) + c(k,e) - c(h.k,e)$

(11) $c(h \lor k,e.i) = c(h,e.i) + c(k,e.i) - c(h.k,e.i)$.

Now, if it should happen that the evidence i drastically alters the relevance of h to k in just the right way our apparently anomalous results can arise. For then, as we shall see in a moment by way of a concrete (fictitious) example, even though condition (8) obtains — i.e., i confirms h and i confirms k — condition (9) may fail. Thus, if

(12) $c(h.k,e.i) > c(h.k,e)$

it may happen that

(13) $c(h \lor k,e.i) < c(h \lor k,e)$

i.e., i disconfirms $h \lor k$. Let us see how this works.

Example 7. Suppose h says that poor old Jones has bacterial pneumonia, and k says that he has viral pneumonia. I am assuming that these are the only varieties of pneumonia, so that $h \lor k$ says simply that he has pneumonia. Let evidence e contain the results of a superficial diagnosis as well as standard medical background knowledge about the disease, on the basis of which we can establish degrees of confirmation for h, k, $h.k$, and $h \lor k$. Suppose, moreover, that the probability on e that Jones has both viral and bacterial pneumonia is quite low, that is, that people do not often get them both simultaneously. For the sake of definiteness, let us

---

[37] To secure this result it is not necessary that $c(h.k,e) = c(h.k,e.i) = 0$; it is sufficient to have $c(h.k,e) = c(h.k,e.i)$, though obviously this condition is not necessary either.

Wesley C. Salmon

introduce some numerical values. Suppose that on the basis of the super-ficial diagnosis it is 98 percent certain that Jones has one or the other form of pneumonia, but the diagnosis leaves it entirely uncertain which type he has. Suppose, moreover, that on the basis of e there is only a 2 percent chance that he has both. We have the following values:

$$c(h,e) \quad = .50 \qquad c(k,e) \quad = .50$$
$$c(h \vee k,e) = .98 \qquad c(h.k,e) = .02.$$

These values satisfy the addition formula (2). Suppose, now, that there is a certain test which indicates quite reliably those rare cases in which the subject has both forms of pneumonia. Let i be the statement that this test was administered to Jones with a positive result, and let this result make it 89 percent certain that Jones has both types. Assume, moreover, that the test rarely yields a positive result if the patient has only one form of pneumonia (i.e., when the positive result occurs for a patient who does not have both types, he usually has neither type). In particular, let

$$c(h,e.i) = .90, \quad c(k,e.i) = .90, \quad c(h.k,e.i) = .89$$
from which it follows that

$$c(h \vee k,e.i) = .91 < c(h \vee k,e) = .98.$$

The test result i thus confirms the hypothesis that Jones has bacterial pneumonia and the hypothesis that Jones has viral pneumonia, but it disconfirms the hypothesis that Jones has pneumonia!

It achieves this feat by greatly increasing the probability that he has both. This increase brings about a sort of clustering together of cases of viral and bacterial pneumonia, concomitantly decreasing the proportion

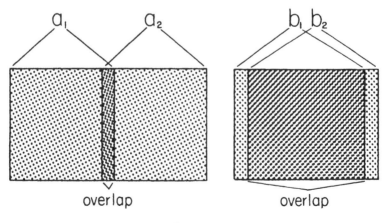

Figure 2

of people with only one of the two types. The effect is easily seen diagrammatically in Figure 2. Even though the rectangles in 2-b are larger than those in 2-a, those in 2-b cover a smaller total area on account of their very much greater degree of overlap. Taking the rectangles to represent the number of cases of each type, we see graphically how the probability of each type of pneumonia can increase simultaneously with a decrease in the overall probability of pneumonia. The evidence i has significantly altered the relevance relation between h and k. Using the multiplication formula (3), we can establish that

$$c(k,h.e) = .04 \qquad c(k,h.e.i) \cong .99$$
$$R(h,k,e) = -0.46 \qquad R(h,k,e.i) \cong .09$$

In the presence of e alone, h is negatively relevant to k; in the presence of i as well, h becomes positively relevant to k. There is nothing outlandish in such changes of relevance in the light of additional evidence. This case thus exemplifies condition (13) by satisfying condition (12).

A similar analysis enables us to understand how an item of evidence can confirm each of two hypotheses, while disconfirming — indeed, even while conclusively refuting — their conjunction. If hypotheses h and k are independent of each other in the presence of evidence $e.i$ and also in the presence of e alone, the following relations obtain:

(14) $c(h.k,e) = c(h,e) \times c(k,e)$

(15) $c(h.k,e.i) = c(h,e.i) \times c(k,e.i)$

so that if

(16) $c(h,e.i) > c(h,e)$ and $c(k,e.i) > c(k,e)$

it follows immediately that

(17) $c(h.k,e.i) > c(h.k,e)$.

Hence, in this special case, if i confirms h and i confirms k, then i must confirm $h.k$.

A different situation obtains if h and k are not independent on both e and $e.i$; in that case we must use the general formulas

(18) $c(h.k,e) = c(h,e) \times c(k,h.e)$

(19) $c(h.k,e.i) = c(h,e.i) \times c(k,h.e.i)$.

Even given that condition (16) still obtains, so that

(20) $c(k,e.i) > c(k,e)$

325

Wesley C. Salmon

it is still possible that

(21) $c(k,h.e.i) < c(k,h.e)$ [38]

which makes it possible, in turn, that

(22) $c(h.k,e.i) < c(h.k,e)$.

Since, according to (20) and (21),

(23) $c(k,e.i) - c(k,e) = R(i,k,e) > 0$
(24) $c(k,h.e.i) - c(k,h.e) = R(i,k,h.e) < 0$

the possibility of i confirming each of two hypotheses while discon-firming their conjunction depends upon the ability of h to make a dif-ference in the relevance of i to k. We said above, however, that the occurrence of the strange confirmation phenomena depends upon the possibility of a change in the relevance of the hypotheses to one another in the light of new evidence. These characterizations are, however, equiv-alent to one another, for the change in relevance of i to k brought about by h is equal to the change in relevance of h to k brought about by i, that is,

(25) $R(h,k,e) - R(h,k,e.i) = R(i,k,e) - R(i,k,h.e)$. [39]

We can therefore still maintain that the apparently anomalous confirma-

---

[38] To establish the compatibility of (20) and (21), perhaps a simple example, in addition to the one about to be given in the text, will be helpful. Let
    $e = X$ is a man. $i = X$ is American.
    $h = X$ is very wealthy. $k = X$ vacations in America.
Under this interpretation, relation (20) asserts: It is more probable that an American man vacations in America than it is that a man (regardless of nationality) vacations in America. Under the same interpretation, relation (21) asserts: It is less probable that a very wealthy American man will vacation in America than it is that a very wealthy man (regardless of nationality) will vacation in America. The interpretation of formula (20) seems like an obviously true statement; the interpretation of (21) seems likely to be true owing to the apparent tendency of the very wealthy to vaca-tion abroad. There is, in any case, no contradiction in assuming that every very wealthy American man vacations on the French Riviera, while every very wealthy man from any other country vacations in America.

[39] This equality can easily be shown by writing out the relevance terms according to their definitions as follows:
    $R(h,k,e) =_{df} c(k,h.e) - c(k,e)$
    $R(h,k,e.i) =_{df} c(k,h.e.i) - c(k,e.i)$
    $R(h,k,e) - R(h,k,e.i) = c(k,h.e) - c(k,e) - c(k,h.e.i) + c(k,e.i)$ (*).
    $R(i,k,e) =_{df} c(k,e.i) - c(k,e)$
    $R(i,k,h.e) =_{df} c(k,h.e.i) - c(k,h.e)$
    $R(i,k,e) - R(i,k,h.e) = c(k,e.i) - c(k,e) - c(k,h.e.i) + c(k,h.e)$ (**).
The right-hand sides of equations (*) and (**) obviously differ only in the arrange-ment of terms.

326

tion situation arises from the ability of new evidence to change relations of relevance between the hypotheses, as was suggested by our initial examination of the general addition and multiplication rules (2) and (3).

Let us illustrate the conjunctive situation with another concrete (though fictitious) example.

Example 8. Suppose that the evidence e tells us that two radioactive atoms A and B decay, each ejecting a particle, and that the probability in each case is 0.7 that it is an alpha particle, 0.2 that it is a negative electron, and 0.1 that it is a positive electron (positron). Assume that the two emissions are independent of one another. Let h be the statement that atom A emits a negative electron; let k be the statement that atom B emits a negative electron. We have the following probabilities:

$$c(h,e) = .2, \quad c(k,e) = .2, \quad c(h.k,e) = .04.$$

Let i be the observation that the two particles approach one another and annihilate upon meeting. Since this occurrence requires the presence of one positive and one negative electron, i entails $\sim(h.k)$. At the same time, since a negative electron must have been present, and since it is just as probable that it was emitted by atom A as atom B, we have

$$c(h,e.i) = .5 \quad \text{and} \quad c(k,e.i) = .5.$$

Hence, evidence i, which refutes the conjunction of the two hypotheses, confirms each one of them.[40]

This occurs because the evidence i makes the hypotheses h and k, which were independent of one another on evidence e alone, into mutually exclusive and exhaustive alternatives, i.e.,

$$c(k,h.e) - c(k,e) = R(h,k,e) = 0$$
$$c(k,h.e.i) - c(k,e.i) = R(h,k,e.i) = -.5.$$

Hypotheses that were totally irrelevant to each other in the absence of evidence i become very strongly relevant in the presence of i. Again, there is nothing especially astonishing about such a change in relevance as a result of new evidence.

Since, as we have seen, all the trouble seems to arise out of a change in the relevance of one hypothesis to the other as a result of new evidence, the most immediate suggestion might be to choose hypotheses h and k whose mutual relevance relations will not change in the light of

---

[40] This case constitutes a counterexample to the Hempel consistency condition H-8.3 discussed in sec. 2 above.

Wesley C. Salmon

the new evidence *i*. We have noted that this constancy of relevance is guaranteed if we begin with hypotheses that are mutually exclusive on evidence *e*; they remain mutually exclusive on any augmented evidence *e.i*. But when we use the conjunction of a test hypothesis *h* with auxiliary hypotheses *a* in order to attempt a test of *h*, we certainly do not want *h* and *a* to be mutually exclusive — that is, we do not want to be in the position of knowing that we must reject our test hypothesis *h* if we are prepared to accept the auxiliaries *a*, even without the addition of any new evidence *i*. It would be more reasonable to insist that the auxiliary hypotheses *a* should themselves be neutral (irrelevant, independent) to the test hypothesis *h*. If that condition is satisfied, we can accept the auxiliary hypotheses *a* and still keep an entirely open mind regarding *h*. We cannot, however, demand that *h* and *a* remain irrelevant to one another after the new evidence *i* has been obtained. The interesting test situation is that in which, given *e*, *h.a* entails some observational consequence *o*. If a result *o'* occurs which is incompatible with *o*, then our hypotheses *h* and *a*, which may have been independent in the presence of *e* alone, are mutually exclusive in the light of the new evidence *o'*. Thus, the very design of that kind of test requires hypotheses whose mutual relevance relations are bound to change in the face of new evidence. Several of our examples (5 — positions at Polly Tech; 6 — celibacy among priests; 8 — electron-positron annihilation) show exactly what can happen when new evidence renders independent hypotheses mutually incompatible.

6. Conclusions

The crude hypothetico-deductive account of scientific inference, according to which hypotheses are confirmed by deducing observational consequences which are then verified by observation, is widely recognized nowadays as an oversimplification (even leaving aside the Duhemian objections). One can hardly improve upon Russell's classic example. From the hypothesis, "Pigs have wings," in conjunction with the observed initial condition, "Pigs are good to eat," we can deduce the consequence, "Some winged things are good to eat." Upon observing that such winged creatures as ducks and turkeys are good to eat, we have a hypothetico-deductive confirmation of the hypothesis, "Pigs have wings."[41] I am inclined to agree with a wide variety of authors who hold that something akin to a Bayesian schema must be involved in the confirmation of scientific hypotheses. If this is correct, it is entirely possible to have positive hypothetico-deductive test results that do not confirm the hypothesis (i.e., that do not add anything to its degree of confirmation on prior evidence). To emphasize this point, Reichenbach aptly described the crude hypothetico-deductive inference as an instance of "the fallacy of incomplete schematization."[42] Recognition of the basic inadequacy of the hypothetico-deductive schema does no violence to the logic of science; it only shows that the methods of science are more complex than this oversimplified schema.

Acknowledging the fact that positive hypothetico-deductive instances may not be confirming instances, I have been discussing the logic of confirmation —that is, I have been investigating the conclusions that can be drawn from the knowledge that this or that evidence confirms this or that hypothesis. By and large, we have found this logic to be poverty-ridden. Although evidence i confirms hypotheses h and k, we have found that we cannot infer that i confirms $h.k$. Evidence i may in fact confirm $h.k$, but to draw that conclusion from the given premises would be another instance of the fallacy of incomplete schematization. Indeed, our investigations have revealed exactly what is missing in the inference. In addition to knowing that i is positively relevant to h and positively relevant to k, we must know what bearing i has on the relevance of h to k. If this is known quantitatively, and if the degrees of relevance of i to h and to k are also known quantitatively, we can ascertain the relevance of i to $h.k$ and to $h \lor k$. Without this quantitative knowledge, we cannot say much of anything. The moral is simple: even if we base our qualitative concept of confirmation (in the relevance sense) upon a quantitative concept of degree of confirmation, the resulting qualitative concept is not very serviceable. It is too crude a concept, and it doesn't carry enough information to be useful. In order to make any substantial headway in understanding the logic of evidential support of scientific hypotheses, we must be prepared to work with at least crude estimates of quantitative values of degree of confirmation

[41] Bertrand Russell, "Dewey's New 'Logic,'" in Paul Arthur Schilpp, ed., The Philosophy of John Dewey (New York: Tudor, 1939), p. 149.
[42] Hans Reichenbach, The Theory of Probability (Berkeley and Los Angeles: University of California Press, 1949), p. 96.

and degree of relevance. Then, in contexts such as the discussion of the Duhemian problem, we must bring the more sophisticated concepts to bear if we hope to achieve greater clarity and avoid logical fallacies. In detailing the shortcomings of the qualitative concept of confirmation, we have, in a way, shown that this sort of confirmation theory is a shambles, but we have done no more violence to the logic of science than to show that its embodies more powerful concepts.

If we are willing, as Carnap has done, to regard degree of confirmation (in the nonrelevance sense) as a probability — that is, as a numerical functor that satisfies the probability calculus — then we can bring the structure of the quantitative probability concept to bear on problems of confirmation. With this apparatus, which gives us the power of Bayes's theorem, we can aspire to a much fuller understanding of relations of confirmation (in both the absolute and the relevance senses).

We can also provide an answer to many who have claimed that confirmation is not a probability concept. Confirmation in the relevance sense is admittedly not a probability; as we have insisted, it is not to be identified with high probability. A quantitative concept of degree of relevance can nevertheless be defined in terms of a concept of degree of confirmation. Degree of relevance, as thus defined, is not a probability; it obviously can take on negative values, which degree of probability cannot do. It is a probability concept, however, in the sense that it is explicitly defined in terms of degree of confirmation which is construed as a probability concept. Thus, even though degree of confirmation in the relevance sense cannot be construed as degree of probability, this fact is no basis for concluding that the concept of probability is an inadequate or inappropriate tool for studying the logic of evidential relations between scientific hypotheses and observational evidence. Moreover, it provides no basis whatever for rejecting the notion that high probabilities as well as high content are what we want our scientific hypotheses eventually to achieve on the basis of experimental testing.

ON INDUCTIVE LOGIC

RUDOLF CARNAP

§1. INDUCTIVE LOGIC

Among the various meanings in which the word 'probability' is used in everyday language, in the discussion of scientists, and in the theories of probability, there are especially two which must be clearly distinguished. We shall use for them the terms 'probability₁' and 'probability₂'. Probability₁ is a logical concept, a certain logical relation between two sentences (or, alternatively, between two propositions); it is the same as the concept of degree of confirmation. I shall write briefly "c" for "degree of confirmation," and "c(h, e)" for "the degree of confirmation of the hypothesis h on the evidence e"; the evidence is usually a report on the results of our observations. On the other hand, probability₂ is an empirical concept; it is the relative frequency in the long run of one property with respect to another. The controversy between the so-called logical conception of probability, as represented e.g. by Keynes[1], and Jeffreys[2], and others, and the frequency conception, maintained e.g. by v. Mises[3] and Reichenbach[4], seems to me futile. These two theories deal with two different probability concepts which are both of great importance for science. Therefore, the theories are not incompatible, but rather supplement each other.[5]

In a certain sense we might regard deductive logic as the theory of L-implication (logical implication, entailment). And inductive logic may be construed as the theory of degree of confirmation, which is, so to speak, partial L-implication. "e L-implies h" says that h is implicitly given with e, in other words, that the whole logical content of h is contained in e. On the other hand, "c(h, e) = 3/4" says that h is not entirely given with e but that the assumption of h is supported to the degree 3/4 by the observational evidence expressed in e.

In the course of the last years, I have constructed a new system of inductive logic by laying down a definition for degree of confirmation and developing a theory based on this definition. A book containing this theory is in preparation. The purpose of the present paper is to indicate briefly and informally the definition and a few of the results found; for lack of space, the reasons for the choice of this definition and the proofs for the results cannot be given here. The book will, of course, provide a better basis than the present informal summary for a critical evaluation of the theory and of the fundamental conception on which it is based.[6]

[1] J. M. Keynes, *A Treatise on Probability*, 1921.

[2] H. Jeffreys, *Theory of Probability*, 1939.

[3] R. v. Mises, *Probability, Statistics, and Truth*, (orig. 1928) 1939.

[4] H. Reichenbach, *Wahrscheinlichkeitslehre*, 1935.

[5] The distinction briefly indicated here, is discussed more in detail in my paper "The Two Concepts of Probability," which will appear in *Philos. and Phenom. Research*, 1945.

[6] In an article by C. G. Hempel and Paul Oppenheim in the present issue of this journal, a new concept of degree of confirmation is proposed, which was developed by the two authors and Olaf Helmer in research independent of my own.

§2. SOME SEMANTICAL CONCEPTS

Inductive logic is, like deductive logic, in my conception a branch of semantics. However, I shall try to formulate the present outline in such a way that it does not presuppose knowledge of semantics.

Let us begin with explanations of some semantical concepts which are important both for deductive logic and for inductive logic.[7]

The system of inductive logic to be outlined applies to an infinite sequence of finite language systems L_N ($N = 1, 2, 3$, etc.) and an infinite language system L_∞. L_∞ refers to an infinite universe of individuals, designated by the individual constants 'a_1', 'a_2', etc. (or 'a', 'b', etc.), while L_N refers to a finite universe containing only N individuals designated by 'a_1', 'a_2', $\cdots$ 'a_N'. Individual variables 'x_1', 'x_2', etc. (or 'x', 'y', etc.) are the only variables occurring in these languages. The languages contain a finite number of predicates of any degree (number of arguments), designating properties of the individuals or relations between them. There are, furthermore, the customary connectives of negation ('$\sim$', corresponding to "not"), disjunction ('$\vee$', "or"), conjunction ('$\cdot$', "and"); universal and existential quantifiers ("for every x," "there is an x"); the sign of identity between individuals '$=$', and 't' as an abbreviation for an arbitrarily chosen tautological sentence. (Thus the languages are certain forms of what is technically known as the lower functional logic with identity.) (The connectives will be used in this paper in three ways, as is customary: (1) between sentences, (2) between predicates (§8), (3) between names (or variables) of sentences (so that, if 'i' and 'j' refer to two sentences, '$i \vee j$' is meant to refer to their disjunction).)

A sentence consisting of a predicate of degree n with n individual constants is called an *atomic sentence* (e.g. 'Pa_1', i.e. 'a_1 has the property P', or 'Ra_3a_5', i.e. 'the relation R holds between a_3 and a_5'). The conjunction of all atomic sentences in a finite language L_N describes one of the possible states of the domain of the N individuals with respect to the properties and relations expressible in the language L_N. If we replace in this conjunction some of the atomic sentences by their negations, we obtain the description of another possible state. All the conjunctions which we can form in this way, including the original one, are called *state-descriptions* in L_N. Analogously, a state-description in L_∞ is a class containing some atomic sentences and the negations of the remaining atomic sentences; since this class is infinite, it cannot be transformed into a conjunction.

In the actual construction of the language systems, which cannot be given here, semantical rules are laid down determining for any given sentence j and any state-description i whether j holds in i, that is to say whether j would be true if i described the actual state among all possible states. The class of those state-descriptions in a language system L (either one of the systems L_N or L_∞) in which j holds is called the *range* of j in L.

The concept of range is fundamental both for deductive and for inductive logic; this has already been pointed out by Wittgenstein. If the range of a

---

[7] For more detailed explanations of some of these concepts see my *Introduction to Semantics*, 1942.

sentence j in the language system L is universal, i.e. if j holds in every state-description (in L), j must necessarily be true independently of the facts; therefore we call j (in L) in this case L-true (logically true, analytic). (The prefix 'L-' stands for "logical"; it is not meant to refer to the system L.) Analogously, if the range of j is null, we call j L-false (logically false, self-contradictory). If j is neither L-true nor L-false, we call it factual (synthetic, contingent). Suppose that the range of e is included in that of h. Then in every possible case in which e would be true, h would likewise be true. Therefore we say in this case that e L-implies (logically implies, entails) h. If two sentences have the same range, we call them L-equivalent; in this case, they are merely different formulations for the same content.

The L-concepts just explained are fundamental for deductive logic and therefore also for inductive logic. Inductive logic is constructed out of deductive logic by the introduction of the concept of degree of confirmation. This introduction will here be carried out in three steps: (1) the definition of regular c-functions (§3), (2) the definition of symmetrical c-functions (§5), (3) the definition of the degree of confirmation c* (§6).

§3. REGULAR C-FUNCTIONS

A numerical function m ascribing real numbers of the interval 0 to 1 to the sentences of a finite language L_N is called a regular m-function if it is constructed according to the following rules:

(1) We assign to the state-descriptions in L_N as values of m any positive real numbers whose sum is 1.

(2) For every other sentence j in L_N, the value $m(j)$ is determined as follows:
  (a) If j is not L-false, $m(j)$ is the sum of the m-values of those state-descriptions which belong to the range of j.
  (b) If j is L-false and hence its range is null, $m(j) = 0$.

(The choice of the rule (2)(a) is motivated by the fact that j is L-equivalent to the disjunction of those state-descriptions which belong to the range of j and that these state-descriptions logically exclude each other.)

If any regular m-function m is given, we define a corresponding function c as follows:

(3) For any pair of sentences e, h in L_N, where e is not L-false, $c(h, e) = \dfrac{m(e \cdot h)}{m(e)}$.

$m(j)$ may be regarded as a measure ascribed to the range of j; thus the function m constitutes a metric for the ranges. Since the range of the conjunction $e \cdot h$ is the common part of the ranges of e and of h, the quotient in (3) indicates, so to speak, how large a part of the range of e is included in the range of h. The numerical value of this ratio, however, depends on what particular m-function has been chosen. We saw earlier that a statement in deductive logic of the form "e L-implies h" says that the range of e is entirely included in that of h. Now we see that a statement in inductive logic of the form "$c(h, e) = 3/4$" says that a certain part—in the example, three fourths—of the range of e is included in

the range of h.[8] Here, in order to express the partial inclusion numerically, it is necessary to choose a regular m-function for measuring the ranges. Any m chosen leads to a particular c as defined above. All functions c obtained in this way are called *regular c-functions*.

One might perhaps have the feeling that the metric m should not be chosen once for all but should rather be changed according to the accumulating experiences.[9] This feeling is correct in a certain sense. However, it is to be satisfied not by the function m used in the definition (3) but by another function m_e dependent upon e and leading to an alternative definition (5) for the corresponding c. If a regular m is chosen according to (1) and (2), then a corresponding function m_e is defined for the state-descriptions in L_N as follows:

(4) Let i be a state-description in L_N, and e a non-L-false sentence in L_N.
(a) If e does not hold in i, $m_e(i) = 0$.

(b) If e holds in i, $m_e(i) = \dfrac{m(i)}{m(e)}$.

Thus m_e represents a metric for the state-descriptions which changes with the changing evidence e. Now $m_e(j)$ for any other sentence j in L_N is defined in analogy to (2) (a) and (b). Then we define the function c corresponding to m as follows:

(5) For any pair of sentences e, h in L_N, where e is not L-false, $c(h, e) = m_e(h)$.

It can easily be shown that this alternative definition (5) yields the same values as the original definition (3).

Suppose that a sequence of regular m-functions is given, one for each of the finite languages L_N ($N = 1, 2$, etc.). Then we define a corresponding m-function for the infinite language as follows:

(6) $m(j)$ in L_∞ is the limit of the values $m(j)$ in L_N for $N \to \infty$.

c-functions for the finite languages are based on the given m-functions according to (3). We define a corresponding c-function for the infinite language as follows:

(7) $c(h, e)$ in L_∞ is the limit of the values $c(h, e)$ in L_N for $N \to \infty$.

The definitions (6) and (7) are applicable only in those cases where the specified limits exist.

We shall later see how to select a particular sub-class of regular c-functions (§5) and finally one particular c-function c* as the basis of a complete system of inductive logic (§6). For the moment, let us pause at our first step, the definition of regular c-functions just given, in order to see what results this definition alone can yield, before we add further definitions. The theory of regular c-functions, i.e. the totality of those theorems which are founded on the definition

[8] See F. Waismann, "Logische Analyse des Wahrscheinlichkeitsbegriffs," *Erkenntnis*, vol. 1, 1930, pp. 228–248.
[9] See Waismann, op. cit., p. 242.

stated, is the first and fundamental part of inductive logic. It turns out that we find here many of the fundamental theorems of the classical theory of probability, e.g. those known as the theorem (or principle) of multiplication, the general and the special theorems of addition, the theorem of division and, based upon it, Bayes' theorem.

One of the cornerstones of the classical theory of probability is the principle of indifference (or principle of insufficient reason). It says that, if our evidence e does not give us any sufficient reason for regarding one of two hypotheses h and h' as more probable than the other, then we must take their probabilities$_1$ as equal: $c(h, e) = c(h', e)$. Modern authors, especially Keynes, have correctly pointed out that this principle has often been used beyond the limits of its original meaning and has then led to quite absurd results. Moreover, it can easily be shown that, even in its original meaning, the principle is by far too general and leads to contradictions. Therefore the principle must be abandoned. If it is and we consider only those theorems of the classical theory which are provable without the help of this principle, then we find that these theorems hold for all regular c-functions. The same is true for those modern theories of probability$_1$ (e.g. that by Jeffreys, op.cit.) which make use of the principle of indifference. Most authors of modern axiom systems of probability$_1$ (e.g. Keynes (op.cit.), Waismann (op.cit.), Mazurkiewicz[10], Hosiasson[11], v. Wright[12]) are cautious enough not to accept that principle. An examination of these systems shows that their axioms and hence their theorems hold for all regular c-functions. Thus these systems restrict themselves to the first part of inductive logic, which, although fundamental and important, constitutes only a very small and weak section of the whole of inductive logic. The weakness of this part shows itself in the fact that it does not determine the value of c for any pair h, e except in some special cases where the value is 0 or 1. The theorems of this part tell us merely how to calculate further values of c if some values are given. Thus it is clear that this part alone is quite useless for application and must be supplemented by additional rules. (It may be remarked incidentally, that this point marks a fundamental difference between the theories of probability$_1$ and of probability$_2$ which otherwise are analogous in many respects. The theorems concerning probability$_2$ which are analogous to the theorems concerning regular c-functions constitute not only the first part but the whole of the logico-mathematical theory of probability$_2$. The task of determining the value of probability$_2$ for a given case is—in contradistinction to the corresponding task for probability$_1$—an empirical one and hence lies outside the scope of the logical theory of probability$_2$.)

[10] St. Mazurkiewicz, "Zur Axiomatik der Wahrscheinlichkeitsrechnung," *C. R. Soc. Science Varsovie*, Cl. III, vol. 25, 1932, pp. 1–4.

[11] Janina Hosiasson-Lindenbaum, "On Confirmation," *Journal Symbolic Logic*, vol. 5, 1940, pp. 133–148.

[12] G. H. von Wright, *The Logical Problem of Induction*, (Acta Phil. Fennica, 1941, Fasc. III). See also C. D. Broad, *Mind*, vol. 53, 1944.

§4. THE COMPARATIVE CONCEPT OF CONFIRMATION

Some authors believe that a metrical (or quantitative) concept of degree of confirmation, that is, one with numerical values, can be applied, if at all, only in certain cases of a special kind and that in general we can make only a comparison in terms of higher or lower confirmation without ascribing numerical values. Whether these authors are right or not, the introduction of a merely comparative (or topological) concept of confirmation not presupposing a metrical concept is, in any case, of interest. We shall now discuss a way of defining a concept of this kind.

For technical reasons, we do not take the concept "more confirmed" but "more or equally confirmed." The following discussion refers to the sentences of any finite language L_N. We write, for brevity, "$MC(h, e, h', e')$" for "h is confirmed on the evidence e more highly or just as highly as h' on the evidence e'".

Although the definition of the comparative concept MC at which we aim will not make use of any metrical concept of degree of confirmation, let us now consider, for heuristic purposes, the relation between MC and the metrical concepts, i.e. the regular c-functions. Suppose we have chosen some concept of degree of confirmation, in other words, a regular c-function c, and further a comparative relation MC; then we shall say that MC is in accord with c if the following holds:

(1) For any sentences h, e, h', e', if $MC(h, e, h', e')$ then $c(h, e) \geq c(h', e')$.

However, we shall not proceed by selecting one c-function and then choosing a relation MC which is in accord with it. This would not fulfill our intention. Our aim is to find a comparative relation MC which grasps those logical relations between sentences which are, so to speak, prior to the introduction of any particular m-metric for the ranges and of any particular c-function; in other words, those logical relations with respect to which all the various regular c-functions agree. Therefore we lay down the following requirement:

(2) The relation MC is to be defined in such a way that it is in accord with all regular c-functions; in other words, if $MC(h, e, h', e')$, then for every regular c, $c(h, e) \geq c(h', e')$.

It is not difficult to find relations which fulfill this requirement (2). First let us see whether we can find quadruples of sentences h, e, h', e' which satisfy the following condition occurring in (2):

(3) For every regular c, $c(h, e) \geq c(h', e')$.

It is easy to find various kinds of such quadruples. (For instance, if e and e' are any non-L-false sentences, then the condition (3) is satisfied in all cases where e L-implies h, because here $c(h, e) = 1$; further in all cases where e' L-implies $\sim h'$, because here $c(h', e') = 0$; and in many other cases.) We could, of course, define a relation MC by taking some cases where we know that the condition (3) is satisfied and restricting the relation to these cases. Then the relation would fulfill the requirement (2); however, as long as there are cases

which satisfy the condition (3) but which we have not included in the relation,
the relation is unnecessarily restricted. Therefore we lay down the following
as a second requirement for MC:

(4) MC is to be defined in such a way that it holds in all cases which satisfy
the condition (3); in such a way, in other words, that it is the most comprehensive relation which fulfills the first requirement (2).

These two requirements (2) and (4) together stipulate that $MC(h, e, h', e')$
is to hold if and only if the condition (3) is satisfied; thus the requirements
determine uniquely one relation MC. However, because they refer to the c-
functions, we do not take these requirements as a definition for MC, for we intend to give a purely comparative definition for MC, a definition which does not
make use of any metrical concepts but which leads nevertheless to a relation
MC which fulfills the requirements (2) and (4) referring to c-functions. This
aim is reached by the following definition (where ' $=_{Df}$' is used as sign of definition).

(5) $MC(h, e, h', e')$ $=_{Df}$ the sentences h, e, h', e' (in L_N) are such that e and e'
are not L-false and at least one of the following three conditions is fulfilled:
(a) e L-implies h,
(b) e' L-implies $\sim h'$,
(c) $e' \cdot h'$ L-implies $e \cdot h$ and simultaneously e L-implies $h \lor e'$.

((a) and (b) are the two kinds of rather trivial cases earlier mentioned; (c) com-
prehends the interesting cases; an explanation and discussion of them cannot
be given here.)

The following theorem can then be proved concerning the relation MC defined
by (5). It shows that this relation fulfills the two requirements (2) and (4).

(6) For any sentences h, e, h', e' in L_N the following holds:
(a) If $MC(h, e, h', e')$, then, for every regular c, $c(h, e) \geqq c(h', e')$.
(b) If, for every regular c, $c(h, e) \geqq c(h', e')$, then $MC(h, e, h', e')$.

(With respect to L_∞, the analogue of (6)(a) holds for all sentences, and that of
(6)(b) for all sentences without variables.)

§5. SYMMETRICAL C-FUNCTIONS

The next step in the construction of our system of inductive logic consists in
selecting a narrow sub-class of the comprehensive class of all regular c-functions.
The guiding idea for this step will be the principle that inductive logic should
treat all individuals on a par. The same principle holds for deductive logic; for
instance, if '$\cdots a \cdots b \cdots$' L-implies '$-b-c-$' (where the first expression in quotation
marks is meant to indicate some sentence containing 'a' and 'b', and the second
another sentence containing 'b' and 'c'), then L-implication holds likewise between corresponding sentences with other individual constants, e.g. between
'$\cdots d \cdots c \cdots$' and '$-c-a-$'. Now we require that this should hold also for inductive logic, e.g. that $c('-b-c-', '\cdots a \cdots b \cdots') = c('-c-a-', '\cdots d \cdots c \cdots')$. It seems

that all authors on probability₁ have assumed this principle—although it has seldom, if ever, been stated explicitly—by formulating theorems in the following or similar terms: "On the basis of observations of s things of which s_1 were found to have the property M and s_2 not to have this property, the probability that another thing has this property is such and such." The fact that these theorems refer only to the number of things observed and do not mention particular things shows implicitly that it does not matter which things are involved; thus it is assumed, e.g., that $c('Pd', 'Pa \cdot Pb \cdot \sim Pc') = c('Pc', 'Pa \cdot Pd \cdot \sim Pb')$.

The principle could also be formulated as follows. Inductive logic should, like deductive logic, make no discrimination among individuals. In other words, the value of c should be influenced only by those differences between individuals which are expressed in the two sentences involved; no differences between particular individuals should be stipulated by the rules of either deductive or inductive logic.

It can be shown that this principle of non-discrimination is fulfilled if c belongs to the class of symmetrical c-functions which will now be defined. Two state-descriptions in a language L_N are said to be *isomorphic* or to have the same structure if one is formed from the other by replacements of the following kind: we take any one-one relation R such that both its domain and its converse domain is the class of all individual constants in L_N, and then replace every individual constant in the given state-description by the one correlated with it by R. If a regular m-function (for L_N) assigns to any two isomorphic state-descriptions (in L_N) equal values, it is called a symmetrical m-function; and a c-function based upon such an m-function in the way explained earlier (see (3) in §3) is then called a *symmetrical c-function*.

§6. THE DEGREE OF CONFIRMATION C*

Let i be a state-description in L_N. Suppose there are n_i state-descriptions in L_N isomorphic to i (including i itself), say i, i', i'', etc. These n_i state-descriptions exhibit one and the same structure of the universe of L_N with respect to all the properties and relations designated by the primitive predicates in L_N. This concept of structure is an extension of the concept of structure or relation-number (Russell) usually applied to one dyadic relation. The common structure of the isomorphic state-descriptions i, i', i'', etc. can be described by their disjunction $i \lor i' \lor i'' \lor \cdots$. Therefore we call this disjunction, say j, a *structure-description* in L_N. It can be shown that the range of j contains only the isomorphic state-descriptions i, i', i'', etc. Therefore (see (2)(a) in §3) $m(j)$ is the sum of the m-values for these state-descriptions. If m is symmetrical, then these values are equal, and hence

(1) $m(j) = n_i \times m(i)$.

And, conversely, if $m(j)$ is known to be q, then

(2) $m(i) = m(i') = m(i'') = \cdots = q/n_i$.

This shows that what remains to be decided, is merely the distribution of m-values among the structure-descriptions in L_N. We decide to give them equal m-values. This decision constitutes the third step in the construction of our inductive logic. This step leads to one particular m-function m^* and to the c-function c^* based upon m^*. According to the preceding discussion, m^* is characterized by the following two stipulations:

(3) (a) m^* is a symmetrical m-function;
    (b) m^* has the same value for all structure-descriptions (in L_N).

We shall see that these two stipulations characterize just one function. Every state-description (in L_N) belongs to the range of just one structure-description. Therefore, the sum of the m^*-values for all structure-descriptions in L_N must be the same as for all state-descriptions, hence 1 (according to (1) in §3). Thus, if the number of structure-descriptions in L_N is m, then, according to (3)(b),

(4) for every structure-description j in L_N, $m^*(j) = \dfrac{1}{m}$.

Therefore, if i is any state-description in L_N and n_i is the number of state-descriptions isomorphic to i, then, according to (3)(a) and (2),

(5) $$m^*(i) = \frac{1}{mn_i}.$$

(5) constitutes a definition of m^* as applied to the state-descriptions in L_N. On this basis, further definitions are laid down as explained above (see (2) and (3) in §3): first a definition of m^* as applied to all sentences in L_N, and then a definition of c^* on the basis of m^*. Our inductive logic is the theory of this particular function c^* as our concept of degree of confirmation.

It seems to me that there are good and even compelling reasons for the stipulation (3)(a), i.e. the choice of a symmetrical function. The proposal of any non-symmetrical c-function as degree of confirmation could hardly be regarded as acceptable. The same can not be said, however, for the stipulation (3)(b). No doubt, to the way of thinking which was customary in the classical period of the theory of probability, (3)(b) would appear as validated, like (3)(a), by the principle of indifference. However, to modern, more critical thought, this mode of reasoning appears as invalid because the structure-descriptions (in contradistinction to the individual constants) are by no means alike in their logical features but show very conspicuous differences. The definition of c^* shows a great simplicity in comparison with other concepts which may be taken into consideration. Although this fact may influence our decision to choose c^*, it cannot, of course, be regarded as a sufficient reason for this choice. It seems to me that the choice of c^* cannot be justified by any features of the definition which are immediately recognizable, but only by the consequences to which the definition leads.

There is another c-function c_W which at the first glance appears not less plausible than c^*. The choice of this function may be suggested by the following consideration. Prior to experience, there seems to be no reason to regard one state-description as less probable than another. Accordingly, it might seem natural to assign equal m-values to all state-descriptions. Hence, if the number of the state-descriptions in L_N is n, we define for any state-description i

(6) $$m_W(i) = 1/n.$$

This definition (6) for m_W is even simpler than the definition (5) for m^*. The measure ascribed to the ranges is here simply taken as proportional to the cardinal numbers of the ranges. On the basis of the m_W-values for the state-descriptions defined by (6), the values for the sentences are determined as before (see (2) in §3), and then c_W is defined on the basis of m_W (see (3) in §3).[13]

In spite of its apparent plausibility, the function c_W can easily be seen to be entirely inadequate as a concept of degree of confirmation. As an example, consider the language L_{101} with 'P' as the only primitive predicate. Let the number of state-descriptions in this language be n (it is 2^{101}). Then for any state-description, $m_W = 1/n$. Let e be the conjunction $Pa_1 \cdot Pa_2 \cdot Pa_3 \cdots Pa_{100}$ and let h be 'Pa_{101}'. Then $e \cdot h$ is a state-description and hence $m_W (e \cdot h) = 1/n$. e holds only in the two state-descriptions $e \cdot h$ and $e \cdot \sim h$; hence $m_W(e) = 2/n$. Therefore $c_W(h, e) = \frac{1}{2}$. If e' is formed from e by replacing some or even all of the atomic sentences with their negations, we obtain likewise $c_W(h, e') = \frac{1}{2}$. Thus the c_W-value for the prediction that a_{101} is P is always the same, no matter whether among the hundred observed individuals the number of those which we have found to be P is 100 or 50 or 0 or any other number. Thus the choice of c_W as the degree of confirmation would be tantamount to the principle never to let our past experiences influence our expectations for the future. This would obviously be in striking contradiction to the basic principle of all inductive reasoning.

§7. LANGUAGES WITH ONE-PLACE PREDICATES ONLY

The discussions in the rest of this paper concern only those language systems whose primitive predicates are one-place predicates and hence designate properties, not relations. It seems that all theories of probability constructed so far have restricted themselves, or at least all of their important theorems, to properties. Although the definition of c^* in the preceding section has been stated in a

[13] It seems that Wittgenstein meant this function c_W in his definition of probability, which he indicates briefly without examining its consequences. In his *Tractatus Logico-Philosophicus*, he says: "A proposition is the expression of agreement and disagreement with the truth-possibilities of the elementary [i.e. atomic] propositions" (*4.4); "The world is completely described by the specification of all elementary propositions plus the specification, which of them are true and which false" (*4.26). The truth-possibilities specified in this way correspond to our state-descriptions. Those truth-possibilities which verify a given proposition (in our terminology, those state-descriptions in which a given sentence holds) are called the truth-grounds of that proposition (*5.101). "If T_r is the number of the truth-grounds of the proposition "r", T_{rs} the number of those truth-grounds of the proposition "s" which are at the same time truth-grounds of "r", then we call the ratio $T_{rs}:T_r$ the measure of the *probability* which the proposition "r" gives to the proposition "s" "(*5.15). It seems that the concept of probability thus defined coincides with the function c_W.

general way so as to apply also to languages with relations, the greater part of our inductive logic will be restricted to properties. An extension of this part of inductive logic to relations would require certain results in the deductive logic of relations, results which this discipline, although widely developed in other respects, has not yet reached (e.g. an answer to the apparently simple question as to the number of structures in a given finite language system).

Let L_N^p be a language containing N individual constants 'a_1', $\cdots$ 'a_N', and p one-place primitive predicates 'P_1', $\cdots$ 'P_p'. Let us consider the following expressions (sentential matrices). We start with '$P_1x \cdot P_2x \cdots P_px$'; from this expression we form others by negating some of the conjunctive components, until we come to '$\sim P_1x \cdot \sim P_2x \cdots \sim P_px$', where all components are negated. The number of these expressions is $k = 2^p$; we abbreviate them by 'Q_1x', $\cdots$ 'Q_kx'. We call the k properties expressed by those k expressions in conjunctive form and now designated by the k new Q-predicates the Q-properties with respect to the given language L_N^p. We see easily that these Q-properties are the strongest properties expressible in this language (except for the L-empty, i.e., logically self-contradictory, property); and further, that they constitute an exhaustive and non-overlapping classification, that is to say, every individual has one and only one of the Q-properties. Thus, if we state for each individual which of the Q-properties it has, then we have described the individuals completely. Every state-description can be brought into the form of such a statement, i.e. a conjunction of N Q-sentences, one for each of the N individuals. Suppose that in a given state-description i the number of individuals having the property Q_1 is N_1, the number for Q_2 is N_2, $\cdots$ that for Q_k is N_k. Then we call the numbers N_1, N_2, $\cdots$ N_k the Q-numbers of the state-description i; their sum is N. Two state-descriptions are isomorphic if and only if they have the same Q-numbers. Thus here a structure-description is a statistical description giving the Q-numbers N_1, N_2, etc., without specifying which individuals have the properties Q_1, Q_2, etc.

Here—in contradistinction to languages with relations—it is easy to find an explicit function for the number m of structure-descriptions and, for any given state-description i with the Q-numbers N_1, $\cdots$ N_k, an explicit function for the number n_i of state-descriptions isomorphic to i, and hence also a function for m*(i).[14]

Let j be a non-general sentence (i.e. one without variables) in L_N^p. Since

[14] The results are as follows.

(1)
$$m = \frac{(N + k - 1)!}{N!(k - 1)!}$$

(2)
$$n_i = \frac{N!}{N_1! N_2! \cdots N_k!}$$

Therefore (according to (5) in §6):

(3)
$$m^*(i) = \frac{N_1! N_2! \cdots N_k!(k - 1)!}{(N + k - 1)!}$$

there are effective procedures (that is, sets of fixed rules furnishing results in a finite number of steps) for constructing all state-descriptions in which j holds and for computing m* for any given state-description, these procedures together yield an effective procedure for computing m*(j) (according to (2) in §3). However, the number of state-descriptions becomes very large even for small language systems (it is k^N, hence, e.g., in L_7^3 it is more than two million.) Therefore, while the procedure indicated for the computation of m*(j) is effective, nevertheless in most ordinary cases it is impracticable; that is to say, the number of steps to be taken, although finite, is so large that nobody will have the time to carry them out to the end. I have developed another procedure for the computation of m*(j) which is not only effective but also practicable if the number of individual constants occurring in j is not too large.

The value of m* for a sentence j in the infinite language has been defined (see (6) in §3) as the limit of its values for the same sentence j in the finite languages. The question arises whether and under what conditions this limit exists. Here we have to distinguish two cases. (i) Suppose that j contains no variable. Here the situation is simple; it can be shown that in this case m*(j) is the same in all finite languages in which j occurs; hence it has the same value also in the infinite language. (ii) Let j be general, i.e., contain variables. Here the situation is quite different. For a given finite language with N individuals, j can of course easily be transformed into an L-equivalent sentence j_N' without variables, because in this language a universal sentence is L-equivalent to a conjunction of N components. The values of m*(j_N') are in general different for each N; and although the simplified procedure mentioned above is available for the computation of these values, this procedure becomes impracticable even for moderate N. Thus for general sentences the problem of the existence and the practical computability of the limit becomes serious. It can be shown that for every general sentence the limit exists; hence m* has a value for all sentences in the infinite language. Moreover, an effective procedure for the computation of m*(j) for any sentence j in the infinite language has been constructed. This is based on a procedure for transforming any given general sentence j into a non-general sentence j' such that j and j', although not necessarily L-equivalent, have the same m*-value in the infinite language and j' does not contain more individual constants than j; this procedure is not only effective but also practicable for sentences of customary length. Thus, the computation of m*(j) for a general sentence j is in fact much simpler for the infinite language than for a finite language with a large N.

With the help of the procedure mentioned, the following theorem is obtained: If j is a purely general sentence (i.e. one without individual constants) in the infinite language, then m*(j) is either 0 or 1.

§8. INDUCTIVE INFERENCES

One of the chief tasks of inductive logic is to furnish general theorems concerning inductive inferences. We keep the traditional term "inference"; however, we do not mean by it merely a transition from one sentence to another (viz.

from the evidence or premiss e to the hypothesis or conclusion h) but the determination of the degree of confirmation $c(h, e)$. In deductive logic it is sufficient to state that h follows with necessity from e; in inductive logic, on the other hand, it would not be sufficient to state that h follows—not with necessity but to some degree or other—from e. It must be specified to what degree h follows from e; in other words, the value of $c(h, e)$ must be given. We shall now indicate some results with respect to the most important kinds of inductive inference. These inferences are of special importance when the evidence or the hypothesis or both give statistical information, e.g. concerning the absolute or relative frequencies of given properties.

If a property can be expressed by primitive predicates together with the ordinary connectives of negation, disjunction, and conjunction (without the use of individual constants, quantifiers, or the identity sign), it is called an *elementary property*. We shall use 'M', 'M'', 'M_1', 'M_2', etc. for elementary properties. If a property is empty by logical necessity (e.g. the property designated by '$P \cdot \sim P$') we call it L-empty; if it is universal by logical necessity (e.g. '$P \lor \sim P$'), we call it L-universal. If it is neither L-empty nor L-universal (e.g. 'P_1', '$P_1 \cdot \sim P_2$'), we call it a *factual property*; in this case it may still happen to be universal or empty, but if so, then contingently, not necessarily. It can be shown that every elementary property which is not L-empty is uniquely analysable into a disjunction (i.e. or-connection) of Q-properties. If M is a disjunction of n Q-properties ($n \geq 1$), we say that the (logical) *width* of M is n; to an L-empty property we ascribe the width 0. If the width of M is w (≥ 0), we call w/k its *relative width* (k is the number of Q-properties).

The concepts of width and relative width are very important for inductive logic. Their neglect seems to me one of the decisive defects in the classical theory of probability which formulates its theorems "for any property" without qualification. For instance, Laplace takes the probability a priori that a given thing has a given property, no matter of what kind, to be $\frac{1}{2}$. However, it seems clear that this probability cannot be the same for a very strong property (e.g. '$P_1 \cdot P_2 \cdot P_3$') and for a very weak property (e.g. '$P_1 \lor P_2 \lor P_3$'). According to our definition, the first of the two properties just mentioned has the relative width $\frac{1}{8}$, and the second $\frac{7}{8}$. In this and in many other cases the probability or degree of confirmation must depend upon the widths of the properties involved. This will be seen in some of the theorems to be mentioned later.

§9. THE DIRECT INFERENCE

Inductive inferences often concern a situation where we investigate a whole population (of persons, things, atoms, or whatever else) and one or several samples picked out of the population. An inductive inference from the whole population to a sample is called a direct inductive inference. For the sake of simplicity, we shall discuss here and in most of the subsequent sections only the case of one property M, hence a classification of all individuals into M and $\sim M$. The theorems for classifications with more properties are analogous but more complicated. In the present case, the evidence e says that in a whole population of n individuals there are n_1 with the property M and $n_2 = n - n_1$ with $\sim M$; hence the relative frequency of M is $r = n_1/n$. The hypothesis h says that a sample of s individuals taken from the whole population will contain s_1 individuals with the property M and $s_2 = s - s_1$ with $\sim M$. Our theory yields in this case the same values as the classical theory.[15]

If we vary s_1, then c^* has its maximum in the case where the relative frequency s_1/s in the sample is equal or close to that in the whole population.

If the sample consists of only one individual c, and h says that c is M, then $c^*(h, e) = r$.

As an approximation in the case that n is very large in relation to s, Newton's theorem holds.[16] If furthermore the sample is sufficiently large, we obtain as an approximation Bernoulli's theorem in its various forms.

It is worthwhile to note two characteristics which distinguish the direct inductive inference from the other inductive inferences and make it, in a sense, more closely related to deductive inferences:

(i) The results just mentioned hold not only for c^* but likewise for all symmetrical c-functions; in other words, the results are independent of the particular m-metric chosen provided only that it takes all individuals on a par.

(ii) The results are independent of the width of M. This is the reason for the agreement between our theory and the classical theory at this point.

§10. THE PREDICTIVE INFERENCE

We call the inference from one sample to another the predictive inference. In this case, the evidence e says that in a first sample of s individuals, there are s_1 with the property M, and $s_2 = s - s_1$ with $\sim M$. The hypothesis h says that in a second sample of s' other individuals, there will be s_1' with M, and $s_2' = s' - s_1'$ with $\sim M$. Let the width of M be w_1; hence the width of $\sim M$ is $w_2 = k - w_1$.[17]

---

[15] The general theorem is as follows:

$$c^*(h, e) = \frac{\binom{n_1}{s_1}\binom{n_2}{s_1}}{\binom{n}{s}}.$$

[16]
$$c^*(h, e) = \binom{s}{s_1} r^{s_1}(1 - r)^{s_2}.$$

[17] The general theorem is as follows:

$$c^*(h, e) = \frac{\binom{s_1 + s_1' + w_1 - 1}{s_1'}\binom{s_2 + s_2' + w_2 - 1}{s_2'}}{\binom{s + s' + k - 1}{s'}}.$$

The most important special case is that where h refers to one individual c only and says that c is M. In this case,

(1) $$c^*(h, e) = \frac{s_1 + w_1}{s + k}.$$

Laplace's much debated rule of succession gives in this case simply the value $\frac{s_1 + 1}{s + 2}$ for any property whatever; this, however, if applied to different properties, leads to contradictions. Other authors state the value s_1/s, that is, they take simply the observed relative frequency as the probability for the prediction that an unobserved individual has the property in question. This rule, however, leads to quite implausible results. If $s_1 = s$, e.g., if three individuals have been observed and all of them have been found to be M, the last-mentioned rule gives the probability for the next individual being M as 1, which seems hardly acceptable. According to (1), c^* is influenced by the following two factors (though not uniquely determined by them):

(i) w_1/k, the relative width of M;
(ii) s_1/s, the relative frequency of M in the observed sample.

The factor (i) is purely logical; it is determined by the semantical rules. (ii) is empirical; it is determined by observing and counting the individuals in the sample. The value of c^* always lies between those of (i) and (ii). Before any individual has been observed, c^* is equal to the logical factor (i). As we first begin to observe a sample, c^* is influenced more by this factor than by (ii). As the sample is increased by observing more and more individuals (but not including the one mentioned in h), the empirical factor (ii) gains more and more influence upon c^* which approaches closer and closer to (ii); and when the sample is sufficiently large, c^* is practically equal to the relative frequency (ii). These results seem quite plausible.[18]

The predictive inference is the most important inductive inference. The kinds of inference discussed in the subsequent sections may be construed as special cases of the predictive inference.

[18] Another theorem may be mentioned which deals with the case where, in distinction to the case just discussed, the evidence already gives some information about the individual c mentioned in h. Let M_1 be a factual elementary property with the width w_1 ($w_1 \geq 2$); thus M_1 is a disjunction of w_1 Q-properties. Let M_2 be the disjunction of w_2 among those w_1 Q-properties ($1 \leq w_2 < w_1$); hence M_2 L-implies M_1 and has the width w_2. e specifies first how the s individuals of an observed sample are distributed among certain properties, and, in particular, it says that s_1 of them have the property M_1 and s_2 of these s_1 individuals have also the property M_2; in addition, e says that c is M_1; and h says that c is also M_2. Then,

$$c^*(h, e) = \frac{s_2 + w_2}{s_1 + w_1}.$$

This is analogous to (1); but in the place of the whole sample we have here that part of it which shows the property M_1.

§11. THE INFERENCE BY ANALOGY

The inference by analogy applies to the following situation. The evidence known to us is the fact that individuals b and c agree in certain properties and, in addition, that b has a further property; thereupon we consider the hypothesis that c too has this property. Logicians have always felt that a peculiar difficulty is here involved. It seems plausible to assume that the probability of the hypothesis is the higher the more properties b and c are known to have in common; on the other hand, it is felt that these common properties should not simply be counted but weighed in some way. This becomes possible with the help of the concept of width. Let M_1 be the conjunction of all properties which b and c are known to have in common. The known similarity between b and c is the greater the stronger the property M_1, hence the smaller its width. Let M_2 be the conjunction of all properties which b is known to have. Let the width of M_1 be w_1, and that of M_2, w_2. According to the above description of the situation, we presuppose that M_2 L-implies M_1 but is not L-equivalent to M_1; hence $w_1 > w_2$. Now we take as evidence the conjunction $e \cdot j$; e says that b is M_2, and j says that c is M_1. The hypothesis h says that c has not only the properties ascribed to it in the evidence but also the one (or several) ascribed in the evidence to b only, in other words, that c has all known properties of b, or briefly that c is M_2. Then

(1) $$c^*(h, e \cdot j) = \frac{w_2 + 1}{w_1 + 1}.$$

j and h speak only about c; e introduces the other individual b which serves to connect the known properties of c expressed by j with its unknown properties expressed by h. The chief question is whether the degree of confirmation of h is increased by the analogy between c and b, in other words, by the addition of e to our knowledge. A theorem[19] is found which gives an affirmative answer to this question. However, the increase of c^* is under ordinary conditions rather small; this is in agreement with the general conception according to which reasoning by analogy, although admissible, can usually yield only rather weak results.

Hosiasson[20] has raised the question mentioned above and discussed it in detail. She says that an affirmative answer, a proof for the increase of the degree of confirmation in the situation described, would justify the universally accepted reasoning by analogy. However, she finally admits that she does not find such a proof on the basis of her axioms. I think it is not astonishing that neither the classical theory nor modern theories of probability have been able to give a satisfactory account of and justification for the inference by analogy. For, as the theorems mentioned show, the degree of confirmation and its increase depend

[19] $$\frac{c^*(h, e \cdot j)}{c^*(h, j)} = 1 + \frac{w_1 - w_2}{w_2(w_1 + 1)}.$$

This theorem shows that the ratio of the increase of c^* is greater than 1, since $w_1 > w_2$.

[20] Janina Lindenbaum-Hosiasson, "Induction et analogie: Comparaison de leur fondement," *Mind*, vol. 50, 1941, pp. 351–365; see especially pp. 361–365.

here not on relative frequencies but entirely on the logical widths of the properties involved, thus on magnitudes neglected by both classical and modern
theories.

The case discussed above is that of simple analogy. For the case of multiple
analogy, based on the similarity of c not only with one other individual but with
a number n of them, similar theorems hold. They show that c^* increases with
increasing n and approaches 1 asymptotically. Thus, multiple analogy is shown
to be much more effective than simple analogy, as seems plausible.

§12. THE INVERSE INFERENCE

The inference from a sample to the whole population is called the inverse
inductive inference. This inference can be regarded as a special case of the
predictive inference with the second sample covering the whole remainder of
the population. This inference is of much greater importance for practical
statistical work than the direct inference, because we usually have statistical
information only for some samples and not for the whole population.

Let the evidence e say that in an observed sample of s individuals there are s_1
individuals with the property M and $s_2 = s - s_1$ with $\sim M$. The hypothesis h
says that in the whole population of n individuals, of which the sample is a part,
there are n_1 individuals with M and n_2 with $\sim M$ ($n_1 \geqq s_1$, $n_2 \geqq s_2$). Let the
width of M be w_1, and that of $\sim M$ be $w_2 = k - w_1$. Here, in distinction to the
direct inference, $c^*(h, e)$ is dependent not only upon the frequencies but also
upon the widths of the two properties.[21]

§13. THE UNIVERSAL INFERENCE

The universal inductive inference is the inference from a report on an observed
sample to a hypothesis of universal form. Sometimes the term 'induction' has
been applied to this kind of inference alone, while we use it in a much wider
sense for all non-deductive kinds of inference. The universal inference is not
even the most important one; it seems to me now that the role of universal
sentences in the inductive procedures of science has generally been overestimated.
This will be explained in the next section.

Let us consider a simple law l, i.e. a factual universal sentence of the form
"all M are M'" or, more exactly, "for every x, if x is M, then x is M'", where M
and M' are elementary properties. As an example, take "all swans are white".
Let us abbreviate '$M \cdot \sim M'$' ("non-white swan") by 'M_1' and let the width of

[21] The general theorem is as follows:

$$c^*(h, e) = \frac{\binom{n_1 + w_1 - 1}{s_1 + w_1 - 1}\binom{n_2 + w_2 - 1}{s_2 + w_2 - 1}}{\binom{n + k - 1}{n - s}} .$$

Other theorems, which cannot be stated here, concern the case where more than two properties are involved, or give approximations for the frequent case where the whole population
is very large in relation to the sample.

M_1 be w_1. Then l can be formulated thus: "M_1 is empty", i.e. "there is no individual (in the domain of individuals of the language in question) with the property M_1" ("there are no non-white swans"). Since l is a factual sentence, M_1 is a factual property; hence $w_1 > 0$. To take an example, let w_1 be 3; hence M_1 is a disjunction of three Q-properties, say $Q \lor Q' \lor Q''$. Therefore, l can be transformed into: "Q is empty, and Q' is empty, and Q'' is empty". The weakest factual laws in a language are those which say that a certain Q-property is empty; we call them Q-laws. Thus we see that l can be transformed into a conjunction of w_1 Q-laws. Obviously l asserts more if w_1 is larger; therefore we say that the law l has the strength w_1.

Let the evidence e be a report about an observed sample of s individuals such that we see from e that none of these s individuals violates the law l; that is to say, e ascribes to each of the s individuals either simply the property $\sim M_1$ or some other property L-implying $\sim M_1$. Let l, as above, be a simple law which says that M_1 is empty, and w_1 be the width of M_1; hence the width of $\sim M_1$ is $w_2 = k - w_1$. For finite languages with N individuals, $c^*(l, e)$ is found to decrease with increasing N, as seems plausible.[22] If N is very large, c^* becomes very small; and for an infinite universe it becomes 0. The latter result may seem astonishing at first sight; it seems not in accordance with the fact that scientists often speak of "well-confirmed" laws. The problem involved here will be discussed later.

So far we have considered the case in which only positive instances of the law l have been observed. Inductive logic must, however, deal also with the case of negative instances. Therefore let us now examine another evidence e' which says that in the observed sample of s individuals there are s_1 which have the property M_1 (non-white swans) and hence violate the law l, and that $s_2 = s - s_1$ have $\sim M_1$ and hence satisfy the law l. Obviously, in this case there is no point in taking as hypothesis the law l in its original forms, because l is logically incom-

[22] The general theorem is as follows:

$$(1) \qquad c^*(l, e) = \frac{\dbinom{s + k - 1}{w_1}}{\dbinom{N + k - 1}{w_1}}.$$

In the special case of a language containing 'M_1' as the only primitive predicate, we have $w_1 = 1$ and $k = 2$, and hence $c^*(l, e) = \dfrac{s + 1}{N + 1}$. The latter value is given by some authors as holding generally (see Jeffreys, op.cit., p. 106 (16)). However, it seems plausible that the degree of confirmation must be smaller for a stronger law and hence depend upon w_1.

If s, and hence N, too, is very large in relation to k, the following holds as an approximation:

$$(2) \qquad c^*(l, e) = \left(\frac{s}{N}\right)^{w_1}.$$

For the infinite language L_∞ we obtain, according to definition (7) in §3:

$$(3) \qquad c^*(l, e) = 0.$$

patible with the present evidence e', and hence $c^*(l, e') = 0$.   That all individuals satisfy l is excluded by e'; the question remains whether at least all unobserved individuals satisfy l.   Therefore we take here as hypothesis the restricted law l' corresponding to the original unrestricted law l; l' says that all individuals not belonging to the sample of s individuals described in e' have the property $\sim M_1$. w_1 and w_2 are, as previously, the widths of M_1 and $\sim M_1$ respectively. It is found that $c^*(l', e')$ decreases with an increase of N and even more with an increase in the number s_1 of violating cases.[23]   It can be shown that, under ordinary circumstances with large N, c^* increases moderately when a new individual is observed which satisfies the original law l.   On the other hand, if the new individual violates l, c^* decreases very much, its value becoming a small fraction of its previous value.   This seems in good agreement with the general conception.

For the infinite universe, c^* is again 0, as in the previous case.   This result will be discussed in the next section.

<h3 style="text-align:center">§14. THE INSTANCE CONFIRMATION OF A LAW</h3>

Suppose we ask an engineer who is building a bridge why he has chosen the building materials he is using, the arrangement and dimensions of the supports, etc.   He will refer to certain physical laws, among them some general laws of mechanics and some specific laws concerning the strength of the materials.   On further inquiry as to his confidence in these laws he may apply to them phrases like "very reliable", "well founded", "amply confirmed by numerous experiences".   What do these phrases mean?   It is clear that they are intended to say something about probability$_1$ or degree of confirmation.   Hence, what is meant could be formulated more explicitly in a statement of the form "$c(h, e)$ is high" or the like.   Here the evidence e is obviously the relevant observational knowledge of the engineer or of all physicists together at the present time.   But what is to serve as the hypothesis h?   One might perhaps think at first that h is the law in question, hence a universal sentence l of the form: "For every space-time point x, if such and such conditions are fulfilled at x, then such and such is the case at x".   I think, however, that the engineer is chiefly interested not in this sentence l, which speaks about an immense number, perhaps an infinite number, of instances dispersed through all time and space, but rather in one instance of l or a relatively small number of instances.   When he says that the law is very reliable, he does not mean to say that he is willing to bet that among the billion of billions, or an infinite number, of instances to which the law applies there is not one counter-instance, but merely that this bridge will not be a counter-instance, or that among all bridges which he will construct during his lifetime, or among those which all engineers will construct during the next one

[23] The theorem is as follows:

$$c^*(l', e') = \frac{\dbinom{s + k - 1}{s_1 + w_1}}{\dbinom{N + k - 1}{s_1 + w_1}}.$$

thousand years, there will be no counter-instance. Thus h is not the law l itself but only a prediction concerning one instance or a relatively small number of instances. Therefore, what is vaguely called the reliability of a law is measured not by the degree of confirmation of the law itself but by that of one or several instances. This suggests the subsequent definitions. They refer, for the sake of simplicity, to just one instance; the case of several, say one hundred, instances can then easily be judged likewise. Let e be any non-L-false sentence without variables. Let l be a simple law of the form earlier described (§13). Then we understand by the *instance confirmation* of l on the evidence e, in symbols "c_i^* (l, e)", the degree of confirmation, on the evidence e, of the hypothesis that a new individual not mentioned in e fulfills the law l.[24]

The second concept, now to be defined, seems in many cases to represent still more accurately what is vaguely meant by the reliability of a law l. We suppose here that l has the frequently used conditional form mentioned earlier: "For every x, if x is M, then x is M'" (e.g. "all swans are white"). By the *qualified-instance confirmation* of the law that all swans are white we mean the degree of confirmation for the hypothesis h' that the next swan to be observed will likewise be white. The difference between the hypothesis h used previously for the instance confirmation and the hypothesis h' just described consists in the fact that the latter concerns an individual which is already qualified as fulfilling the condition M. That is the reason why we speak here of the qualified-instance confirmation, in symbols "c_{qi}^*".[25] The results obtained concerning instance confirmation and qualified-instance confirmation[26] show that the values of these two functions are independent of N and hence hold for all finite and infinite universes. It has been found that, if the number s_1 of observed counter-instances

[24] In technical terms, the definition is as follows:
$c_i^*(l, e) = {}_{Df} c^*(h, e)$, where h is an instance of l formed by the substitution of an individual constant not occurring in e.

[25] The technical definition will be given here. Let l be 'for every x, if x is M, then x is M'. Let l be non-L-false and without variables. Let 'c' be any individual constant not occurring in e; let j say that c is M, and h' that c is M'. Then the qualified-instance confirmation of l with respect to 'M' and 'M'' on the evidence e is defined as follows: $c_{qi}^*('M', 'M'', e) = {}_{Df} c^*(h', e \cdot j)$.

[26] Some of the theorems may here be given. Let the law l say, as above, that all M are M'. Let 'M_1' be defined, as earlier, by '$M \cdot \sim M'$' ("non-white swan") and 'M_2' by '$M \cdot M'$' ("white swan"). Let the widths of M_1 and M_2 be w_1 and w_2 respectively. Let e be a report about s observed individuals saying that s_1 of them are M_1 and s_2 are M_2, while the remaining ones are $\sim M$ and hence neither M_1 nor M_2. Then the following holds:

(1)
$$c_i^*(l, e) = 1 - \frac{s_1 + w_1}{s + k} \cdot$$

(2)
$$c_{qi}^*('M', 'M'', e) = 1 - \frac{s_1 + w_1}{s_1 + w_1 + s_2 + w_2} \cdot$$

The values of c_i^* and c_{qi}^* for the case that the observed sample does not contain any individuals violating the law l can easily be obtained from the values stated in (1) and (2) by taking $s_1 = 0$.

is a fixed small number, then, with the increase of the sample s, both c_i^* and c_{qi}^* grow close to 1, in contradistinction to c^* for the law itself. This justifies the customary manner of speaking of "very reliable" or "well-founded" or "well confirmed" laws, provided we interpret these phrases as referring to a high value of either of our two concepts just introduced. Understood in this sense, the phrases are not in contradiction to our previous results that the degree of confirmation of a law is very small in a large domain of individuals and 0 in the infinite domain (§13).

These concepts will also be of help in situations of the following kind. Suppose a scientist has observed certain events, which are not sufficiently explained by the known physical laws. Therefore he looks for a new law as an explanation. Suppose he finds two incompatible laws l and l', each of which would explain the observed events satisfactorily. Which of them should he prefer? If the domain of individuals in question is finite, he may take the law with the higher degree of confirmation. In the infinite domain, however, this method of comparison fails, because the degree of confirmation is 0 for either law. Here the concept of instance confirmation (or that of qualified-instance confirmation) will help. If it has a higher value for one of the two laws, then this law will be preferable, if no reasons of another nature are against it.

It is clear that for any deliberate activity predictions are needed, and that these predictions must be "founded upon" or "(inductively) inferred from" past experiences, in some sense of those phrases. Let us examine the situation with the help of the following simplified schema. Suppose a man X wants to make a plan for his actions and, therefore, is interested in the prediction h that c is M'. Suppose further, X has observed (1) that many other things were M and that all of them were also M', let this be formulated in the sentence e; (2) that c is M, let this be j. Thus he knows e and j by observation. The problem is, how does he go from these premisses to the desired conclusion h? It is clear that this cannot be done by deduction; an inductive procedure must be applied. What is this inductive procedure? It is usually explained in the following way. From the evidence e, X infers inductively the law l which says that all M are M'; this inference is supposed to be inductively valid because e contains many positive and no negative instances of the law l; then he infers h ("c is white") from l ("all swans are white") and j ("c is a swan") deductively. Now let us see what the procedure looks like from the point of view of our inductive logic. One might perhaps be tempted to transcribe the usual description of the procedure just given into technical terms as follows. X infers l from e inductively because $c^*(l, e)$ is high; since $l \cdot j$ L-implies h, $c^*(h, e \cdot j)$ is likewise high; thus h may be inferred inductively from $e \cdot j$. However, this way of reasoning would not be correct, because, under ordinary conditions, $c^*(l, e)$ is not high but very low, and even 0 if the domain of individuals is infinite. The difficulty disappears when we realize on the basis of our previous discussions that X does not need a high c^* for l in order to obtain the desired high c^* for h; all he needs is a high c_{qi}^* for l; and this he has by knowing e and j. To put it in another way, X need not take the roundabout way through the law l at all, as is usually believed; he can instead go from his observational knowledge $e \cdot j$ directly to the prediction h. That is to say, our inductive logic makes it possible to determine $c^*(h, e \cdot j)$ directly and to find that it has a high value, without making use of any law. Customary thinking in every-day life likewise often takes this short-cut, which is now justified by inductive logic. For instance, suppose somebody asks Mr. X what color he expects the next swan he will see to have. Then X may reason like this: he has seen many white swans and no non-white swans; therefore he presumes, admittedly not with certainty, that the next swan will likewise be white; and he is willing to bet on it. He does perhaps not even consider the question whether all swans in the universe without a single exception are white; and if he did, he would not be willing to bet on the affirmative answer.

We see that the use of laws is not indispensable for making predictions. Nevertheless it is expedient of course to state universal laws in books on physics, biology, psychology, etc. Although these laws stated by scientists do not have a high degree of confirmation, they have a high qualified-instance confirmation and thus serve us as efficient instruments for finding those highly confirmed singular predictions which we need for guiding our actions.

§15. THE VARIETY OF INSTANCES

A generally accepted and applied rule of scientific method says that for testing a given law we should choose a variety of specimens as great as possible. For instance, in order to test the law that all metals expand by heat, we should examine not only specimens of iron, but of many different metals. It seems clear that a greater variety of instances allows a more effective examination of the law. Suppose three physicists examine the law mentioned; each of them makes one hundred experiments by heating one hundred metal pieces and observing their expansion; the first physicist neglects the rule of variety and takes only pieces of iron; the second follows the rule to a small extent by examining iron and copper pieces; the third satisfies the rule more thoroughly by taking his one hundred specimens from six different metals. Then we should say that the third physicist has confirmed the law by a more thoroughgoing examination than the two other physicists; therefore he has better reasons to declare the law well-founded and to expect that future instances will likewise be found to be in accordance with the law; and in the same way the second physicist has more reasons than the first. Accordingly, if there is at all an adequate concept of degree of confirmation with numerical values, then its value for the law, or for the prediction that a certain number of future instances will fulfill the law, should be higher on the evidence of the report of the third physicist about the positive results of his experiments than for the second physicist, and higher for the second than for the first. Generally speaking, the degree of confirmation of a law on the evidence of a number of confirming experiments should depend not only on the total number of (positive) instances found but also on their variety, i.e. on the way they are distributed among various kinds.

Ernest Nagel[27] has discussed this problem in detail. He explains the difficulties involved in finding a quantitative concept of degree of confirmation that

[27] E. Nagel, *Principles of the Theory of Probability*. Int. Encycl. of Unified Science, vol. I, No. 6, 1939; see pp. 68–71.

would satisfy the requirement we have just discussed, and he therefore expresses his doubt whether such a concept can be found at all. He says (pp. 69f): "It follows, however, that the degree of confirmation for a theory seems to be a function not only of the absolute number of positive instances but also of the kinds of instances and of the relative number in each kind. It is not in general possible, therefore, to order degrees of confirmation in a linear order, because the evidence for theories may not be comparable in accordance with a simple linear schema; and a fortiori degrees of confirmation cannot, in general, be quantized." He illustrates his point by a numerical example. A theory T is examined by a number E of experiments all of which yield positive instances; the specimens tested are taken from two non-overlapping kinds K_1 and K_2. Nine possibilities $P_1, \cdots P_9$ are discussed with different numbers of instances in K_1 and in K_2. The total number E increases from 50 in P_1 to 200 in P_9. In P_1, 50 instances are taken from K_1 and none from K_2; in P_9, 198 from K_1 and 2 from K_2. It does indeed seem difficult to find a concept of degree of confirmation that takes into account in an adequate way not only the absolute number E of instances but also their distribution among the two kinds in the different cases. And I agree with Nagel that this requirement is important. However, I do not think it impossible to satisfy the requirement; in fact, it is satisfied by our concept c^*.

This is shown by a theorem in our system of inductive logic, which states the ratio in which the c^* of a law l is increased if s new positive instances of one or several different kinds are added by new observations to some former positive instances. The theorem, which is too complicated to be given here, shows that c^* is greater under the following conditions: (1) if the total number s of the new instances is greater, *ceteris paribus*; (2) if, with equal numbers s, the number of different kinds from which the instances are taken is greater; (3) if the instances are distributed more evenly among the kinds. Suppose a physicist has made experiments for testing the law l with specimens of various kinds and he wishes to make one more experiment with a new specimen. Then it follows from (2), that the new specimen is best taken from one of those kinds from which so far no specimen has been examined; if there are no such kinds, then we see from (3) that the new specimen should best be taken from one of those kinds which contain the minimum number of instances tested so far. This seems in good agreement with scientific practice. [The above formulations of (2) and (3) hold in the case where all the kinds considered have equal width; in the general and more exact formulation, the increase of c^* is shown to be dependent also upon the various widths of the kinds of instances.] The theorem shows further that c^* is much more influenced by (2) and (3) than by (1); that is to say, it is much more important to improve the variety of instances than to increase merely their number.

The situation is best illustrated by a numerical example. The computation of the increase of c^*, for the nine possible cases discussed by Nagel, under certain plausible assumptions concerning the form of the law l and the widths of the properties involved, leads to the following results. If we arrange the nine possibilities in the order of ascending values of c^*, we obtain this: P_1, P_3, P_7, P_9;

P_3, P_4, P_5, P_6, P_8. In this order we find first the four possibilities with a bad distribution among the two kinds, i.e. those where none or only very few (two) of the instances are taken from one of the two kinds, and these four possibilities occur in the order in which they are listed by Nagel; then the five possibilities with a good or fairly good distribution follow, again in the same order as Nagel's. Even for the smallest sample with a good distribution (viz., P_2, with 100 instances, 50 from each of the two kinds) c^* is considerably higher—under the assumptions made, more than four times as high—than for the largest sample with a bad distribution (viz. P_9, with 200 instances, divided into 198 and 2). This shows that a good distribution of the instances is much more important than a mere increase in the total number of instances. This is in accordance with Nagel's remark (p. 69): "A large increase in the number of positive instances of one kind may therefore count for less, in the judgment of skilled experimenters, than a small increase in the number of positive instances of another kind."

Thus we see that the concept c^* is in satisfactory accordance with the principle of the variety of instances.

§16. THE PROBLEM OF THE JUSTIFICATION OF INDUCTION

Suppose that a theory is offered as a more exact formulation—sometimes called a "rational reconstruction"—of a body of generally accepted but more or less vague beliefs. Then the demand for a justification of this theory may be understood in two different ways. (1) The first, more modest task is to validate the claim that the new theory is a satisfactory reconstruction of the beliefs in question. It must be shown that the statements of the theory are in sufficient agreement with those beliefs; this comparison is possible only on those points where the beliefs are sufficiently precise. The question whether the given beliefs are true or false is here not even raised. (2) The second task is to show the validity of the new theory and thereby of the given beliefs. This is a much deeper going and often much more difficult problem.

For example, Euclid's axiom system of geometry was a rational reconstruction of the beliefs concerning spatial relations which were generally held, based on experience and intuition, and applied in the practices of measuring, surveying, building, etc. Euclid's axiom system was accepted because it was in sufficient agreement with those beliefs and gave a more exact and consistent formulation for them. A critical investigation of the validity, the factual truth, of the axioms and the beliefs was only made more than two thousand years later by Gauss.

Our system of inductive logic, that is, the theory of c^* based on the definition of this concept, is intended as a rational reconstruction, restricted to a simple language form, of inductive thinking as customarily applied in everyday life and in science. Since the implicit rules of customary inductive thinking are rather vague, any rational reconstruction contains statements which are neither supported nor rejected by the ways of customary thinking. Therefore, a comparison is possible only on those points where the procedures of customary inductive thinking are precise enough. It seems to me, that on these points sufficient agreement is found to show that our theory is an adequate reconstruction;

this agreement is seen in many theorems, of which a few have been mentioned in this paper.

An entirely different question is the problem of the validity of our or any other proposed system of inductive logic, and thereby of the customary methods of inductive thinking. This is the genuinely philosophical problem of induction. The construction of a systematic inductive logic is an important step towards the solution of the problem, but still only a preliminary step. It is important because without an exact formulation of rules of induction, i.e. theorems on degree of confirmation, it is not clear what exactly is meant by "inductive procedures", and therefore the problem of the validity of these procedures cannot even be raised in precise terms. On the other hand, a construction of inductive logic, although it prepares the way towards a solution of the problem of induction, still does not by itself give a solution.

Older attempts at a justification of induction tried to transform it into a kind of deduction, by adding to the premisses a general assumption of universal form, e.g. the principle of the uniformity of nature. I think there is fairly general agreement today among scientists and philosophers that neither this nor any other way of reducing induction to deduction with the help of a general principle is possible. It is generally acknowledged that induction is fundamentally different from deduction, and that any prediction of a future event reached inductively on the basis of observed events can never have the certainty of a deductive conclusion; and, conversely, the fact that a prediction reached by certain inductive procedures turns out to be false does not show that those ɩnductive procedures were incorrect.

The situation just described has sometimes been characterized by saying that a theoretical justification of induction is not possible, and hence, that there is no problem of induction. However, it would be better to say merely that a justification in the old sense is not possible. Reichenbach[28] was the first to raise the problem of the justification of induction in a new sense and to take the first step towards a positive solution. Although I do not agree with certain other features of Reichenbach's theory of induction, I think it has the merit of having first emphasized these important points with respect to the problem of justification: (1) the decisive justification of an inductive procedure does not consist in its plausibility, i.e., its accordance with customary ways of inductive reasoning, but must refer to its success in some sense; (2) the fact that the truth of the predictions reached by induction cannot be guaranteed does not preclude a justification in a weaker sense; (3) it can be proved (as a purely logical result) that induction leads in the long run to success in a certain sense, provided the world is "predictable" at all, i.e. such that success in that respect is possible. Reichenbach shows that his rule of induction R leads to success in the following sense: R yields in the long run an approximate estimate of the relative frequency in the whole of any given property. Thus suppose that we observe the relative frequencies of a property M in an increasing series of samples, and that we determine on the basis of each sample with the help of the rule R the probability

---

[28] Hans Reichenbach, *Experience and Prediction*, 1938, §§38 ff., and earlier publications.

q that an unobserved thing has the property M, then the values q thus found approach in the long run the relative frequency of M in the whole. (This is, of course, merely a logical consequence of Reichenbach's definition or rule of induction, not a factual feature of the world.)

I think that the way in which Reichenbach examines and justifies his rule of induction is an important step in the right direction, but only a first step. What remains to be done is to find a procedure for the examination of any given rule of induction in a more thoroughgoing way. To be more specific, Reichenbach is right in the assertion that any procedure which does not possess the character-istic described above (viz. approximation to the relative frequency in the whole) is inferior to his rule of induction. However, his rule, which he calls "the" rule of induction, is far from being the only one possessing that characteristic. The same holds for an infinite number of other rules of induction, e.g., for Laplace's rule of succession (see above, §10; here restricted in a suitable way so as to avoid contradictions), and likewise for the corresponding rule of our theory of c^* (as formulated in theorem (1), §10). Thus our inductive logic is justified to the same extent as Reichenbach's rule of induction, as far as the only criterion of justification so far developed goes. (In other respects, our inductive logic covers a much more extensive field than Reichenbach's rule; this can be seen by the theorems on various kinds of inductive inference mentioned in this paper.) However, Reichenbach's rule and the other two rules mentioned yield different numerical values for the probability under discussion, although these values converge for an increasing sample towards the same limit. Therefore we need a more general and stronger method for examining and comparing any two given rules of induction in order to find out which of them has more chance of success. I think we have to measure the success of any given rule of induction by the total balance with respect to a comprehensive system of wagers made according to the given rule. For this task, here formulated in vague terms, there is so far not even an exact formulation; and much further investigation will be needed before a solution can be found.

University of Chicago, Chicago, Ill.

The New Riddle of Induction

1. The Old Problem of Induction

AT the close of the preceding lecture, I said that today I should examine how matters stand with respect to the problem of induction. In a word, I think they stand ill. But the real difficulties that confront us today are not the traditional ones. What is commonly thought of as the Problem of Induction has been solved, or dissolved; and we face new problems that are not as yet very widely understood. To approach them, I shall have to run as quickly as possible over some very familiar ground.

The problem of the validity of judgments about future or unknown cases arises, as Hume pointed out, because such judgments are neither reports of experience nor logical consequences of it. Predictions, of course, pertain to what has not yet been observed. And they cannot be logically inferred from what has been observed; for what *has* happened imposes no logical restrictions on what *will* happen. Although Hume's dictum that there are no necessary connections of matters of fact has been challenged at times, it has withstood all attacks. Indeed, I should be inclined not merely to agree that there are no necessary connections of matters of fact, but to ask whether there are any necessary connections at all[1]—but that is another story.

Hume's answer to the question how predictions are related to past experience is refreshingly non-cosmic. When an event of one kind frequently follows upon an

359

event of another kind in experience, a habit is formed that leads the mind, when confronted with a new event of the first kind, to pass to the idea of an event of the second kind. The idea of necessary connection arises from the felt impulse of the mind in making this transition.

Now if we strip this account of all extraneous features, the central point is that to the question "Why one prediction rather than another?", Hume answers that the elect prediction is one that accords with a past regularity, because this regularity has established a habit. Thus among alternative statements about a future moment, one statement is distinguished by its consonance with habit and thus with regularities observed in the past. Prediction according to any other alternative is errant.

How satisfactory is this answer? The heaviest criticism has taken the righteous position that Hume's account at best pertains only to the source of predictions, not their legitimacy; that he sets forth the circumstances under which we make given predictions—and in this sense explains why we make them—but leaves untouched the question of our license for making them. To trace origins, runs the old complaint, is not to establish validity: the real question is not why a prediction is in fact made but how it can be justified. Since this seems to point to the awkward conclusion that the greatest of modern philosophers completely missed the point of his own problem, the idea has developed that he did not really take his solution very seriously, but regarded the main problem as unsolved and perhaps as insoluble. Thus we come to speak of 'Hume's problem' as though he propounded it as a question without answer.

All this seems to me quite wrong. I think Hume grasped the central question and considered his answer to be passably effective. And I think his answer is reasonable and relevant, even if it is not entirely satisfactory. I shall explain presently. At the moment, I merely want to record a protest against the prevalent notion that the problem of justifying induction, when it is so sharply dissociated from the problem of describing how induction takes place, can fairly be called Hume's problem.

I suppose that the problem of justifying induction has called forth as much fruitless discussion as has any halfway respectable problem of modern philosophy. The typical writer begins by insisting that some way of justifying predictions must be found; proceeds to argue that for this purpose we need some resounding universal law of the Uniformity of Nature, and then inquires how this universal principle itself can be justified. At this point, if he is tired, he concludes that the principle must be accepted as an indispensable assumption; or if he is energetic and ingenious, he goes on to devise some subtle justification for it. Such an invention, however, seldom satisfies anyone else; and the easier course of accepting an unsubstantiated and even dubious assumption much more sweeping than any actual predictions we make seems an odd and expensive way of justifying them.

2. Dissolution of the Old Problem

Understandably, then, more critical thinkers have suspected that there might be something awry with the problem we are trying to solve. Come to think of it, what precisely would constitute the justification we seek? If the problem is to explain how we know that certain predictions will turn out to be correct, the sufficient answer is that we don't know any such thing. If the problem is to *find* some way of distinguishing antecedently between true and false predictions, we are asking for prevision rather than for philosophical explanation. Nor does it help matters much to say that we are merely trying to show that or why certain predictions are *probable*. Often it is said that while we

E

cannot tell in advance whether a prediction concerning a given throw of a die is true, we can decide whether the prediction is a probable one. But if this means determining how the prediction is related to actual frequency distributions of future throws of the die, surely there is no way of knowing or proving this in advance. On the other hand, if the judgment that the prediction is probable has nothing to do with subsequent occurrences, then the question remains in what sense a probable prediction is any better justified than an improbable one.

Now obviously the genuine problem cannot be one of attaining unattainable knowledge or of accounting for knowledge that we do not in fact have. A better understanding of our problem can be gained by looking for a moment at what is involved in justifying non-inductive inferences. How do we justify a *de*duction? Plainly, by showing that it conforms to the general rules of deductive inference. An argument that so conforms is justified or valid, even if its conclusion happens to be false. An argument that violates a rule is fallacious even if its conclusion happens to be true. To justify a deductive conclusion therefore requires no knowledge of the facts it pertains to. Moreover, when a deductive argument has been shown to conform to the rules of logical inference, we usually consider it justified without going on to ask what justifies the rules. Analogously, the basic task in justifying an inductive inference is to show that it conforms to the general rules of *in*duction. Once we have recognized this, we have gone a long way towards clarifying our problem.

Yet, of course, the rules themselves must eventually be justified. The validity of a deduction depends not upon conformity to any purely arbitrary rules we may contrive, but upon conformity to valid rules. When we speak of *the* rules of inference we mean the valid rules—or better, *some* valid rules, since there may be alternative sets of equally valid rules. But how is the validity of rules to be determined? Here again we encounter philosophers who insist that these rules follow from some self-evident axiom, and others who try to show that the rules are grounded in the very nature of the human mind. I think the answer lies much nearer the surface. Principles of deductive inference are justified by their conformity with accepted deductive practice. Their validity depends upon accordance with the particular deductive inferences we actually make and sanction. If a rule yields inacceptable inferences, we drop it as invalid. Justification of general rules thus derives from judgments rejecting or accepting particular deductive inferences.

This looks flagrantly circular. I have said that deductive inferences are justified by their conformity to valid general rules, and that general rules are justified by their conformity to valid inferences. But this circle is a virtuous one. The point is that rules and particular inferences alike are justified by being brought into agreement with each other. *A rule is amended if it yields an inference we are unwilling to accept; an inference is rejected if it violates a rule we are unwilling to amend.* The process of justification is the delicate one of making mutual adjustments between rules and accepted inferences; and in the agreement achieved lies the only justification needed for either.

All this applies equally well to induction. An inductive inference, too, is justified by conformity to general rules, and a general rule by conformity to accepted inductive inferences. Predictions are justified if they conform to valid canons of induction; and the canons are valid if they accurately codify accepted inductive practice.

A result of such analysis is that we can stop plaguing ourselves with certain spurious questions about induction. We no longer demand an explanation for guarantees that we do not have, or seek keys to knowledge that we cannot

363

obtain. It dawns upon us that the traditional smug insistence upon a hard-and-fast line between justifying induction and describing ordinary inductive practice distorts the problem. And we owe belated apologies to Hume. For in dealing with the question how normally accepted inductive judgments are made, he was in fact dealing with the question of inductive validity.[2] The validity of a prediction consisted for him in its arising from habit, and thus in its exemplifying some past regularity. His answer was incomplete and perhaps not entirely correct; but it was not beside the point. The problem of induction is not a problem of demonstration but a problem of defining the difference between valid and invalid predictions.

This clears the air but leaves a lot to be done. As principles of deductive inference, we have the familiar and highly developed laws of logic; but there are available no such precisely stated and well-recognized principles of inductive inference. Mill's canons hardly rank with Aristotle's rules of the syllogism, let alone with *Principia Mathematica*. Elaborate and valuable treatises on probability usually leave certain fundamental questions untouched. Only in very recent years has there been any explicit and systematic work upon what I call the constructive task of confirmation theory.

3. The Constructive Task of Confirmation Theory

The task of formulating rules that define the difference between valid and invalid inductive inferences is much like the task of defining any term with an established usage. If we set out to define the term "tree", we try to compose out of already understood words an expression that will apply to the familiar objects that standard usage calls trees, and that will not apply to objects that standard usage refuses to call trees. A proposal that plainly violates either condition is rejected; while a definition that meets these tests may be adopted and used to decide cases that are not already settled by actual usage. Thus the interplay we observed between rules of induction and particular inductive inferences is simply an instance of this characteristic dual adjustment between definition and usage, whereby the usage informs the definition, which in turn guides extension of the usage.

Of course this adjustment is a more complex matter than I have indicated. Sometimes, in the interest of convenience or theoretical utility, we deliberately permit a definition to run counter to clear mandates of common usage. We accept a definition of "fish" that excludes whales. Similarly we may decide to deny the term "valid induction" to some inductive inferences that are commonly considered valid, or apply the term to others not usually so considered. A definition may modify as well as extend ordinary usage.[3]

Some pioneer work on the problem of defining confirmation or valid induction has been done by Professor Hempel.[4] Let me remind you briefly of a few of his results. Just as deductive logic is concerned primarily with a relation between statements—namely the consequence relation—that is independent of their truth or falsity, so inductive logic as Hempel conceives it is concerned primarily with a comparable relation of confirmation between statements. Thus the problem is to define the relation that obtains between any statement S_1 and another S_2 if and only if S_1 may properly be said to confirm S_2 in any degree.

With the question so stated, the first step seems obvious. Does not induction proceed in just the opposite direction from deduction? Surely the evidence-statements that inductively support a general hypothesis are consequences of it. That a given piece of copper conducts electricity follows from and confirms the statement that all copper conducts electricity. Since the consequence relation is already well defined by deductive logic, will we not be on firm ground in saying that confirmation embraces the converse relation? The laws of deduction in reverse will then be among the laws of induction.

Let's see where this leads us. We naturally assume further that whatever confirms a given statement confirms also whatever follows from that statement.[5] But if we combine this assumption with our proposed principle, we get the embarrassing result that every statement confirms every other. Surprising as it may be that such innocent beginnings lead to such an intolerable conclusion, the proof is very easy. Start with any statement S_1. It is a consequence of, and so by our present criterion confirms, the conjunction of S_1 and any statement whatsoever—call it S_2. But the confirmed conjunction, $S_1 \cdot S_2$, of course has S_2 as a consequence. Thus every statement confirms all statements.

The fault lies in careless formulation of our first proposal. While the statements that confirm a general hypothesis are consequences of it, not all its consequences confirm it. This may not be immediately evident; for indeed we do in some sense furnish support for a statement when we establish one of its consequences. We settle one of the questions about it. Consider the heterogeneous conjunction:

8497 is a prime number and the other side of the moon is flat and Elizabeth the First was crowned on a Tuesday.

To show that any one of the three component statements is true is to support the conjunction by reducing the net undetermined claim. But support[6] of this kind is not confirmation; for establishment of one component endows the whole statement with no credibility that is transmitted to other component statements. Confirmation of a hypothesis occurs only when an instance imparts to the hypothesis some credibility that is conveyed to other instances. Appraisal of hypotheses, indeed, is incidental to prediction, to the judgment of new cases on the basis of old ones.

Our formula thus needs tightening. This is readily accomplished, as Hempel points out, if we observe that a hypothesis is genuinely confirmed only by those of its consequences that are instances of it in the strict sense of being derivable from it by instantiation. In other words, a singular statement confirms the hypothesis secured by generalizing from the singular statement—where generalizing means replacing the argument-constants in the singular statement by variables, and prefixing universal quantifiers governing these variables. The predicate constants remain fixed. Less technically, the hypothesis says of all things what the evidence statement says of one thing (or of one pair or other n-ad of things). This obviously covers the confirmation of the conductivity of all copper by the conductivity of a given piece; and it excludes confirmation of our heterogeneous conjunction by any of its components. And, when taken together with the principle that what confirms a statement confirms all its consequences, this criterion does not yield the untoward conclusion that every statement confirms every other.

New difficulties promptly appear from other directions, however. One is the infamous paradox of the ravens. The statement that a given object, say this piece of paper, is neither black nor a raven confirms the hypothesis that all non-black things are non-ravens. But this hypothesis is logically equivalent to the hypothesis that all ravens are black. Hence we arrive at the unexpected conclusion that the statement that a given object is neither black nor a raven confirms the hypothesis that all ravens are black. The prospect of being able to investigate ornithological theories without going out in the rain is so attractive that we know there must be a catch in it. The trouble this time, however, lies not in faulty definition, but in tacit and illicit reference to evidence not stated in our example. Taken by itself, the statement that the given object is neither black nor a raven confirms the hypothesis that everything that is not a raven is not black as well as the hypothesis that everything that is not black is not a raven. We tend to ignore the former hypothesis because we know it to be false from abundant other evidence—from all the familiar things that are not ravens but are black. But we are required to assume that no such evidence is available. Under this circumstance, even a much stronger hypothesis is also obviously confirmed: that nothing is either black or a raven. In the light of this confirmation of the hypothesis that there are no ravens, it is no longer surprising that under the artificial restrictions of the example, the hypothesis that all ravens are black is also confirmed. And the prospects for indoor ornithology vanish when we notice that under these same conditions, the contrary hypothesis that no ravens are black is equally well confirmed.

On the other hand, our definition does err in not forcing us to take into account all the *stated* evidence. The unhappy results are readily illustrated. If two compatible evidence statements confirm two hypotheses, then naturally the conjunction of the evidence statements should confirm the conjunction of the hypotheses.[7] Suppose our evidence consists of the statements E_1 saying that a given thing b is black, and E_2 saying that a second thing c is not black. By our present definition, E_1 confirms the hypothesis that everything is black, and E_2 the hypothesis that everything is non-black. The conjunction of these perfectly compatible evidence statements will then confirm the self-contradictory hypothesis that everything is both black and non-black. Simple as this anomaly is, it requires drastic modification of our definition. What given evidence confirms is not what we arrive at by generalizing from separate items of it, but—roughly speaking—what we arrive at by generalizing from the total stated evidence. The central idea for an improved definition is that, within certain limitations, what is asserted to be true for the narrow universe of the evidence statements is confirmed for the whole universe of discourse. Thus if our evidence is E_1 and E_2, neither the hypothesis that all things are black nor the hypothesis that all things are non-black is confirmed; for neither is true for the evidence-universe consisting of b and c. Of course, much more careful formulation is needed, since some statements that are true of the evidence-universe—such as that there is only one black thing—are obviously not confirmed for the whole universe. These matters are taken care of by the studied formal definition that Hempel develops on this basis; but we cannot and need not go into further detail here.

No one supposes that the task of confirmation-theory has been completed. But the few steps I have reviewed—chosen partly for their bearing on what is to follow—show how things move along once the problem of definition displaces the problem of justification. Important and long-unnoticed questions are brought to light and answered; and we are encouraged to expect that the many remaining questions will in time yield to similar treatment.

But our satisfaction is shortlived. New and serious trouble begins to appear.

4. The New Riddle of Induction

Confirmation of a hypothesis by an instance depends rather heavily upon features of the hypothesis other than its syntactical form. That a given piece of copper conducts electricity increases the credibility of statements asserting that other pieces of copper conduct electricity, and thus confirms the hypothesis that all copper conducts electricity. But the fact that a given man now in this room is a third son does not increase the credibility of statements asserting that other men now in this room are third sons, and so does not confirm the hypothesis that all men now in this room are third sons. Yet in both cases our hypothesis is a generalization of the evidence statement. The difference is that in the former case the hypothesis is a *law-like* statement; while in the latter case, the hypothesis is a merely contingent or accidental generality. Only a statement that is *lawlike*—regardless of its truth or falsity or its scientific importance—is capable of receiving confirmation from an instance of it; accidental statements are not. Plainly, then, we must look for a way of distinguishing lawlike from accidental statements.

So long as what seems to be needed is merely a way of excluding a few odd and unwanted cases that are inadvertently admitted by our definition of confirmation, the problem may not seem very hard or very pressing. We fully expect that minor defects will be found in our definition and that the necessary refinements will have to be worked out patiently one after another. But some further examples will show that our present difficulty is of a much graver kind.

Suppose that all emeralds examined before a certain time *t* are green.[8] At time *t*, then, our observations support the hypothesis that all emeralds are green; and this is in accord with our definition of confirmation. Our evidence statements assert that emerald *a* is green, that emerald *b* is green, and so on; and each confirms the general hypothesis that all emeralds are green. So far, so good.

Now let me introduce another predicate less familiar than "green". It is the predicate "grue" and it applies to all things examined before *t* just in case they are green but to other things just in case they are blue. Then at time *t* we have, for each evidence statement asserting that a given emerald is green, a parallel evidence statement asserting that that emerald is grue. And the statements that emerald a is grue, that emerald b is grue, and so on, will each confirm the general hypothesis that all emeralds are grue. Thus according to our definition, the prediction that all emeralds subsequently examined will be green and the prediction that all will be grue are alike confirmed by evidence statements describing the same observations. But if an emerald subsequently examined is grue, it is blue and hence not green. Thus although we are well aware which of the two incompatible predictions is genuinely confirmed, they are equally well confirmed according to our present definition. Moreover, it is clear that if we simply choose an appropriate predicate, then on the basis of these same observations we shall have equal confirmation, by our definition, for any prediction whatever about other emeralds—or indeed about anything else.[9] As in our earlier example, only the predictions subsumed under lawlike hypotheses are genuinely confirmed; but we have no criterion as yet for determining lawlikeness. And now we see that without some such criterion, our definition not merely includes a few unwanted cases, but is so completely ineffectual that it virtually excludes nothing. We are left once again with the intolerable result that anything confirms anything. This difficulty cannot be set aside as an annoying detail to be taken care of in due course. It has to be met before our definition will work at all.

Nevertheless, the difficulty is often slighted because on the surface there seem to be easy ways of dealing with it. Sometimes, for example, the problem is thought to be much like the paradox of the ravens. We are here again, it is pointed out, making tacit and illegitimate use of information outside the stated evidence: the information, for example, that different samples of one material are usually alike in conductivity, and the information that different men in a lecture audience are usually not alike in the number of their older brothers. But while it is true that such information is being smuggled in, this does not by itself settle the matter as it settles the matter of the ravens. There the point was that when the smuggled information is forthrightly declared, its effect upon the confirmation of the hypothesis in question is immediately and properly registered by the definition we are using. On the other hand, if to our initial evidence we add statements concerning the conductivity of pieces of other materials or concerning the number of older brothers of members of other lecture audiences, this will not in the least affect the confirmation, according to our definition, of the hypothesis concerning copper or of that concerning other lecture audiences. Since our definition is insensitive to the bearing upon hypotheses of evidence so related to them, even when the evidence is fully declared, the difficulty about accidental hypotheses cannot be explained away on the ground that such evidence is being surreptitiously taken into account.

A more promising suggestion is to explain the matter in terms of the effect of this other evidence not directly upon the hypothesis in question but *in*directly through other hypotheses that *are* confirmed, according to our definition, by such evidence. Our information about other materials does by our definition confirm such hypotheses as that all pieces of iron conduct electricity, that no pieces of rubber do, and so on; and these hypotheses, the explanation runs, impart to the hypothesis that all pieces of copper conduct electricity (and also to the hypothesis that none do) the character of lawlikeness—that is, amenability to confirmation by direct positive instances when found. On the other hand, our information about other lecture audiences *dis*confirms many hypotheses to the effect that all the men in one audience are third sons, or that none are; and this strips any character of lawlikeness from the hypothesis that all (or the hypothesis that none) of the men in *this* audience are third sons. But clearly if this course is to be followed, the circumstances under which hypotheses are thus related to one another will have to be precisely articulated.

The problem, then, is to define the relevant way in which such hypotheses must be alike. Evidence for the hypothesis that all iron conducts electricity enhances the lawlikeness of the hypothesis that all zirconium conducts electricity, but does not similarly affect the hypothesis that all the objects on my desk conduct electricity. Wherein lies the difference? The first two hypotheses fall under the broader hypothesis—call it "H"—that every class of things of the same material is uniform in conductivity; the first and third fall only under some such hypothesis as—call it "K"—that every class of things that are either all of the same material or all on a desk is uniform in conductivity. Clearly the important difference here is that evidence for a statement affirming that one of the classes covered by H has the property in question increases the credibility of any statement affirming that another such class has this property; while nothing of the sort holds true with respect to K. But this is only to say that H is lawlike and K is not. We are faced anew with the very problem we are trying to solve: the problem of distinguishing between lawlike and accidental hypotheses.

The most popular way of attacking the problem takes its cue from the fact that accidental hypotheses seem typically to involve some spatial or temporal restriction, or reference to some particular individual. They seem to concern the people in some particular room, or the objects on some particular person's desk; while lawlike hypotheses characteristically concern all ravens or all pieces of copper whatsoever. Complete generality is thus very often sup-

posed to be a sufficient condition of lawlikeness; but to
define this complete generality is by no means easy.
Merely to require that the hypothesis contain no term
naming, describing, or indicating a particular thing or
location will obviously not be enough. The troublesome
hypothesis that all emeralds are grue contains no such
term; and where such a term does occur, as in hypotheses
about men in *this room*, it can be suppressed in favor of
some predicate (short or long, new or old) that contains no
such term but applies only to exactly the same things.
One might think, then, of excluding not only hypotheses
that actually contain terms for specific individuals but
also all hypotheses that are equivalent to others that do
contain such terms. But, as we have just seen, to exclude
only hypotheses of which *all* equivalents are free of such
terms is to exclude nothing. On the other hand, to exclude
all hypotheses that have *some* equivalent containing such a
term is to exclude everything; for even the hypothesis

All grass is green

has as an equivalent

All grass in London or elsewhere is green.

The next step, therefore, has been to consider ruling
out predicates of certain kinds. A syntactically universal
hypothesis is lawlike, the proposal runs, if its predicates
are 'purely qualitative' or 'non-positional'.[10] This will
obviously accomplish nothing if a purely qualitative
predicate is then conceived either as one that is equivalent
to some expression free of terms for specific individuals,
or as one that is equivalent to no expression that contains
such a term; for this only raises again the difficulties just
pointed out. The claim appears to be rather that at least
in the case of a simple enough predicate we can readily
determine by direct inspection of its meaning whether or
not it is purely qualitative. But even aside from obscurities

in the notion of 'the meaning' of a predicate, this claim seems to me wrong. I simply do not know how to tell whether a predicate is qualitative or positional, except perhaps by completely begging the question at issue and asking whether the predicate is 'well-behaved'—that is, whether simple syntactically universal hypotheses applying it are lawlike.

This statement will not go unprotested. "Consider", it will be argued, "the predicates 'blue' and 'green' and the predicate 'grue' introduced earlier, and also the predicate 'bleen' that applies to emeralds examined before time t just in case they are blue and to other emeralds just in case they are green. Surely it is clear", the argument runs, "that the first two are purely qualitative and the second two are not; for the meaning of each of the latter two plainly involves reference to a specific temporal position." To this I reply that indeed I do recognize the first two as well-behaved predicates admissible in lawlike hypotheses, and the second two as ill-behaved predicates. But the argument that the former but not the latter are purely qualitative seems to me quite unsound. True enough, if we start with "blue" and "green", then "grue" and "bleen" will be explained in terms of "blue" and "green" and a temporal term. But equally truly, if we start with "grue" and "bleen", then "blue" and "green" will be explained in terms of "grue" and "bleen" and a temporal term; "green", for example, applies to emeralds examined before time t just in case they are grue, and to other emeralds just in case they are bleen. Thus qualitativeness is an entirely relative matter and does not by itself establish any dichotomy of predicates. This relativity seems to be completely overlooked by those who contend that the qualitative character of a predicate is a criterion for its good behavior.

Of course, one may ask why we need worry about such

unfamiliar predicates as "grue" or about accidental hypotheses in general, since we are unlikely to use them in making predictions. If our definition works for such hypotheses as are normally employed, isn't that all we need? In a sense, yes; but only in the sense that we need no definition, no theory of induction, and no philosophy of knowledge at all. We get along well enough without them in daily life and in scientific research. But if we seek a theory at all, we cannot excuse gross anomalies resulting from a proposed theory by pleading that we can avoid them in practice. The odd cases we have been considering are the clinically pure cases that, though seldom encountered in practice, nevertheless display to best advantage the symptoms of a widespread and destructive malady.

We have so far neither any answer nor any promising clue to an answer to the question what distinguishes lawlike or confirmable hypotheses from accidental or nonconfirmable ones; and what may at first have seemed a minor technical difficulty has taken on the stature of a major obstacle to the development of a satisfactory theory of confirmation. It is this problem that I call the new riddle of induction.

5. The Pervasive Problem of Projection

At the beginning of this lecture, I expressed the opinion that the problem of induction is still unsolved, but that the difficulties that face us today are not the old ones; and I have tried to outline the changes that have taken place. The problem of justifying induction has been displaced by the problem of defining confirmation, and our work upon this has left us with the residual problem of distinguishing between confirmable and non-confirmable hypotheses. One might say roughly that the first question was "Why does a positive instance of a hypothesis give any grounds for predicting further instances?"; that the newer question was "What is a positive instance of a hypothesis?"; and that the crucial remaining question is "What hypotheses are confirmed by their positive instances?"

The vast amount of effort expended on the problem of induction in modern times has thus altered our afflictions but hardly relieved them. The original difficulty about induction arose from the recognition that anything may follow upon anything. Then, in attempting to define confirmation in terms of the converse of the consequence relation, we found ourselves with the distressingly similar difficulty that our definition would make any statement confirm any other. And now, after modifying our definition drastically, we still get the old devastating result that any statement will confirm any statement. Until we find a way of exercising some control over the hypotheses to be admitted, our definition makes no distinction whatsoever between valid and invalid inductive inferences.

The real inadequacy of Hume's account lay not in his descriptive approach but in the imprecision of his description. Regularities in experience, according to him, give rise to habits of expectation; and thus it is predictions conforming to past regularities that are normal or valid. But Hume overlooks the fact that some regularities do and some do not establish such habits; that predictions based on some regularities are valid while predictions based on other regularities are not. Every word you have heard me say has occurred prior to the final sentence of this lecture; but that does not, I hope, create any expectation that every word you will hear me say will be prior to that sentence. Again, consider our case of emeralds. All those examined before time t are green; and this leads us to expect, and confirms the prediction, that the next one will be green. But also, all those examined are grue; and

F

377

this does not lead us to expect, and does not confirm the prediction, that the next one will be grue. Regularity in greenness confirms the prediction of further cases; regularity in grueness does not. To say that valid predictions are those based on past regularities, without being able to say *which* regularities, is thus quite pointless. Regularities are where you find them, and you can find them anywhere. As we have seen, Hume's failure to recognize and deal with this problem has been shared even by his most recent successors.

As a result, what we have in current confirmation theory is a definition that is adequate for certain cases that so far can be described only as those for which it is adequate. The theory works where it works. A hypothesis is confirmed by statements related to it in the prescribed way provided it is so confirmed. This is a good deal like having a theory that tells us that the area of a plane figure is one-half the base times the altitude, without telling us for what figures this holds. We must somehow find a way of distinguishing lawlike hypotheses, to which our definition of confirmation applies, from accidental hypotheses, to which it does not.

Today I have been speaking solely of the problem of induction, but what has been said applies equally to the more general problem of projection. As pointed out earlier, the problem of prediction from past to future cases is but a narrower version of the problem of projecting from any set of cases to others. We saw that a whole cluster of troublesome problems concerning dispositions and possibility can be reduced to this problem of projection. That is why the new riddle of induction, which is more broadly the problem of distinguishing between projectible and non-projectible hypotheses, is as important as it is exasperating.

Our failures teach us, I think, that lawlike or projective

hypotheses cannot be distinguished on any merely syntactical grounds or even on the ground that these hypotheses are somehow purely general in meaning. Our only hope lies in re-examining the problem once more and looking for some new approach. This will be my course in the final lecture.

NOTES: III

[1] (page 63) Although this remark is merely an aside, perhaps I should explain for the sake of some unusually sheltered reader that the notion of a necessary connection of ideas, or of an absolutely analytic statement, is no longer sacrosanct. Some, like Quine and White, have forthrightly attacked the notion; others, like myself, have simply discarded it; and still others have begun to feel acutely uncomfortable about it.

[2] (page 68) A hasty reader might suppose that my insistence here upon identifying the problem of justification with a problem of description is out of keeping with my parenthetical insistence in the preceding lecture that the goal of philosophy is something quite different from the mere description of ordinary or scientific procedure. Let me repeat that the point urged there was that the organization of the explanatory account need not reflect the manner or order in which predicates are adopted in practice. It surely must describe practice, however, in the sense that the extensions of predicates as explicated must conform in certain ways to the extensions of the same predicates as applied in practice. Hume's account is a description in just this sense. For it is an attempt to set forth the circumstances under which those inductive judgments are made that are normally accepted as valid; and to do that is to state necessary and sufficient conditions for, and thus to define, valid induction. What I am maintaining above is that the problem of justifying induction is not something over and above the problem of describing or defining valid induction.

[3] (page 69) For a fuller discussion of definition in general see chapter i of *The Structure of Appearance*.

[4] (page 69) The basic article is 'A Purely Syntactical Definition of Confirmation', cited in Note I.9. A much less technical account is given in 'Studies in the Logic of Confirmation', *Mind*, n.s., vol. 54 (1945), pp. 1–26 and 97–121. Later

380

work by Hempel and others on defining *degree* of confirmation does not concern us here.

[5] (page 70) I am not here asserting that this is an indispensable requirement upon a definition of confirmation. Since our commonsense assumptions taken in combination quickly lead us to absurd conclusions, some of these assumptions have to be dropped; and different theorists may make different decisions about which to drop and which to preserve. Hempel gives up the converse consequence condition, while Carnap (*Logical Foundations of Probability*, Chicago and London, 1950, pp. 474–6) drops both the consequence condition and the converse consequence condition. Such differences of detail between different treatments of confirmation do not affect the central points I am making in this lecture.

[6] (page 70) Any hypothesis is 'supported' by its own positive instances; but support — or better, direct factual support — is only one factor in confirmation. This factor has been separately studied by John G. Kemeny and Paul Oppenheim in 'Degree of Factual Support', *Philosophy of Science*, vol. 19 (1952), pp. 307–24. As will appear presently, my concern in these lectures is primarily with certain other important factors in confirmation, some of them quite generally neglected.

[7] (page 72) The status of the conjunction condition is much like that of the consequence condition — see Note III.5. Although Carnap drops the conjunction condition also (p. 394), he adopts for different reasons the requirement we find needed above: that the total available evidence must always be taken into account (pp. 211–13).

[8] (page 74) Although the example used is different, the argument to follow is substantially the same as that set forth in my note 'A Query on Confirmation', cited in Note I.15.

[9] (page 75) For instance, we shall have equal confirmation, by our present definition, for the prediction that roses subsequently examined will be blue. Let "emerose" apply just to emeralds examined before time t, and to roses examined later. Then all emeroses so far examined are grue, and this confirms the hypothesis that all emeroses are grue and hence

the prediction that roses subsequently examined will be blue. The problem raised by such antecedents has been little noticed, but is no easier to meet than that raised by similarly perverse consequents.

[10] (page 78) Carnap took this course in his paper 'On the Application of Inductive Logic', *Philosophy and Phenomenological Research*, vol. 8 (1947), pp. 133–47, which is in part a reply to my 'A Query on Confirmation', cited in Note I.15. The discussion was continued in my note 'On Infirmities of Confirmation Theory', *Philosophy and Phenomenological Research*, vol. 8 (1947), pp. 149–51; and in Carnap's 'Reply to Nelson Goodman', same journal, same volume, pp. 461–2.

Acknowledgments

Ramsey, Frank Plumpton. "Truth and Probability." In *The Foundations of Mathematics and other Logical Essays* (Paterson: Littlefield, Adams and Company, 1960): 156–98. Reprinted with the permission of Littlefield, Adams and Company.

Skyrms, Brian. "Degrees of Belief." In *Pragmatics and Empiricism* (New Haven: Yale University Press, 1984): 20–36. Reprinted with the permission of Yale University Press.

Skyrms, Brian. "Learning from Experience." In *Pragmatics and Empiricism* (New Haven: Yale University Press, 1984): 37–62. Reprinted with the permission of Yale University Press.

Popper, K.R. "The Propensity Interpretation of the Calculus of Probability, and the Quantum Theory." In *Observation and Interpretation: A Symposium of Philosophers and Physicists*, edited by S. Körner (London: Butterworths Scientific Publications): 65–70. Reprinted with the permission of the author.

Carnap, Rudolf. "Statistical and Inductive Probability." In *Readings in the Philosophy of Science*, edited by Baruch A. Brody (Englewood Cliffs, N.J.: Prentice Hall, 1970): 440–50. Reprinted with the permission of Prentice Hall.

van Fraassen, Bas C. "Indifference: The Symmetries of Probability." In *Laws and Symmetry* (Oxford: Oxford University Press, 1989): 293–317. Reprinted with the permission of Oxford University Press.

Lewis, David. "A Subjectivist's Guide to Objective Chance." *Philosophical Papers*, Vol. 2 (Oxford: Oxford University Press, 1986): 83–113. Reprinted with the permission of Oxford University Press.

Lewis, David. "Postscripts to 'A Subjectivist's Guide to Objective Chance.'" *Philosophical Papers*, Vol. 2 (Oxford: Oxford University Press, 1986): 114–132. Reprinted with the permission of Oxford University Press.

Howson, Colin. "Theories of Probability." *British Journal for the Philosophy of Science* 46 (1995): 1–32. Reprinted with the permission of Oxford University Press.

Salmon, Wesley. "Vindication of Induction." In *Current Issues in the Philosophy of Science*, edited by Herbert Feigl and Grover Maxwell (New York: Holt, Reinhart and Winston, 1961): 245–56. Reprinted with the permission of Holt, Rinehart and Winston.

Strawson, P.F. "The 'Justification' of Induction." In *Introduction to Logical Theory*

(London: Methuen, 1952): 248–63. Reprinted with the permission of Methuen, Division of Routledge.

Hempel, Carl G. "Studies in the Logic of Confirmation," Parts I and II. *Mind* 54 (1945): 1–26, 97–121. Reprinted with the permission of the Oxford University Press.

Salmon, Wesley C. "Confirmation and Relevance." In *Minnesota Studies in the Philosophy of Science*, Vol. 6, *Induction, Probability, and Confirmation*, edited by Grover Maxwell and Robert M. Anderson Jr. (Minneapolis: University of Minnesota Press, 1975): 3–36. Reprinted with the permission of the University of Minnesota Press.

Carnap, Rudolf. "On Inductive Logic." *Philosophy of Science* 12 (1945): 72–97. Reprinted with the permission of the University of Chicago Press.

Goodman, Nelson. "The New Riddle of Induction." In *Fact, Fiction, and Forecast* (Cambridge: Harvard University Press, 1955): 63–86. Reprinted with the permission of Harvard University Press.